GREEN INTELLIGENCE

JOHN WARGO

Green

CREATING
ENVIRONMENTS
THAT PROTECT
HUMAN HEALTH

Intelligence

YALE UNIVERSITY PRESS   NEW HAVEN & LONDON

Set in FontShop Scala and Scala Sans type by Duke & Company, Devon, Pennsylvania.
Printed in the United States of America by Sheridan Books, Ann Arbor, Michigan.

Library of Congress Cataloging-in-Publication Data

Wargo, John, 1950–
Green intelligence : creating environments that protect human health / John Wargo.
p.   cm.
Includes bibliographical references and index.
ISBN 978-0-300-11037-1 (cloth : alk. paper)
1. Environmental quality. 2. Environmental degradation. 3. Environmental policy. 4. Nuclear accidents—Environmental aspects. 5. Air pollution. 6. Water pollution. 7. Pesticides—Environmental aspects. I. Title.
GE140.W365 2009
363.7—dc22   2009007814

A catalogue record for this book is available from the British Library.

This paper meets the requirements of ANSI/NISO Z39.48-1992 (Permanence of Paper). It contains 30 percent postconsumer waste (PCW) and is certified by the Forest Stewardship Council (FSC).

10 9 8 7 6 5 4 3 2 1

For Linda, Ellie, Kate, and Adam

CONTENTS

FIGURES

ONE UNEXPECTED SIDE EFFECT of twentieth-century prosperity has been a change in the chemistry of the human body. Each day most people are exposed to thousands of chemicals in mixtures that were never experienced by previous generations. Many of these substances are recognized by the governments of the United States and European Union to be carcinogens, neurotoxins, reproductive and developmental toxins, or endocrine disruptors that mimic or block human hormones. In 1999, the U.S. Centers for Disease Control and Prevention (CDC) began testing human tissue among populations across the country to detect the presence of environmental contaminants, and reported that most individuals carry in their bodies a mixture of metals, pesticides, solvents, fire retardants, waterproofing agents, and by-products of fuel combustion. Children often carry higher concentrations than adults, with the amounts of contaminants also varying according to gender and ethnicity.[1]

Our petroleum- and chemical-dependent economy is the primary cause of these exposures. Every year hundreds of billions of pounds of chemicals are released into the environment as commercial products while trillions of additional pounds of pollution are discharged into the atmosphere, surface and groundwater, oceans, and land as by-products of fuel combustion or as wastes. Often the distinction between commercial chemicals and pollutants is only a matter of time as once sought-after products lose their utility, are discarded, and slowly degrade, releasing their ingredients into the surrounding areas. Our global economy concentrates raw chemicals, mixes them in millions of products that are distributed through markets, and reconcentrates remaining wastes in landfills or incinerators, where once again they are dispersed unpredictably into the air, soil, and water. The human exposures that occur at every step along the way are often unrecognized or ignored by individuals, corporations, and governments. And most in society have little comprehension of the chemical mixtures they experience in everyday life and the dangers posed to their health.

During the last half-century, society's growing chemical imprint has been accompanied by an increase in the prevalence of many illnesses. These include respiratory diseases such as childhood asthma, neurological impairments, declining sperm counts, fertility failure, immune dysfunction, breast and prostate cancers, and developmental disorders among the young. Some of these illnesses have been caused or exacerbated by exposure to commercial chemicals and pollutants. There is little doubt, for example, that tobacco, lead, mercury, radionuclides, solvents, vehicle exhaust, combustion by-products, dioxins, PCBs, and many pesticides have caused extensive human illness. Significantly, these chemicals were once thought to be safe at doses now known to be hazardous; as with other substances, the perception of danger grew as governments tested chemicals more thoroughly.

Three trends help to explain the growing chemical burden on the environment and human health: population growth, longevity, and economic expansion. The U.S. population surpassed 300 million in 2006, having doubled in both the first and second halves of the twentieth century. Meanwhile, life expectancy has increased from 47 to 78 years since 1900, doubling the duration of adulthood. And the economy is now forty-five times larger than it was in 1950. Consequently, people in the United States are now living longer with higher incomes while consuming more goods and energy, all as they create more pollution and waste per capita than any previous generation or any other nation. By 2008 those in the wealthiest countries consumed thirty-two times more energy and raw materials than those in the poorest. And inexpensive oil, gas, and coal have fueled many of these trends.[2]

The United States arguably has the most extensive body of environmental law and regulation in the world and many believe it was intended to prevent the pollution, exposures, and health loss just described. Between 1970 and 1996 several dozen major federal statutes and tens of thousands of regulations were adopted. Congress intended these regulations to limit emissions of hazardous chemicals and their residues in surface and groundwater, food, soils, consumer products, the air, and the oceans.

Legislators may have had high aspirations, but they were naive to assume that the government can control the global economic forces driving the chemical revolution we are experiencing. Moreover, there has never been a master plan for environmental law; instead Congress has adopted statutes in a piecemeal way, responding to compelling and often surprising stories of environmental contamination, damage, and health loss by adopting new laws to govern hazardous materials. Infamous examples include the periodic Cuyahoga River fires in Ohio in the 1950s and 1960s; the discovery of a school built above a chemical dump at Love Canal in New York in the 1970s; the spraying of PCBs on the back roads of Times Beach in Missouri in the mid 1980s; the Union Carbide disaster in Bhopal in 1984; the Chernobyl nuclear plant explosion in the Ukrainian Republic in 1987; and the *Exxon Valdez* oil spill in Alaska in 1989. The result of this reactionary, rather than preventive, approach is an odd patchwork of poorly coordinated regulations rather than a comprehensive body of law to manage chemicals we all experience in daily life. Equally important, U.S. law has come to depend on technical risk assessments that demand ever more evidence to prove chemical danger, rather than requirements that companies prove chemicals' safety prior to their production and sale.

The United States has banned some products, chemicals, and technologies, demonstrating that the law can improve environmental quality and reduce human exposures to health-threatening chemicals. There are now bans on nuclear weapons testing in the atmosphere, lead from paints and gasoline, DDT and other persistent chlorinated insecticides, ozone-gobbling CFCs formerly used as aerosol propellants, and PCBs in electronics. The CDC's human-tissue testing program provides clear evidence that bans led quickly to reduced levels of DDT, lead, and PCBs in the population. Bans, however, have been rare: of the nearly 100,000 chemicals now traded internationally, the Environmental Protection Agency (EPA) has prohibited fewer than 200 since its creation in 1970.

Rather than curtail a chemical's production or use, the EPA has normally attempted to limit the chemical's release to the environment, or to set ceilings for how much chemical residue is allowed to linger in water, air, soils, food, and occasionally, consumer products. Tens of thousands of separate regulations establish chemical residue limits. But these choices create many more challenges than real solutions. They require expensive and time-consuming environmental monitoring. They demand analyses of the danger posed by each chemical in question, along with their nearly infinite mixtures—tests that have rarely been done. And they initiate protracted negotiations between government and the regulated community that often last decades, with debates about quality of evidence and acceptable concentration limits continuing for years in Congress, regulatory agencies, and the courts. Consequently, the government has at times decided only to warn the public about chemical hazards, as it has done by posting signs along rivers and lakes to advise pregnant women about the dangers to fetuses of mercury in fish.

Indeed, most progress in reducing pollution and waste occurred during the last half of the twentieth century only if sources were easy to identify. Power plants, incinerators, vehicles, sewage treatment plants, and landfills thus became targets for EPA regulators and environmental activists. The problems we now face are far more subtle, decentralized, and imperceptible. Today everyone is a consumer and producer of dangerous chemical mixtures. We buy gas for a vehicle and burn it, producing exhaust, or purchase pesticides for our home or lawn, without understanding the broader implications of these simple, everyday activities.

Today the chemical burden on the environment and human health is enormous and growing at an ever faster pace. The U.S. Chemical Abstract Service maintains a registry of over 32 million organic and inorganic chemicals, 14 million of which are commercially available. Nearly 100,000 of these are registered in the United States and the European Union and are traded internationally. Pesticide manufacturers alone, which constitute merely 5 percent of the synthetic chemical industry, market nearly 20,000 different compounds. The amounts of chemicals released in the United States are also astonishing. In the year 2000, three hundred chemicals were produced in amounts exceeding one billion pounds annually and about 2,700 others exceeded a million pounds each year. But the story doesn't end there. The mixing of these chemicals to form millions of different consumer products and their consequent entry into the waste stream have produced chemical exposures that far exceed our understanding as well as the ability of governments to assure safety.[3]

One important guardian of the health of the U.S. population, the EPA, has since 1976 been required by the Toxic Substances Control Act (TSCA) to maintain an inventory of potentially toxic substances; but the agency cannot demand premarket testing or regulate production unless it has compelling evidence of significant environmental or health risk. This requirement places the burden on government to conduct the testing needed to justify regulation, an impossibility given the staggering number of untested chemicals and combinations. When this law went into effect, 62,000 chemicals already in commerce were listed but immediately exempted or grandfathered from any data submission requirements. Since that time, 45,000 additional chemicals have been introduced to commerce, yet nearly half of these were reported to the EPA after companies began to sell them. The effect is that among U.S. chemicals produced in highest volumes, 90 percent are exempted from federal review under TSCA. Moreover, in 1998 the EPA found that basic toxicity information was available for only 7 percent of these, while none was available for 43 percent. For chemicals produced at lower volumes, the agency had even less information.

"High production volume" (HPV) chemicals are those produced in amounts exceeding one million pounds per year. Data necessary to judge basic health risks are available for fewer than 15 percent of these. Given its absence of regulatory authority, the EPA in 1998 decided to create a "voluntary partnership" with manufacturers, called the High Production Volume Challenge Program, to request basic data. A decade later, the government's understanding of these compounds' hazards has improved little, while the overall knowledge base has probably declined given that thousands of additional untested chemicals were introduced to global markets during this period. Part of the problem is the ineffectiveness of the voluntary program. As of 2008, chemical companies had refused to provide basic studies for no fewer than two hundred HPV chemicals. The EPA also reported that 95 percent of the information submitted by manufacturers is classified as "confidential business information" and therefore is not accessible to the public—nor to state, local, or foreign governments. These data are often crucial for understanding and preventing dangerous exposures to people living near industrial plants or waste facilities, workers transporting chemicals, and those who respond to emergencies such as fires and explosions involving such chemicals. The dangers posed to workers and the public after the World Trade Center collapse in 2001 illustrate well the importance of bringing to light companies' knowledge of product ingredients and chemical hazards.[4]

Markets for chemicals and goods have grown beyond the capacity of any government to monitor and control. There are too many chemicals, too many products, and too many shipments moving around the world to make surveillance and enforcement efficient and effective. Ideally consumer impressions of safety would grow from premarket chemical testing and a vigilant and protective government that would prevent imperceptible hazards from reaching the marketplace. Instead, consumers unwittingly believe themselves to be safe from danger based on the advertising claims of manufacturers, which are rarely supported by independent testing.

Certainly the challenge has overwhelmed the limited knowledge, authority, and resources of public agencies charged to manage environmental quality. Legislators never predicted the enormous investment required to create the knowledge we need to protect the public's health from environmental insults. Without it our Gross Domestic Product, a primary indicator of national economic health, has fully neglected the costs of chemical testing, regulation, environmental contamination and restoration, and the associated burden of disease. If in fact market prices did reflect these expenses, our society would certainly consume, pollute, and waste far less and we would all be living with fewer industrial chemicals flowing through our veins.[5]

To explore what might be done to confront these challenges, I chose to compare four histories of other dangerous technologies and practices that have changed the chemistry of the planet and our bodies. These include two cases concerning the government's pursuit of national security—atmospheric testing of nuclear weapons, and contaminated military training sites—as well as those of pesticides and vehicle emissions, which evolved from private companies' profit-seeking. Each example includes the discovery that children have been exposed more intensely than others, and each problem has been potent, persistent, and transnational if not global in scale. The four histories illustrate different types of risk, why some are more threatened than others, and possibilities for reducing future exposure.

The U.S. nuclear weapons program unintentionally produced the very first paradigm for understanding global environmental problems such as climate change, ozone depletion, and mercury contamination in marine food chains. The discovery of the hazards of nuclear testing profoundly influenced the evolution of both environmental science and environmental law. Nuclear weapons were intended to cause massive loss of human life and environmental ruin, but the discovery by the Atomic Energy Commission (AEC) of global fallout and its effects on humans was unanticipated. The pattern of discovery that radionuclides persist, move through the atmosphere, follow complex ecological pathways that lead to human exposures, and produce life-threatening health effects became a model for later efforts to understand and manage pollution and hazardous chemicals.

The history of Vieques, a small island off the east coast of Puerto Rico, illustrates another kind of environmental degradation that can occur in the name of national defense. The U.S. government in 1941 chose Puerto Rico as an Atlantic Basin counterpart to the Pearl Harbor naval base and acquired land for training and conventional weapons testing. The island has long been home to nearly nine thousand residents who have lived between a munitions training range on the east end and weapons storage bunkers to their west. U.S. and allied forces bombed and shelled the east end of the island for fifty years, releasing nearly 150 million pounds of munitions, chemical warfare agents, depleted uranium, pesticides, solvents, and fuels. In 2003 the U.S. Navy closed the facility and left the island, leaving behind an enormous amount of chemical waste as well as a cratered, toxic moonscape created by the explosions. These hazardous substances continue to seep into the coastal environment as they have for nearly half a century.

Vieques is hardly unique as a recipient of government abuse. The U.S. Congress in 2005 estimated that 50,000 sites under the authority of the Departments of Defense, Energy and Interior are severely contaminated. At the time, the cost of simply containing the site hazards, not restoring the sites to their original conditions, was estimated to be at least $337 billion. Some bases lie on remote islands such as Pacific atolls or the Aleutian Islands in Alaska, while others include urban settings such as Hunter's Point Naval Shipyard in San Francisco. Most of these sites are being transferred to private developers for homes, schools, offices, and parks without being fully restored to their premilitary and far more natural condition. The Defense Department has often left a toxic chemical soup behind for surrounding communities that have neither the expertise nor the resources to identify potential associated threats to human health.

The history of pesticides suggests how society might protect itself from more dispersed hazardous exposures. Each year 3 billion pounds of pesticides are sold in the United States, and nearly 400 million acres of the national landscape are chemically treated. Pesticides are intended to kill various insects, bacteria, fungi, algae, parasites, and other species that can harm economic productivity and human health. Their intentional release to the environment causes these deliberately toxic substances to linger in air, foods, water, soils, and human tissues. They are added to commercial products such as plastics, fabrics, paints, fuels, wood products, swimming pools and spas, personal care products, and some pharmaceuticals. Many pesticides may legally be sprayed in indoor environments including homes, offices, restaurants, schools, hospitals, hotels, and vehicles. Some are encapsulated within tiny plastic beads that slowly dissolve and can cling to insects, but also to clothing, skin, and foods. Nearly every home in the nation contains some registered pesticide. Among all chemicals, pesticides have been studied the longest, as scientists have sought to understand their health and environmental effects and to manage human exposure and associated illnesses. Despite these efforts, everyone in the nation still carries pesticide residues in their tissues and tens of thousands of people report their exposures to poison control centers each year. Understanding the successes and failures of pesticide law could help to guide effective control of plastics and the larger chemical universe.

Vehicle emissions are inhaled by most people every day of their lives. In 2008, more than 240 million motor vehicles were registered in the United States, one for every person in the nation eligible to drive. This passion for movement has had an important effect on air quality, human health, climate change, and the nation's dependence on foreign oil. By 2007, U.S. vehicles were consuming twice the amount of petroleum produced annually in the nation. Each year U.S. drivers burn almost 175 billion gallons of fuel as they travel 2.5 trillion miles. They accomplish this remarkable act while sitting less than ten feet from exhaust pipes that emit highly hazardous chemicals to the air. I recently sat on a freeway between Los Angeles and San Diego while traffic was either crawling or stalled in both directions in all fourteen lanes. Several thousand drivers idled or crept along for hours while inhaling each other's emissions. This is considered a normal "rush hour" in Southern California. Although most who experience it feel exasperated by the wasted time, many are unaware of the well-documented toll it is taking on their health.

Why do people accept vehicle exhaust as part of their daily routine? Several mistaken beliefs underlie our willingness to overlook the dangers, including that any individual's contribution to the problem is inconsequential; changes in fuel chemistry will have little effect on health; gradual improvements in new model vehicle engines have improved air quality; pollution is dangerous only if visible; living, working, or exercising near highways creates no special threat to health; chemical concentrations should be averaged across large regions and long time periods to judge the severity of the health threat; pollution only threatens respiratory health; and a population's risk of losing health is uniformly shared. The chapters on air quality challenge all of these premises, and highlight the growing seriousness of air pollution.

I close the book with an overview of the plastics industry that illuminates the chemical challenges to health we all face today. Plastics have entered every aspect of our lives, and their chemical residues now can be found in air, water, soils, and the tissues of nearly every human tested. Government studies confirm that children carry the highest concentrations of some plastic compounds that either mimic or block human hormones. These compounds are known to induce reproductive and developmental disorders in several "model" species—that is, animals that are often tested to infer human health risks. But the sometimes alarming information that is known about these compounds is eclipsed by what we simply do not know and in many cases, cannot control. Chemical production and manufacturing occur in so many parts of the world that no national government has the ability to control sales, product ingredients, dangerous exposures, or wastes. Plastic manufacturers introduce nearly 100 billion pounds of resins to markets each year, with fewer than 5 percent of products containing them recycled annually in the United States. Collectively these trends demonstrate the longstanding and growing environmental experiment on human health being conducted by the plastics industry, one predominantly overlooked but nevertheless authorized and often subsidized by the federal government.

The antagonists in this book include the Atomic Energy Commission (AEC), which was eventually absorbed by the Department of Energy (DOE); the Department of Defense (DOD); and the plastics, pesticide, and vehicle industries. Each employed startlingly similar patterns of environmental neglect and strategies of public deception to their advantage. Officials in each organization thought carefully and creatively about how they could shape favorable public perceptions of their products or programs while neglecting the environment and often deceiving the public. Whether government bureaucrat or corporate leader, each attempted to control scientific inquiry and findings, sometimes by choosing to avoid research on obvious questions or by concealing discoveries that threatened their missions. They explained results in stories that announced the benefits and safety of their actions or products while they dismissed dangers as improbable, easily recognized, and manageable. These optimistic narratives dominated public discourse and formed favorable public impressions of technologies, products, or practices that eventually left many exposed and ill. In all cases the antagonists' concentration of authority, wealth, and expertise overwhelmed competing interpretations of environmental threats to human health.

Secrecy plays a special role in these histories. The primary intent of trade secrecy and classified information is to prevent public access to information and to protect competitive advantage either in markets or international relations. The effects are menacing; secrecy skews the balance of power toward government or corporate leaders while ensuring public ignorance. As these environmental histories illustrate, secrecy also creates false impressions of naturalness, wildness, purity, and safety, leading generations to experience contaminated environments without their knowledge or consent.

Most secrets have limited lifespans, and in each case the credibility of government and corporate narratives disintegrated gradually as independent scientists challenged claims of safety with new evidence, and sometimes developed plausible alternative interpretations of data disclosed by governments or corporations. The effect was a new understanding of the sources of danger, patterns of human exposure, and the seriousness of health effects. These insights led government officials to stop testing nuclear weapons in the atmosphere, ban numerous chlorinated pesticides, and eliminate lead from paints and gasoline. But decades often passed before convincing challenges to conventional wisdom led to regulatory action. Between the time these hazards were released and the government responded, widespread and serious environmental damage and loss of health often occurred. If thorough testing of chemicals or technologies had been required prior to their deployment, and if the government had reacted by controlling the testing, production, sales, and environmental release of these toxins, the chemistry of the planet and our bodies would be very different.

But is the antidote to failed intelligence about environmental hazards simply more information? Not necessarily. Even if perfect knowledge did exist, many conditions could still inhibit environmental law and policy from offering practical protection to all. A depressed economy, litigation, imprecise legal authority, bureaucratic incompetence, insufficient budgets, lack of political will, and competing social problems such as national security all may diminish resources and public support to protect health and environmental quality. Yet lack of knowledge is the most fundamental obstacle to effective safeguards. If information does not exist, or is secret, distorted, or misunderstood, then there is little hope of slowing or reversing the current trend toward releasing ever greater quantities and varieties of substances that are hazardous to humans and other species.

In an ideal world, everyone would understand the distinction between the dangerous and the benign. We would be able to recognize hazardous chemicals in our foods, air, water, soil, buildings, and consumer goods; and we would understand their relative dangers, as well as how risks increase with the intensity of exposure. As it is, however, the gulf between what we know and what we need to know to protect society from environmental hazards is enormous and will be extraordinarily expensive to remedy. The underlying causes of our ignorance include illiteracy, secrecy, deception, privacy, language barriers, and perhaps most importantly, the control and ownership of science by powerful institutions. The origin of the problem is that governments and corporations have each failed to produce and disseminate knowledge of hazards in ways that make public comprehension and effective protection possible. Without this intelligence, the global chemical experiment on public health will remain wildly out of control.

What can we learn from environmental history that might help steer us toward a more health-sustaining future? What principles should guide governments, corporations, and individuals to create an environment less threatening to human health? And while waiting for more responsible leadership, how can you reduce your personal exposure to serious environmental dangers? These are the questions that stimulated this book; and the answers, however unsettling, must be confronted if we are to have any chance of creating a health-supportive environment for future generations.

PART ONE THE LEGACY OF NUCLEAR TESTING

CHAPTER ONE

Perfecting the Art of Terror

ON APRIL 8, 1951, dignitaries sat in Adirondack lawn chairs at the Officers' Beach Club Patio, near the shore of Runit Island in the Pacific Ocean. Dressed in light summer clothing and darkened ski goggles, they waited patiently, staring out to sea. Precisely at 10:00 A.M., a blinding white light filled the sky, followed by a yellow and red fireball that lifted a fiery swirling cloud behind it. The force of the explosion pulled millions of tons of water, mud, and plant and animal life into the atmosphere, turning much of it into fine radioactive particles and mist. At its peak, nearly twenty miles high, the cloud spilled over the top of its stem in every direction, forming the rounded image of a mushroom. The flash and fireball lasted only fifteen seconds, and were followed quickly by a thunderous roar and shockwave. After that, only the sound of waves broke the silence. Winds gradually reshaped the cloud, pushing it slowly toward the horizon. Most onlookers believed the danger had passed, dissipating into the atmosphere. No one recognized the radiation exposure they had experienced or its implications for their future health.

The stage had been set for this event nearly six years earlier, when the first above-ground nuclear test occurred at Alamogordo, New Mexico, in July 1945. The United States, seeking a location for additional tests, believed the small and isolated Marshall Islands would be an ideal experimental test range. The area's nearly 2,000 islands lie in 3 million square miles of ocean, with a total land area of less than 1,000 square miles. Relatively few people would need to be relocated, security would be assured due to the area's isolation, and human exposure to the radioactive isotopes released by the atmospheric explosions might be minimized.

BIKINI

The Bikini Atoll, within the Marshall Islands, became a focal point for the experiments. Nearly 2,500 miles from Hawaii, the atoll, a collection of twenty-three narrow islands that surround a central lagoon, was home to only 167 people on 3.4 square miles of land. At first the Navy intended to use Bikini only for two experiments, one in the atmosphere and another underwater. The program was expanded, however, when the initial experiments unexpectedly demonstrated the nation's vulnerability to attack as well as the scientists' limited ability to predict and control the fate of radioactive particles.

Although the fishing industry protested plans to expand the experiments, worrying that radiation would harm aquatic life, and others objected to radiation experiments on animals, no one complained about the removal of Bikini residents from their homeland to create the test site. Their relocation, however, was to have a profound effect on their identities as individuals and community members. The islanders depended on fishing, the collection of native fruits, and the cultivation of vegetables. With no television, radio, and very little contact with the more "civilized" world, much of their identity emanated from their close connection to the islands, the sea, and especially the lagoon. In addition, the islanders' social status, wealth, and power in the community were all tightly bound to the distribution of island property rights.

In February 1946, Benjamin Wyatt, the military governor of the Marshall Islands, flew to Bikini to ask the residents to move temporarily from their homeland so that the United States could conduct experiments. Wyatt told Chief Juda that his people should leave their home "so it could be used for the good of mankind and to end all world wars." The governor then compared the residents to "the children of Israel whom the Lord saved from their enemy and led into the Promised Land." While agonizing over this difficult "choice," Chief Juda and the islanders did not know that President Truman had already approved the plan, and as the islanders began their exodus the Navy started blasting through the atoll's coral reef so that warships could enter the lagoon.[1]

On March 7, 1946, aware that tens of thousands of troops had already moved to other Marshall Islands for the operation, the Bikini residents packed their belongings and boarded a Navy transport for their new island home of Rongerik, believing that they would be back within a year or at most, two.[2] But Rongerik is half the size of Bikini, and its soils are infertile, causing crop failure. The new residents struggled with nutritional deficiencies, a situation made worse by a rumor that an evil witch had poisoned fish in the lagoon. Scientists later found that fish inhabiting the lagoon did contain compounds toxic to humans, which may explain why the island had been uninhabited.

The following spring a fire destroyed a third of the coconut trees on Rongerik. One man recalled later that residents at the time were so desperate for food that they began to eat the pulp wood of coconut trees as well as rotting fruit that had washed up on shore. In 1948 an anthropologist sent by the Navy to the island found the residents living on raw flour diluted with water. All were suffering from malnutrition, and some were near starvation. They pleaded with the military officials to let them go home following the first test of a 20 kiloton atomic bomb in July 1946. Yet only months after leaving their homeland, the AEC knew that radiation levels in the island's soils, water, and plant and animal life were all far higher than predicted and that the islanders would remain in exile for many years.[3]

The Navy moved the Bikinians from Rongerik to a tent camp on a nearby naval base, Kwajalein, where they remained for six more months before being relocated again to the island of Kili. Kili was larger and had more coconut trees than Rongerik, but it had no lagoon for fishing and the heavy surf that pounded the shores kept the fishermen from venturing out in their small boats for nearly half the year. Eventually the Navy provided a forty-foot whaleboat, and the Trust Territory donated an older copra trade boat, while making additional investments in Kili's agriculture. The whaleboat sank in 1951 in rough seas, however, and in 1957 a typhoon eventually destroyed the copra boat while also damaging much of the island's agriculture. These setbacks added to the deep depression felt by many Bikinians. More than a decade had passed since they had left their atoll, and many desperately yearned to return home.

Although Bikini residents were promised that only one or two blasts would occur on their homeland, eventually more nuclear energy would be released on the island by the United States than at any other site in human history. The cumulative force exploded at Bikini exceeded that of 75 megatons of TNT, or nearly four times the energy released by all 928 tests at the Nevada Test Site that lies 65 miles northwest of Las Vegas.[4]

Operation Crossroads

Operation Crossroads began in 1946, only a year following the Japanese surrender. Its two tests, Able and Baker, were designed to explore the ability of naval vessels to survive a nuclear attack. In contrast to earlier secret tests, these experiments were highly publicized and broadcast live by radio around the world. The target fleet included ninety-five vessels— battleships, two aircraft carriers, twelve destroyers, eight submarines, and transports and landing craft—made up of craft captured from the Japanese or Germans as well as damaged American vessels. All were chosen and placed near the blast site to test the resilience of construction materials and ship designs to the force of atomic blasts. Sheep in cages were placed on the ships to test their ability to survive both the blast and the subsequent radiation. Able was dropped by plane, while Baker was suspended beneath the surface of the lagoon (Figure 1.1).

The operation sparked considerable conflict among scientists over its need and possible effects. Lee DuBridge, president of the California Institute of Technology, wrote in the *Bulletin of the Atomic Scientists*: "Wouldn't science and engineering be far better off if the 100 million dollars or so which tests will cost could be devoted to laboratory research under controlled conditions? . . . Who can say that a sudden rain storm would not precipitate dangerous quantities of this material onto one or more of the ships packed with observers? . . . The surface burst will raise a great cloud of water spray and where will it be carried? . . . Might not a cloud of this lethal dust be carried hundreds of miles and deposited on unsuspecting inhabitants?"[5]

The Bravo Experiment

Castle Bravo was the first megaton hydrogen bomb detonated above Bikini Atoll just before dawn on March 1, 1954. It exploded with a force of

FIGURE 1.1. Nuclear test Baker, Bikini Atoll, 1946. The U.S. Navy tested the resilience to a nuclear attack by placing nearly a hundred ships in the immediate vicinity of the blast site, all within the atoll. Live test animals were also placed on the vessels to understand radiation dangers. *Source:* U.S. Atomic Energy Commission.

15 megatons, twice what was intended, and nearly eight hundred times that released at Hiroshima. Nearly 270 miles away, the blast lit the sky for several minutes, and was soon followed by the sound of thunder.[6]

The enormity of the Bravo explosion shocked everyone who witnessed it. The blast produced hurricane-force winds and stripped trees of vegetation. Radioactive ash and debris fell from the sky within hours of detonation, and just as DuBridge had anticipated several years earlier, soon covered the decks of vessels within thirty miles of the atoll. All Navy ships were ordered to retreat to at least fifty miles from ground zero. When government scientists returned to the site, they were stunned to find an area twelve miles in diameter, about the size of Washington, D.C., and some of its suburbs, completely obliterated: the island, coral reefs, and the lagoon bottom had all been vaporized and sucked into the atmosphere via the enormous fireball and mushroom cloud. At the center of the blast was a mile-wide crater beneath the sea. Within several months of the test, a report on fallout hazards from "very high yield nuclear weapons" was produced by Armed Forces Special Weapons Project, yet its contents would be classified as "secret" for the next forty-five years.[7] Despite many previous nuclear tests, Bravo was the first in which scientists carefully identified areas that had received different intensities of radiation and how much was received. The contours on Figure 1.2 represent intensity of radiation, measured in roentgens; the distance from the blast site and the arrival time of the radioactive dust are also noted. The potentially lethal zone—that is, the area receiving 500 roentgens or more—is shown to be approximately 170 miles long (nearly the distance between New York City and Boston), and 50 miles wide, far beyond the zone of visible devastation. The results were unsettling, because New York City at the time spanned only 365 square miles, and Moscow only 117 square miles.[8]

Only two of eight previous surface blasts had been monitored to permit the re-creation of radiation dose rate contours like those developed from the Bravo test.[9] The largest bomb ever detonated in world history was the Soviets' Tzar Bomba, which was detonated on the Arctic island of Novaya Zemlya in 1961. The 50-megaton device produced blast damage nearly 600 miles away, and potentially lethal fallout over nearly 15,000 square miles. The Soviets claimed that they were capable of delivering a 100-megaton device as well, but that its enormous weight, and their inability to move the delivery aircraft far enough from the blast to keep it from becoming a suicide mission for the pilots and crew, kept them from testing it.[10]

For the Bravo experiment, winds blowing west were predicted to push radioactive debris away from the inhabited islands and atolls that lie to the east. Yet after the blast, winds became variable, and fallout patterns less predictable. The elliptical shape was caused by winds blowing first from the south and then the west that deposited significant radioactive debris over nearly 5,000 square miles. Four inhabited islands between one hundred and three hundred miles east of Bikini received unexpectedly high levels of fallout. On Rongerik, one nearby atoll, twenty-eight servicemen huddled inside a protective structure staring at their radiation meter. The needle moved quickly beyond the scale's maximum value as the radioactive ash began floating gently to the ground, leading them to stay indoors, and to put on additional clothing. They were evacuated thirty-four hours following the blast. Two days later, the indigenous residents were also evacuated.[11]

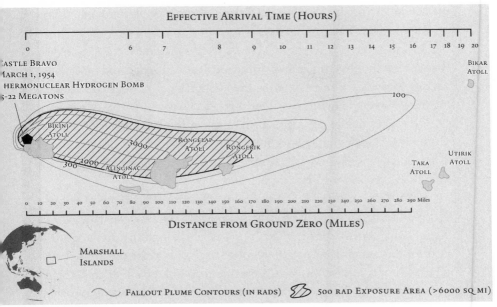

EFFECTIVE ARRIVAL TIME (HOURS)

DISTANCE FROM GROUND ZERO (MILES)

MARSHALL
ISLANDS

FALLOUT PLUME CONTOURS (IN RADS)   500 RAD EXPOSURE AREA (>6000 SQ MI)

FIGURE 1.2. Spread of radiation from Bravo thermonuclear test, 1954. Bravo was
the largest atmospheric test conducted by the United States. It distributed potentially
lethal doses of radiation within a zone defined by the 500 rad contour, roughly 170
miles long and 50 miles wide. Many within the zone experienced less intense radiation;
exposures varied considerably by shelter. *Source:* U.S. Atomic Energy Commission.

The island of Rongelap was also unexpectedly poisoned by the Bravo
experiment. Within hours of the detonation, nearly 250 of its inhabitants received radiation doses high enough to burn their skin, induce
radiation sickness, and cause illnesses such as thyroid disease. Another
one hundred Rongelap residents were luckily off-island during the blast,
and eventually provided scientists with an unexposed "control group" to
monitor health effects.

Nearly an inch and a half of radioactive ash fell on Rongelap on the
day of the blast. Children ran and played in the warm "snow," rolling in it,
throwing it in the air, inhaling it, and tasting it. Within hours, they realized something was very wrong as burns appeared on their skin and many
were overwhelmed by nausea and diarrhea. For many, hair loss quickly
followed. A one-year-old on Rongelap at the time of the test eventually died
of myelogenous leukemia. And nineteen of the twenty-two children on

the island during the blast required surgical removal of thyroid nodules, several of which were malignant.

After the test, Rongelap residents were evacuated to a U.S. naval base several hundred miles away for medical examination and treatment. Among Rongelap women exposed to radiation, 13 of 32 pregnancies during the following four years were terminated due to "nonviable offspring." Children ages one to five years at the time of the blast experienced abnormally low growth in bone density, weight, and stature. In fact, among thirty-one exposed children, twenty-five experienced several pounds of weight loss during the first two months.

By 1957, U.S. officials had declared that the former residents of Rongelap could return to the island, where a new village was constructed for them.[12] During the following twenty years, the Rongelap population, reassured about their safety, gathered food as usual on the more northern islands. In 1979, however, the Department of Energy conducted a new survey and found that the northern islands were so heavily contaminated with radiation that they were unsafe to visit, let alone to provide food. They also concluded that eleven islands or atolls other than those found contaminated in 1954 had in fact received "intermediate" levels of radiation. The AEC's earlier declaration of the islands' safety had justified the avoidance of medical monitoring, which had in turn prevented any correlation between levels of radiation exposure and illness. Although it is difficult to know if the earlier avoidance of testing was intentional, it is certain that if radiation had been found, the government's moral and financial responsibility would have been enormous.

Bravo taught that shelter and evacuation were necessary to survive a nuclear conflict. Early warning and rapid movement to well-shielded shelters soon became core elements of civil defense strategies. The AEC reported that radiation exposure during the first three hours following a blast would be greater than levels likely to be experienced during the following week; similarly, exposure during the first week would on average be higher than that anticipated during the remainder of one's life, if spent in the area. These were crude estimates, because the physicists mistakenly assumed that radiation would be distributed evenly; only later would they become aware of the phenomenon we know today as fallout "hotspots."[13]

The suggested remedy was to take shelter as quickly as possible: subways, underground bunkers buried at least three feet deep, and basements could all offer significant protection. Earth-covered shelters could reduce radiation by a factor of 1,000. AEC scientists recommended that residents stay in their shelters for a week to ten days, then quickly evacuate.

When announcing Bravo's test results, the AEC stated that 70 million people in ninety-two cities lived in "targeted zones" and should plan for evacuation. But how could cities such as New York, Boston, Washington, and Los Angeles evacuate millions of people within the warning period of just a few hours? AEC experts cautioned that evacuation would not be efficient or effective, and could lead to serious exposures in traffic jams. They also warned that communications would likely be poor following an attack and one could easily flee in the direction of the radioactive cloud rather than away from it, especially because in the event of a Soviet nuclear attack, more than one strike would probably occur. If Boston were hit, for example, some would flee south or east toward Connecticut or Rhode Island, where they might encounter a radioactive dust cloud floating northeast from a strike in New York City. The sobering estimates of destruction prompted President Eisenhower to propose spending $100 billion over ten years on national highway construction to ease congestion and increase evacuation rates from urban areas.[14] At the height of the Cold War, the United States simply did not have the capacity to evacuate its population centers.

Details of Bravo's destruction were so deeply unsettling that over the next five years, the Atomic Energy Commission released only fragments of reports about it. Even a report to Congress was delayed due to "international political sensitivity."[15] Most of the scientific studies of radiation were conducted by academic institutes funded by the AEC. And many of the scientific articles published in the open literature followed AEC reports that had been classified—and unchallenged—for several years. Together, these limitations assured strong agency influence over public knowledge of the extent of destruction and contamination that would likely follow an attack. And it demonstrated the potent role that classified information could play to insulate political and military leaders from public criticism and accountability.

Bravo made the world seem smaller and less secure to many. Three years after the test, in June 1957, the Congressional Joint Committee on Atomic Energy held eight days of hearings during which the effects of a thermonuclear war were finally described to the public as "local fall out." "Local fallout can cover . . . thousands of square miles from megaton explosions.

... Multi-weapon attacks (say 200 to 300 bombs of megaton size, 2,500–3,000 megatons total yield), can blanket half or more of the continental United States with lethal (death producing) or near-lethal radiation levels from local fallout alone. This of course could be applied to any nation. There would be great difficulties of forecasting in advance such information as the weather, the exact locations of detonations, and the exact characteristics of weapons, it is to be expected, that there would be great difficulties in forecasting local fallout patterns associated with nuclear war."[16]

Given the darkness of these forecasts, the AEC worked hard to maintain positive public sentiment toward the weapons testing program. Political acceptance of continued testing at the Nevada Test Site in many ways depended on wide acceptance of narratives told by government scientists about the hazards experienced in the Pacific Proving Grounds. The enormity of Bravo's devastation thus posed the first serious threat to the future of the U.S. nuclear weapons program.

By 1958, twenty-three nuclear weapons had been tested on or near the fragile Bikini Atoll; its lagoon, surrounding waters, and coral reefs had taken an enormous toll. Even so, in 1967 Chief Juda pleaded with the Trust Territory Commissioner to allow his people to return from Kili to their homeland. The commissioner remembered vividly being introduced to "Chief Juda, who very emotionally and persuasively, and almost tearfully, pleaded with me to either get them back to Bikini or, failing that, to get them a better place than Kili."[17] That same year, the AEC found that the island was "safe enough" based on the opinions of "eight of the most highly qualified experts available."[18]

"Restoration" and the Islanders' Return

In 1968, drawing on information from these and other AEC reports, President Johnson made the decision to restore the island and allow the former residents to return following their twenty-two-year odyssey. The AEC spent $5 million to remove all vegetation, plow up the soils, and replant nearly 50,000 coconut trees and other crops to encourage resettlement. Afterward, the commission concluded that there was virtually no detectable radiation on the island, and that the islanders' health would in no way be threatened. To convince skeptical Bikinians that the food was safe if one simply followed their expert dietary advice, AEC scientists themselves consumed island fruits, fish, and coconuts. Many former residents remained unconvinced and refused the incentives, prompting additional offers of free housing and food to Marshallese government officials.

The Bikinians gathered in 1968 to decide whether to return home, and although many remained doubtful of government claims of safety, only two or three of the three hundred exiles voted to remain on Kili. Some even participated in efforts to restore and rebuild Bikini Island. Soils were ploughed, trees cut and buried, others were planted, and nearly eighty new homes were constructed. In 1975 some islanders wished to build homes on the island's interior, prompting yet another round of radiation studies that found radiation hazards far higher than earlier recorded. Favorite foods including breadfruit and pandanas had absorbed dangerous levels of radioactivity from the soil, but coconuts were reported to be safe.

Contradicting the AEC's earlier conclusions, in 1975 Lawrence Livermore Laboratory scientists reported that "all living patterns involving Bikini Island exceed federal (radiation) guidelines." The AEC scientists had somehow miscalculated the persistence and movement of the isotopes in the soil, water, and native plants—especially coconuts, a food treasured and consumed often by the islanders. That year the Energy Department—which had been created by patching together the AEC and other federal agencies—tested the urine of 100 residents and found cesium-137 levels to be eleven times higher than those of nonresidents. Worried U.S. officials suggested that Bikini residents should no longer eat foods grown on the island, and that consuming imported foods would eventually reduce the residents' radiation burdens.[19]

Many of the Marshall Islands have shallow and porous soils that permit radionuclides to move quickly and deeply into the groundwater following rains. The AEC expected that by plowing the soils, removing contaminated vegetation, and replanting food crops, the radionuclides would be sufficiently removed from the island's ecosystems. "There was no hint in 1969 that there would be a problem with coconuts, vegetables and water," testified one Interior Department official at a 1978 U.S. Congressional hearing.[20]

Soon Livermore Laboratory scientists also found that well water on Bikini exceeded the U.S. standard for strontium-90, a regulation adopted after the passage of the Safe Drinking Water Act of 1976. Starting in the early 1950s, the AEC knew that rain carried radionuclides from the atmosphere to the ground, and that radiation was detectable in surface water that often flowed to groundwater. During the thirty-year period following the first atmospheric nuclear tests in 1946, however, no U.S. standard had been created to limit radionuclides in drinking water in Bikini or anywhere else. In fact, during the entire atmospheric testing era (1945–1963), U.S. water supplies were not tested for the presence of strontium-90.

When the energy officials studied indigenous plants on Bikini more carefully, they discovered that coconuts efficiently absorbed both cesium-137 and strontium-90. The rapid rise in the islanders' body burdens of these radionuclides could be explained in part by their penchant for coconuts. So the government ordered the residents to limit their intake to one coconut per day, although it recommended avoiding the fruit completely. Islanders were also directed to eat surplus government foods, with the exception of fish or fowl.[21] Most complied with the new restrictions, although during the drought of 1977 many resumed drinking coconut milk and their body burdens of cesium-137 and strontium-90 rose once more. Consequently, in 1978 the Bikinians were forced to leave their homeland for the second time in thirty-two years. Two years later Energy Department scientists again declared it was safe to go home, but only under certain conditions: if instead of Bikini Island they inhabited another small part in the atoll; if 50 percent of their food was grown from off-island sources; and if they spent no more than 10 percent of their time on Bikini Island itself.

The problem of the government's haphazard radiation monitoring was compounded by scientists' not understanding that radionuclides travel from soil and water to humans through the plants and animals they eat—and by their presumption that some radiation exposure would prove harmless. They thus tried to limit islanders' access to certain foods, or keep them from living in hot zones, so that they could contain their exposure beneath some "bright line" of safety. But to be successful, such an approach would have required that fallout patterns, as well as variations in exposure and consequent health effects, be understood with precision.

Robert Conrad of the Brookhaven National Laboratory led the AEC's Marshall Islands' medical program for twenty-three years following the Pacific tests. During this time, he repeatedly assured the islanders of their health and the negligible effect from radiation.[22] Some who needed the most reassurance were the nearly 160 residents of Utrik, a small atoll 280 miles to the east of Bikini who were exposed to fallout following Bravo's explosion. Fallout rained down on Utrik for twenty-two hours following the explosion and residents remained for three more days before being evacuated. Even so, they were allowed to return several months later: the AEC believed they were too far from the blast to have been affected, a confidence that led the agency to neglect to measure radiation or monitor health effects in Utrik in the years following the blast. When questioned, the residents demonstrated that they understood almost nothing about the nuclear testing program, including its effects on the environment or human health, making them especially vulnerable to the comforting claims of the government scientists. Conrad eventually reconsidered his earlier dismissal of health effects as well as his definition of a "safe dose." As he confessed, "Thyroid nodules have been increasing in the Utrik people and this was quite unpredicted, and we had some of the best experts in the U.S. . . . It turns out we were wrong." Another Energy Department official reflected on the bureaucratic bungling, "We made a mess there and we're going to stay and clean it up. Let's face it, the people of Bikini were screwed by history, but it wasn't deliberate."[23]

ENEWETAK

Like Bikini, the Enewetak Atoll was chosen to be the site of nuclear tests between 1947 and 1958, including Operation Mike, a thermonuclear device that evaporated one of the islands, leaving a crater in the ocean more than a mile in diameter. The United States has spent over $100 million to clean up radioactivity that has lingered after forty-two nuclear devices were detonated on or over the atoll. On Runit Island, which lies within the atoll, a 15-kiloton bomb exploded but did not produce a nuclear blast, spreading plutonium over much of the island site. Radiation levels are expected to preclude resettlement for at least 24,000 years. The Defense Nuclear Agency, recognizing the intensity of the problem, decided not to clean it up, and instead made the island a repository for the most highly contaminated debris removed from other nearby islands. The crater left in the lagoon and the dome constructed above the plutonium waste are easily visible in public-domain satellite images of the area.

In creating the nuclear waste depository, nearly 100,000 cubic yards of radioactive soils were scraped from the islands, mixed with concrete, and placed within an atomic bomb crater on Runit Island. Workers were given no special gear or equipment to prevent inhalation of plutonium particles from dusts and soils that blew around the contaminated sites. The northern islands were more heavily bombed, and are still more radioactive. On the southern islands, new houses and community facilities were constructed and coconut trees planted, in anticipation of returning local populations.[24]

The Enewetak people were relocated and remained in exile for thirty-three years on Ujelang Atoll. Their wish to go home prompted intense debate over acceptable levels of cancer risk. Two scientists from Brookhaven National Laboratory, which was funded by the AEC and later the Energy Department, concluded that the islanders' cancer risk would be low.[25] But Rosalie Bertell, a consultant to the Nuclear Regulatory Commission, disagreed with the finding and challenged: "The population of Enewetak has the right to know that a value judgment has been made for them, namely that induction of cancer is their only concern. They may, if informed about hypothyroidism, aplastic anemia, premature aging, benign tumors and other such disorders, make different judgments. They [the AEC] 'reduced' the radiation dose of the inhabitants by averaging in the population less exposed. This is like telling one member of a family his or her risk of lung cancer is lowered if the other nonsmoking members of the family are included and an 'average' [family] risk is given. It is a scientifically ridiculous approach to public health."[26]

Energy Department officials had contended for twenty-five years that the only islands with lingering and dangerous levels of radiation included Bikini, Rongelap, Enewetak, and Utirik. Then in 1978 it reported that eleven other islands or atolls had received significant fallout from the larger bomb tests. Many residents have faced unusual medical problems that are associated with radiation exposure, including thyroid disease, cancer, mental retardation, miscarriages, and gross physical birth defects. The populations are relatively small as are the number of those ill; moreover, the failure to test for radiation immediately following the explosions makes it difficult to understand the role that weapons testing may have played in the islanders' diseases. Federal officials presumed no persistent radioactive particles and therefore that no serious exposures had occurred. This in turn justified their choice not to monitor nuclides in the environment or the health of inhabitants.[27]

By 1958, the United States had detonated more than a hundred nuclear weapons in the Pacific: 66 in the Marshall Islands, as well as 24 near

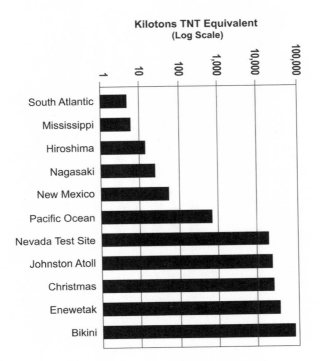

FIGURE 1.3. Explosive power of U.S. nuclear weapons tests by site. Bikini Atoll experienced more explosive force than any other site used by the United States during the eighteen-year-long atmospheric nuclear testing era. The chart also demonstrates that more explosive force was released on Pacific islands than on the U.S. mainland. *Source:* U.S. Department of Energy.

Kiritimati Island and 12 at Johnston Atoll, both south of Hawaii.[28] When testing stopped, Bikini Atoll had become a wasteland. The cumulative explosive force of devices detonated on, above, or beneath the atoll was 5,900 times higher than that released by the Hiroshima bomb. Ten of the Bikini tests were each 100 times more destructive than the Hiroshima bomb, while two were nearly 1,000 times greater (Figures 1.3, 1.4). Largely due to their own haphazard environmental testing, and the secrecy that precluded other scientists from adding to the effort, AEC scientists long misunderstood the persistence and fate of the radionuclides produced by these nuclear weapons tests.

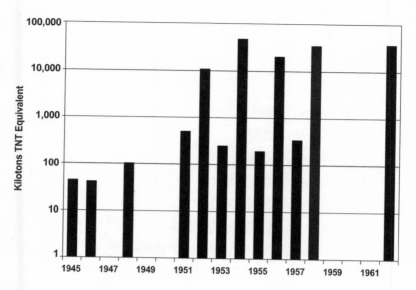

FIGURE 1.4. Explosive power of U.S. nuclear weapons tests by year. The years 1951–1958 were the most active period of atmospheric nuclear testing. Note the annual rise and decline in experimentation during the period, as well as the pause between 1958 and 1961 when the United States and Soviet Union agreed to a brief testing moratorium. *Source:* U.S. Department of Energy.

The government's failure to test the environment created a false sense of its safety and kept the Bikinians' hope for return alive. Nearly three decades after first giving up their homeland atoll "for the good of mankind," and then being removed a second time due to radiation poisoning, Chief Juda reflected on the Navy commander's persuasion that the Bikinians were akin to "the children of Israel whom the Lord saved from their enemy and led into the Promised Land." Instead, he lamented, "We are sadly more akin to the Children of Israel when they left Egypt and wandered through the desert for 40 years. We left Bikini and have wandered through the ocean for 32 years and we will never return to our Promised Land."[29]

CHAPTER TWO

The Strontium-90 Odyssey

EARLY IN 1955, following the Bravo nuclear test, John Bugher chaired a secret meeting convened by the Atomic Energy Commission. Scientists had gathered to review the effects of the first thermonuclear bomb tests in the Pacific, and Bugher's opening comments conveyed a sense of urgency. "The picture we had two years ago of being able to continue in a reasonable measured pace with respect to problems in environmental contamination has obviously not been possible to maintain. That is, events have overtaken us, and we must [accelerate our work;] weapons have taken an enormous leap in energy release during the intervening time."[1]

Bugher then asked Willard Libby to elaborate. Libby was a nuclear physicist at the University of Chicago who specialized in understanding what happened to radioactive particles after nuclear bombs exploded. He had been appointed in 1953 as one of the five AEC commissioners, and led Project Sunshine, a commission effort to understand the environmental and health consequences of nuclear testing and warfare.[2] Curiously, Libby had never witnessed an atomic explosion at this point in his career, and he claimed publicly that he never wished to.[3] Dubbed "Wild Bill of the Atom" by some due to his outspoken advocacy for nuclear testing, he believed that as weapons became more powerful, the probability of warfare declined. Project Sunshine received whatever research money its leaders requested as long as it fulfilled two requirements: rapid progress and complete secrecy.

When the project began, neither Libby nor anyone else knew what happened to radioactive particles following nuclear explosions. Nor did scientists understand the threat they posed to health or how human susceptibility varied by age or intensity of exposure. Yet this knowledge was crucial to protect against unnecessary contamination and human exposure, and importantly to plan to manage the effects of nuclear war or a serious accidental release of radiation. By 1953, however, Libby had come to believe that strontium-90, one of nearly two hundred radionuclides produced by nuclear explosions, posed the most serious threat to human health—in part because of its very long, twenty-seven-year half-life. (That is, after twenty-seven years, the isotope would still be emitting half of its original radioactive energy.)

Libby was asked at Bugher's meeting to provide an overview of critical gaps in the scientists' understanding of the threat posed by fallout. He quickly pointed out that little was known about human contact and emphasized the importance and difficulty of sampling human tissue for the presence of strontium-90. There could be no better metric of human exposure to fallout than the testing of human tissues. But how could they obtain enough samples to know with confidence whether fallout varied significantly within the United States, and around the world? Libby had come up with a rather gruesome plan, which he shared with his colleagues: sampling tissues from the recently deceased. "We hired an expensive law firm to look up the law of body snatching. . . . It is not very encouraging. It shows you how very difficult it is going to be to do it legally . . . if anybody knows how to do a good job of body snatching they will really be serving their country."[4]

By 1953, the government-sponsored scientists had already begun to test strontium-90 levels in cadaver bones. Most bone samples were collected by the researchers' personal contacts within medical schools or public morgues. Some were stillborn samples, others were legitimately donated, and still others were transferred to New York or Chicago with unknown histories. Libby and his colleagues at Lamont Observatory in New York were especially interested in bones from fetuses and babies, because during the atmospheric testing era, skeletal formation in utero occurred under constant maternal exposure to strontium-90. How would fetal, infant, and maternal concentrations of strontium-90 differ, and how might that variance be explained? Was fallout more severe in some parts of the country than others, and if so, why? Could pregnant women, fetuses, and children somehow be protected from strontium-90 by varying or restricting their diets? Given the intensity of the AEC's atmospheric testing program, and the very real possibility of nuclear warfare, these questions haunted Libby and others at the AEC.

The process of tissue collection proved complicated. The secrecy that surrounded all AEC activities meant that the scientists' purpose could only be revealed to those with high-level security clearances. The AEC commissioners and staff struggled over what to tell the doctors and other officials without security clearances, especially those who would need to authorize the shipment of body parts. They finally decided to lie, telling them the government was conducting a study of radium in human bone to better identify a threshold of exposure that would induce bone cancer. As Robert Dudley from the AEC wrote to the laboratories that would conduct the analyses: "In order to keep the AEC out of the picture where possible, we intend to have the samples shipped directly to you."[5] "As for the emphasis on infants, we can say that such samples are easy to obtain here, and that we would like to keep our foreign collections comparable."[6]

By 1960 Project Sunshine scientists had successfully obtained tissues from nearly 15,000 bodies found in New York, San Francisco, Houston, Chicago, Japan, India, South Africa, Brazil, Colombia, Peru, Chile, Bolivia, and numerous European nations. Permission to take the body parts was neither sought nor obtained from next of kin before bones were "cleaned," packaged in formalin, shipped to New York or Chicago, crushed, and finally analyzed by the laboratory technicians. Sometimes "pooled samples" were created from some regions, meaning that body parts from different individuals living in the same area were combined. No federal policy then existed to require permission, despite the warnings of the law firm secured by Libby about the illegality of body snatching.[7]

Libby understood that Project Sunshine was crucial to the success of the AEC weapons testing program. Public support for the arms race depended on popular belief that radiation threats from testing were insignificant, especially when compared with the anticipated devastation of a nuclear attack. Yet the Project Sunshine scientists were quickly patching together data that demonstrated the presence and movement of strontium-90 in rain, soil, water, plants, animals used for food, and humans— evidence that could threaten the AEC's efforts if widely understood. If the public realized that radioactive particles from U.S. weapons became concentrated in children's bones, the weapons development program could be jeopardized.

Libby and other AEC commissioners relied on the authority given them by Congress to prevent scientific secrets from leaking to the Soviets. Instead of releasing their discoveries to the public, they crafted a story that played down the risks and prevalence of radiation exposure, and exaggerated the need to build more and larger weapons.

Indeed the highest levels of strontium-90 were found in the bones of one-year-old children; these toddlers had levels four times higher than those found in adults. The finding was explained by the rapid rate of human skeletal growth during development; human bones grow fastest in utero and during early childhood, with another surge occurring during adolescence (Figure 2.1). Adult bone that had formed before atmospheric testing contained far lower levels. But given the competing testing programs of the United States and Soviet Union, where was the isotope coming from?

STRONTIUM-90 IN THE ATMOSPHERE

The drive to identify the most important sources of human exposure to strontium-90 was fueled by two concerns: a desire to continue weapons testing with far less risk to public health, and a wish to protect the public from radiation from contaminated foods and water in the event of a nuclear war or serious nuclear accident. By the mid-1950s, it was clear that fallout from the explosions was causing widespread contamination: strontium-90, iodine-131, and cesium-137 were found wherever the elements were sought.

But the primitive nature of the environmental testing conducted during the early 1950s, as well as a poor ability to forecast weather patterns, meant that it was nearly impossible to predict the timing and patterns of fallout with the precision needed to prevent dangerous exposures. For example, after the 1945 Trinity test at Alamogordo, New Mexico, Warren Shields, head of radiological safety for the Manhattan Project, recommended to General Leslie Groves, head of the Manhattan Project, that some future tests be conducted a mere 150 miles from human populations.[8] The larger blasts in the Marshall Islands had led scientists to believe the maximum distance might need to be 600 miles. Moreover, even

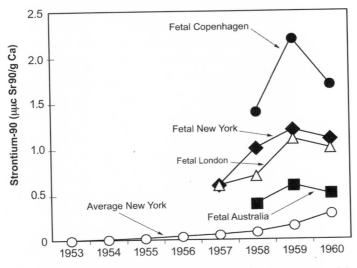

FIGURE 2.1. Fetal bone levels of strontium-90, 1953–1960. Unsuccessful pregnancies provided federal scientists with an opportunity to study fetal bone concentrations of strontium-90. Surprisingly, levels in fetuses far exceeded those in adults, a fact eventually explained by the movement of the radionuclide across placental tissues, its uptake by bone, and rapid rates of fetal bone formation. *Source:* U.S. Atomic Energy Commission.

these misguided recommendations were not followed. During a search for the best location for a continental nuclear weapons testing facility, President Truman, after considering sites in Maine and the Carolinas (among other locations), chose the Nevada Test Site, which is located only 65 miles northwest of Las Vegas.

The enormity and intensity of a thermonuclear blast are difficult to fathom. When a bomb greater than one megaton explodes, the extreme heat immediately following the blast lifts millions of tons of radioactive debris into the stratosphere at a rate of hundreds of miles per hour, with the largest weapons producing clouds that extend as high as thirty miles.

The threat to human health posed by this radioactive debris depends on how quickly the particles settle on the earth's surface. The largest particles tended to settle to earth a few hours after the test blasts, whereas smaller particles would remain suspended, moving with the weather. Prevailing winds tended to blow the dusts to the north and east of the Nevada

Test Site; while the test site was active, portions of Custer County, Idaho, received the greatest exposure to atomic radiation in U.S. history.

The isotopes' rate of decay affects the toxicity of the debris. Iodine-131 has a half-life of about eight days, cesium-137's half-life is 30.2 years, and strontium-90's is 27 years. The longer decay rates of cesium-137 and strontium-90 mean that they pose the most significant long-term threat to human health.[9]

By 1953 government scientists recognized that wind direction varied by altitude, and that radioactive material from a single blast would move along different pathways at different heights and speeds. This meant that a precise understanding of exposure patterns would require monitoring carefully and simultaneously both the climate and the distribution of radioactive contamination. But until 1954 responsibility for monitoring this fallout with the primitive methods of the era rested with the U.S. Army and the Los Alamos National Laboratory, hardly impartial actors.[10] Before each test the government would set out near the blasts roughly a hundred glass plates each one foot square and covered with a gummed substance to capture the particles falling to earth. Several hundred samples placed in close proximity to the test did not provide even a general understanding of national or global patterns of fallout, but it did provide a rough guide to some particles' direction and speed.

Also missing was an understanding of the relation between fallout and rainfall; for the first decade of atmospheric testing all tests (with the exception of several Pacific blasts) were conducted during fair, dry weather. That strontium-90 would be washed from the atmosphere by rain was well recognized by Libby, who had found strontium-90 in Chicago's rainfall by 1952. By 1953 scientists at the Lamont Observatory in the Palisades had identified the radioisotope in New York rainfall. Soon scientists were discovering radionuclides in rain wherever they tested, although the concentrations varied dramatically across space and over time, due to the ever changing influence of wind, humidity, and temperature on the movement of the dusts.

Since rainfall is the predominant means by which strontium-90 is removed from the atmosphere, one might expect surface contamination to be greater in the wettest areas if the element was evenly distributed in the atmosphere. Between 1958 and 1960, in the Northern Hemisphere, ground deposition of strontium-90 was highest between latitudes 40 to

60 degrees north.[11] Within these latitudes lie both Shell Mera, a community at the base of a mountain in Equador, and nearby Guayaquil. But foods grown in Shell Mera, with its average rainfall of nearly three meters per year, had nearly ten times the strontium-90 of foods grown in nearby Guayaquil, which has an average rainfall of only twenty inches per year. Similarly compelling evidence showed that several indigenous groups living in the Upper Amazon Valley, which has rainfall levels of more than five meters a year, had six times the amount of strontium-90 in their diets than did the average American during that time.[12]

The amount of strontium-90 in soils was also recorded during the Cold War era. Not surprisingly to scientists today, levels in the United States were higher than those recorded in other nations, with the greatest concentrations documented in 1958 and 1959 coming from areas in New York City; Coral Gables, Florida; and Tulsa, Oklahoma, all far from the Nevada test site. Fallout levels in New York City doubled between 1954 and 1958, when dozens of atmospheric tests were conducted by the United States and Soviet Union.[13] Another soil-related hot spot turned out to be mountain peaks. New Hampshire peaks that were normally surrounded by clouds had captured higher levels of radionuclides, which were measurable in the soil, vegetation, and even the snowpack.

Elsewhere, too, snow trapped radioactive dust from the atmosphere, causing snowpack, glaciers, and polar ice sheets to become reservoirs for the radionuclides. Buffalo, New York, was called a "radioactive sewer" in 1953 due to the contamination in its lake-effect snowfall. In the northern United States, when the winter snowpack melted in the spring, it released its store of radionuclides into soils, streams, rivers, lakes, and the oceans, where it entered the marine food chains and concentrated in some fish. These seasonal pulses from snowmelt also influenced human exposure, since most Americans received their drinking and wash water from surface water, and radionuclides were not removed by available filtration methods.

While prospecting for uranium in Belle Fourche, South Dakota, in July 1957, Edward Lindstad was surprised when his Geiger counter began clicking furiously. Initially annoyed with what he thought was his malfunctioning meter, Lindstad nevertheless began to pay attention to the exceptionally high readings and notified local health officials, who in turn contacted the Atomic Energy Commission. A chemistry professor at the South Dakota School of Mines also detected the airborne particles at around the same time.

Lindstad had recorded radioactive fallout from the Diablo atomic test at the Nevada Test Site two days earlier. The explosion had occurred nearly a thousand miles to the southwest of Belle Fourche. The radioactive cloud of dust and debris had drifted over Colorado, then moved toward Wyoming before enough moisture had accumulated in the atmosphere to produce the thundershowers that then swept across South Dakota.

The incident at Belle Fourche was not the first time that AEC officials had been surprised by the extent and location of radioactive fallout. Following a 1953 atmospheric blast, only a light fallout had been expected at the test site, but civil defense authorities measured levels high enough at St. George, Utah, 135 miles away, that residents were advised to remain indoors. Pedestrian and motor vehicle traffic were halted, "forcing 5,000 people to take shelter for three hours" while a radioactive cloud passed overhead. Cars, trucks, and buses were then checked for radioactive particles on their tires as they passed through roadblocks. If levels were high enough, the vehicles were required to pull into certain gas stations where the radioactive particles would be washed away.[14]

The Belle Fourche incident prompted a national debate over the severe hazard that U.S. citizens had experienced without warning. The *Washington Post* reported: "The AEC says it has no record of any other case of heavy fallout but admits it has not been following the situation closely. AEC officials argue that it would take a 'tremendous' amount of manpower and money to keep a close check on atomic test effects all over the country." These excuses implied that no one knew where "hot spots" were. And the AEC had little incentive to find them, given the public anxiety their identity created.[15]

The specter of hot spots captured the public's attention and became the subject of hearings held by the Joint Congressional Committee on Atomic Energy in 1957 and 1959. These hearings provided a remarkable summary of the AEC's history of understanding strontium-90 deposition as well as the commission's internal debates over the biological effects of radiation. Through the hearings, the public learned that Lindstad's accidental discovery had not been completely unexpected: AEC scientists knew that during 1957, no fewer than fifteen nuclear explosions in Nevada had produced radiation that had drifted over the Dakotas. Two additional shots fired during late July had also produced radiation that followed the same pathway, but because during July the area between the Dakotas and Nevada tended to have little rainfall, by the time the radioactive clouds reached North Dakota they were still fully loaded with the isotopes. The strong Midwestern thunderstorms that would often carry the radiation to the earth at that point also could produce an intense and rapid washout. Strontium-90 levels in the Red River jumped to nine times the "maximum permissible level" for drinking water following the July tests. If the contaminated cloud did not meet a rainstorm, it would continue to drift eastward with the prevailing winds, explaining why fallout levels in Wisconsin, Michigan, New York, and Connecticut were also periodically high.[16]

The Joint Congressional Committee on Atomic Energy again held hearings in 1959 to review the state of knowledge about fallout. These hearings were far more revealing than those held in 1957. For the first time, maps were released to the public demonstrating variable pathways of radioactive fallout tracked from the Nevada tests. More than a dozen maps are printed in the transcript and demonstrate how difficult it was to forecast the movement of debris as it moved across the nation and where it might fall to earth. Figures 2.2 and 2.3 re-create two of these, and illustrate the route of radioactive dust movement from tests conducted during the summer of 1957. The maps demonstrate that the dusts moved in different directions at different elevations, and that speed varied with height, although they oversimplify the particles' complex three-dimensional movement through the environment.[17]

The government's limited ability to predict fallout patterns is illustrated by one official at the AEC who concluded in 1956, "Our ignorance in this field is so great that we cannot say with any certainty that we have not already put so much strontium-90 into the stratosphere that harmful fallout is now inevitable."[18] Moreover, assumptions were made that turned out to be incorrect. Libby hypothesized that the hydrogen bombs would force the majority of their radiation into the stratosphere, which would act like a "reservoir," slowly releasing the particles to produce uniform concentrations around the world. This presumption of uniform contamination led Libby and the AEC to average estimates of human exposure and health hazard, even in the face of conflicting data from the U.S. Weather Bureau.[19] In response, Weather Bureau scientists collected their own data

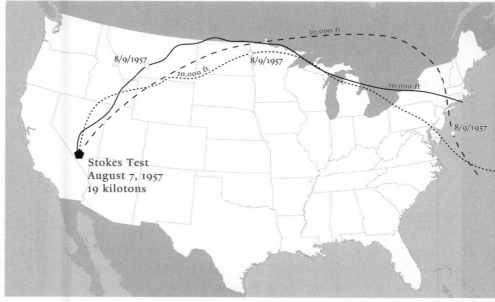

FIGURE 2.2. Pathways of radioactive debris from Nevada nuclear test Stokes, 1957. The Atomic Energy Commission realized by 1957 that radioactive debris traveled long distances in the atmosphere. These maps demonstrate the different pathways followed at various elevations within the United States. Debris at higher altitudes traveled farther than that which remained lower in the atmosphere (see the data for August 9), and deposition was most intense where the dust clouds intersected low pressure systems, especially thunderstorms. *Source:* U.S. Atomic Energy Commission.

for presentation to the Joint Congressional Committee on Atomic Energy in 1957, when they announced that northern-tier states in fact received two to three times more fallout than southern-tier states.[20]

Lawrence Kulp, a geophysicist at Columbia University's Lamont Geological Observatory in New York and a central academic participant in the AEC's Project Sunshine along with Libby, estimated in 1957 that only 10 percent of the stratospheric strontium-90 would settle to the troposphere each year, where it would then drop with rain to the earth's surface. In addition, Willard Libby estimated a seven-year half-life for the isotope. The estimates seemed reassuring; it is less dangerous to have the particles' energy dissipated before reaching the ground, and according to this scenario, only roughly half of the stratospheric strontium-90 would settle out every five years. Indeed, the tests of megaton weapons conducted

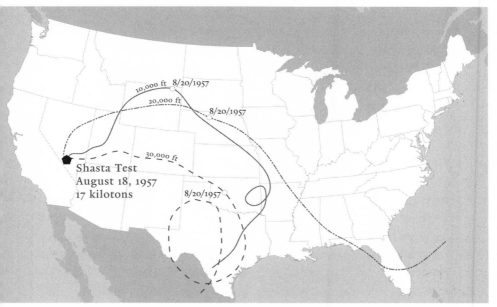

FIGURE 2.3. Pathways of radioactive debris from Nevada nuclear test Shasta, 1957. The Shasta test of 1957 demonstrated how debris could swirl unpredictably with the winds. Debris floated at 30,000 feet over Texas for nearly a week. *Source:* U.S. Atomic Energy Commission.

by the United States and the Soviet Union between 1958 and 1959, and then again in 1962, produced an accumulation of strontium-90 in the atmosphere that took nearly eighteen months to begin to decline after aboveground testing ceased in 1963.[21]

Although the Congressional Joint Committee on Atomic Energy recognized the existence of "hot spots" of radioactivity, where fallout had led to intense contamination of soils and plants, the AEC admitted as late as 1960 that it had little capacity to discern these more highly contaminated patches and instead was still relying on average exposure estimates. As the commission reported about its budget for that year: "A massive sampling program designed to give detailed coverage of every case of higher-than-average level of radioactivity is beyond our present fiscal and physical capability."[22]

What happens to fallout when it settles in the oceans? Libby and others expected that the seas would have an infinite capacity to dilute the radionuclides to safe levels. Instead, they discovered that soluble fission products normally linger in the top one hundred meters of the ocean's waters, where, significantly, many species of commercial fish spend their early development and absorb radioactive elements while feeding.[23] Because of such findings, Project Sunshine scientists began to focus on the importance of the human food supply as the most worrisome pathway for human exposure. Since the diet of Americans varies by age, ethnicity, season, and region of the country, food intake surveys became an important research tool to explain variability in human exposure.

DISCOVERING THE SOURCES OF HUMAN EXPOSURE

As the scientists expanded their monitoring network, they gradually realized that radionuclides encircled the earth and that no area was escaping strontium-90 deposition. While involved with Project Sunshine, for example, Kulp was finding the nuclide in nearly every bone sample tested, often with highest concentrations found in the youngest children, but he could not identify its sources. In addition, it seemed that hot spots of contamination were caused not only by wind and weather, which the scientists already knew, but also by the diets of the animals and people in various environments. Alaska experienced higher fallout levels than many parts of the Northern Hemisphere due to Soviet atmospheric testing in Siberia and the Arctic, and fallout sometimes reached the state from the Nevada Test Site. Concentration of strontium-90 in Alaskan caribou meat was found in 1959 to be ten times higher than in other meats because the animals liked to eat lichen and mosses, which had readily absorbed the radionuclides. In turn, the concentration of strontium-90 in urine among Alaskan indigenous groups that ate large amounts of caribou meat was four times the average levels of Americans living in the lower forty-eight states.[24]

Strontium-90, iodine-131, and cesium-137 may all be absorbed by plants, animals, and humans. Once deposited by rainfall or as dry particles, they tend to remain in the upper two inches of soil, unless lands are plowed or the soil is somehow disturbed. Most strontium-90 is biologically available, meaning that it is easily absorbed by plants either directly from exposed surfaces, or taken up by their roots. Its movement into grains, vegetables, and fruits depends on many factors, including the chemical properties of the soil, the species of plant, former deposition of strontium-90 on the site, patterns of rainfall, and soil characteristics.

The Lamont team demonstrated that under some conditions liming the soil to enrich it with calcium prevented some plants from absorbing strontium-90. If levels in pasture grasses could be reduced, grazing animals would be less exposed—that is, unless fallout that landed directly on the exposed surfaces of plants contributed significantly to the total strontium-90 ingested, which turned out to be true.[25]

Radionuclides in the U.S. food supply became a serious concern during the late 1950s and 1960s—at least among those who knew there was a problem. As the frequency of testing and the size of weapons increased during 1957 and 1958, General Mills and other major food processors became concerned about the quality of their raw grains, fruits, and vegetables. Fallout was present in more samples and concentrations were increasing. Although both public and corporate officials claimed that the tainted foods were not considered "an immediate hazard to health," the possibility that a public scare would cause a "large-scale dislocation of the economy" (and harm the food-processing industry) caused alarm.[26] President Eisenhower was informed of the problem as well when he was advised that nuclear tests in 1959 had contaminated wheat in nine states, doubling the human dose. Wheat is an ingredient of many processed foods such as breads, cereals, pasta, crackers, cookies, and cakes, and many of these are consumed in large amounts by children.[27]

MOVING UP THE FOOD CHAIN

Both Libby and Kulp knew by 1953 that animals absorbed strontium-90 from grasses and grains. Its discovery within animal bone led them to worry that small bone fragments, common in ground beef, could be an important source of strontium-90 in the human diet. Strontium-90 tends to follow the same metabolic pathway as calcium due to its similar chemical structure. The human body prefers calcium to strontium-90 and "discriminates" against its absorption in the intestine, causing some of the isotope to be eliminated by the kidneys in urine. Consequently, those exposed to strontium-90 generally absorb far less into their bones than what they actually ingest. Kulp reasoned that if enough bone material could be analyzed, scientists could back-calculate exposure from fallout with considerable precision.[28]

Project Sunshine scientists increasingly focused attention on dietary patterns, knowing that calcium is a fundamental building block of human

life and the skeleton. The Columbia research team began to explore concentrations in foods consumed in New York, and found the highest levels
in milk, followed by fresh vegetables, bakery products, flour, fresh fruit,
and potatoes.

By 1960 milk accounted for 61 percent of the average New Yorker's
strontium-90 exposure from food. A comparative food survey of low-
income families in Puerto Rico found the highest strontium-90 concentrations in plantains, beans, milk, mangos, rice, and bread. Average total
intake of strontium-90 among Puerto Ricans for the same period was
less than 30 percent the level of New Yorkers.[29] Yet neither study took
into account differences in diet among certain susceptible groups such
as pregnant women, infants, and children. Nor did they consider where
the food had come from (with the exception of milk, which tends to be
consumed near the producing dairy farm).

As far back as the 1940s, the presence in milk of environmental
contaminants such as pesticides and veterinary drug residues was well
known. DDT was first detected in dog's milk in 1945, and penicillin was
found in human and animal milk by that time.[30] But when did the government know that radioactive iodine, cesium-137, and strontium-90 would
move into the animal and human milk supply? Healy's detection of Soviet
atmospheric testing in 1949 demonstrated that the government knew the
potential for global circulation and deposition of radionuclides. Government scientists also understood the potential for iodine-131 to concentrate
in the thyroid of mammals, and especially in humans, as early as 1945.
The absence of milk testing by the government, and of special precautionary measures for milk related to the schedule of tests, is difficult to
explain. Were such oversights due to purposeful neglect, or was the safety
of the milk supply simply one among the many ecological questions that
scientists had not yet asked?

By 1954, scientists were reasonably certain that milk was an important
source of radionuclides for humans. Scientists from the three laboratories
working on Project Sunshine reported in July of that year: "It is reasonable
to expect human intake of I-131 to be largely through milk. Data obtained by
Brody at the University of Missouri indicate 10 percent iodine transmission
into cows' milk. Experiments at Hanford [a plutonium enrichment plant
in Washington State] on sheep suggest that approximately 20 percent of
the iodine ingested by these animals was transmitted to their milk."[31]

They also concluded that "the extent of Sr [strontium-90] incorpora-
tion into milk is an important problem, since milk is the most important
source of Ca [calcium] in the American diet."[32] Nearly a decade later,
AEC scientists at Hanford conducted an experiment by placing iodine-131
in dairy cattle feed. They found that radionuclides concentrated in the
cattle's thyroid and were excreted in milk. The milk was then consumed
by human volunteers to determine the rate of accumulating iodine-131
in the thyroid gland, and the potential of nonradioactive iodine to block
thyroid uptake of this isotope.[33] The test was essential: if milk was contaminated, certainly children would be at special risk due to their high
milk intake, and the relatively small size of their thyroids, where the
iodine-131 concentrates.

But why did it take so long to test whether the nuclides would concentrate in the milk of cows that grazed on contaminated pastures? It is
now well understood that the most important pathway of human exposure
to iodine-131 is from tainted cows' milk. The Department of Energy now
claims that it did not understand the iodine-131 exposure pathway at the
time of these experimental releases.[34] Yet iodine contamination near the
Hanford facility was recognized in 1945, and stack releases of iodine-131
were well measured, as were worker exposure levels.[35]

Moreover, by 1953, still years before the formal experiment on human
volunteers, the Lamont scientists were testing for the isotope in cows'
milk, the grasses they ate, and their underlying soils.[36] They learned that
cows fed in pastures acquired strontium-90 from alfalfa and excreted it
in their milk. The cattle that were fed grains in feed lots tended to have
lower levels of strontium-90 than those that grazed, especially if the feed
had been grown outside of areas receiving the greatest fallout. These
zones tended to be ellipses that began at the Nevada Testing Site and extended northwest toward Utah, Colorado, Wyoming, parts of Idaho, and
the Dakotas. Yet these hot zones were not easily predicted, and deposition
patterns varied from test to test due to shifting weather patterns.

Libby reported to his colleagues that during the summer of 1954, he
began to explore techniques to purify milk. "Last summer in Berkeley I
took a week off and worked on the problem of purifying milk. We took
milk and loaded it with strontium isotope and calcium isotope for tracer
and removed the alkaline earths with ion exchange resin, first by using a
column, and then by just stirring the powder. Stirring the powder worked

very well. It was possible to remove the calcium and strontium from the milk essentially completely, by adding, I would say a couple of tablespoonfuls of resin to a pint of milk. . . . Milk is the main way strontium gets into us, and I think it certainly should be followed up. Of course you have to fortify it with calcium afterwards."[37] Several days after his vacation, Libby, then a commissioner of the AEC, recommended an engineering and economic study of national milk purification.[38]

In June 1957, Herman Kalckar, a scientist at the National Institutes of Health, proposed a worldwide project to test children's baby teeth for strontium-90.

> The Public Health Service in every nation and especially in the large nations such as the US, USSR and China should organize, perhaps on a country-wide basis, a colossal collection of milk teeth . . . and do radioactive measurements on this material. . . . A possible occurrence in the rise of radioactivity in children's teeth structure should soon be detected. The results should be conveyed to the public without unnecessary interpretations which might either raise complacency or fear, but rather in a spirit so as to encourage a sober, active concern. Any family, be it in America or in Russia, who saw for themselves how the teeth of their younger children grew more radioactive would soon, out of concern for the fact that their own children carry the main burden, develop a realistic restraint on the question of "atomic might." Moreover, they would have a natural incentive for taking an active constructive part in the constant search of our society for affecting atomic disarmament.[39]

Libby responded favorably: "I think your idea of using children's teeth for strontium 90 measurement is a good one. However, I would not encourage publicity in connection with the program. We have found that in collecting human samples publicity is not particularly helpful."[40] One wonders whether if the AEC had collected children's teeth rather than human bones from around the world, it might have developed a more accurate understanding of strontium-90 exposure and body burdens.

In 1959 the Consumers Union collected milk from forty-eight cities in the United States and two in Canada. The group found levels of

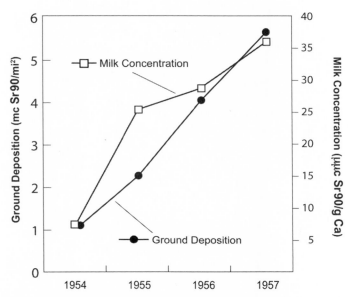

FIGURE 2.4. Concentrations of strontium-90 on the ground and in milk, New York, 1954–1957. Scientists with the Atomic Energy Commission correlated a rise in strontium-90 concentrations in ground detectors and milk in New York. They did not understand how long radioactive particles would continue to fall from the sky, and became concerned by the data showing that as testing intensity increased during the mid-1950s, so did the radioactivity of some foods such as milk. *Source:* U.S. Atomic Energy Commission.

strontium-90 higher than previously reported, but more importantly, they found considerable geographic variability.[41] As Figure 2.4 shows, strontium-90 levels measured in New York around the same time were elevated as well.

Government scientists' delays in investigating further the clear links between contaminated milk and human exposure to radioactive isotopes must have been motivated by something other than protecting the public or cost savings. When compared to the cost of the weapons development program, the cost of food testing was virtually zero, and could not possibly have justified the low budget and poor-quality milk sampling program that prevailed through the era. Perhaps the scientists were concerned that more precise predictions of fallout, and more compelling connections between nuclear blasts and the contamination of people—

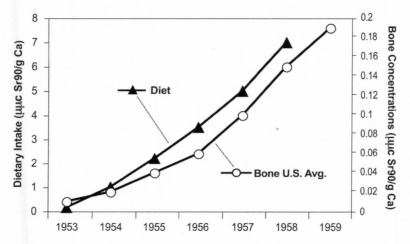

FIGURE 2.5. Strontium-90 dietary intake and bone concentrations, 1953–1959. Government scientists found rising bone concentrations of strontium-90 during the mid-1950s. These levels were correlated with increased weapons testing and growing food contamination. *Source:* U.S. Atomic Energy Commission.

especially children (see, for example, Figure 2.5)—would empower those local groups that opposed testing.

HUMAN BREAST MILK

The AEC's New York office first tested human milk in 1953 for the presence of strontium-90. The result was the first credible evidence that mothers transferred radionuclides to nursing children. This important finding, however, went unreported in the scientific literature and the national press; the AEC scientists buried their discovery in a poorly labeled table published within a 1957 Joint Congressional hearing report.[42] In the test, samples had been taken from mothers in Los Angeles and San Francisco, areas unlikely to have received a great deal of strontium-90, since prevailing winds from the south and west tended to blow the Nevada Test Site fallout away from California to the north and east. Fallout from the Pacific Testing Grounds regularly reached California, but as the contamination in Belle Fourche, South Dakota, had demonstrated, fallout depended primarily upon rainfall, and much of California is arid.[43]

Iodine-131's shorter half-life of eight days makes it less of a threat

than strontium-90 or cesium-137. Medical experiments conducted by
AEC-funded scientists in 1951 demonstrated that iodine-131 passed easily
from breast milk to infants' thyroids. They also demonstrated that stable
iodine can saturate the thyroid and thereby prevent women's absorption
of iodine-131, protecting the nursing infant.[44]

The response by the AEC to the threat to human breast milk was
predictable given its tradition of secrecy and experimentation. When the
commission found strontium-90 in cow's milk, it sought ways to reduce
human exposure without informing the public. It experimented with
methods to decontaminate milk using resins; it fed cattle dried feed grown
when the fallout was believed to have been less severe, while prohibiting
cattle from grazing in open fields; and it diluted contaminated milk with
pure milk. The commission also experimented with calcium fortification
of foods, to block bone absorption of strontium-90, and with the application of calcium to soils, to prevent crop absorption. Not until 1962 was
the issue of childhood exposure to contaminated breast milk addressed
directly, in a paper by Allan Lough of the Atomic Energy Commission:
"It has been observed that in the human lactating female only about 1.3
per cent of ingested strontium 90 appears in the milk . . . Observations
in human infants have shown, however, that at very early ages there is
much less discrimination against strontium. Indeed the ratio of strontium
to calcium in the bones of very young children who died in early 1961 is
about four times as great as in adults who died in the same period. . . .
The population group exposed to the highest concentration of strontium
90 in bone is made up of persons born within two years after the test
series which have been completed."[45]

The AEC's firm grip on the scientific information surrounding radioactive
contamination, including by strontium-90, finally began to waver in the
early 1960s. In 1962, tests of hydrogen bombs by the United States and
Soviet Union forced more radioactive debris into the atmosphere than
did all previous nuclear tests combined. In response to the resulting high
levels of iodine-131 circulating around the globe, and the associated risks
of thyroid cancer and other diseases, Britain twice considered prohibiting
milk from the marketplace in 1961 and 1962. During this period, milk in
Utah and Minnesota was voluntarily removed from markets, and cattle
were prohibited from grazing in contaminated fields; instead they were

fed grains and hay grown during less contaminated years. Surprisingly, the "hottest" spots in the nation at that time included Salt Lake City, Utah; Minot, North Dakota; and Palmer, Alaska. The fallout's unpredictability was unsettling, and may have resulted from inconsistent testing as much as variability in fallout patterns.

Researchers with the Department of Agriculture worked feverishly to find ways to undo the damage to the nation's dairy supplies. Vermiculite, a natural mineral, was added to cattle feed, and found to scavenge some of the radiation. Two test cows, Madame Curie and Sophie Tucker, were fed strontium-90 and then had their milk filtered through a resinous mixture of metallic salts, which effectively removed 98 percent of the isotope. But chemical treatments could introduce other contaminants into the nation's diet and so were not ideal.[46]

By 1962, President Kennedy had become alarmed that the nation was consuming two to three billion pounds fewer dairy products than the year before, despite a growing population of children. He announced that milk would be served at every White House meal, as a way of reassuring mothers that milk "offers no hazards." But almost simultaneously, levels of radioactive strontium-90, cesium, and iodine-131 were found in dairy products at unprecedented levels—and the news began to be noticed. As the levels of the radioactive isotopes continued to rise through most of 1963 and 1964—due to fallout from extensive thermonuclear tests in 1962 and 1963—the Public Health Service, not the AEC, was responsible for testing and reporting residue levels in foods, further eroding the AEC's control over the alarming information.[47]

The radioactive food supply challenged the future of the AEC's testing program. Yet the AEC encouraged public acceptance of continued testing with its narrative of insignificant risk, and with arguments that became ever more tortured. Yes, AEC officials reasoned, the food supply does contain fallout from weapons testing, but the amounts discovered after "extensive" environmental testing do not constitute a significant hazard; in fact, the risks were well understood and known to be less than the danger posed by natural radioactivity. Besides, the Soviets were conducting their own tests, and so must share considerable responsibility for global fallout. The agency also argued that atmospheric testing was necessary to learn how best to protect citizens in the event of nuclear war, and that the risk of misunderstanding fallout in the event of a Soviet strike would be far greater than that posed by insignificant residues in foods. Finally, the agency was "unable to disclose" details of its testing for independent verification, because, in its view, doing so would provide the Soviets with a competitive advantage in the arms race and the design of war strategies. While these arguments were persuasive to some, many others became angry after learning about the extent of environmental contamination, human exposures, and the special threats posed to fetuses and children.[48]

CHAPTER THREE

Experiments on Humans

WHEN GENERAL LESLIE GROVES asked the DuPont Corporation to become the primary contractor for the Hanford Works nuclear weapons production facility, Walter Carpenter, the company's president, hesitated and then demanded that a $20 million claims fund be established to protect against unforeseeable radiation-induced harm to workers. Carpenter eventually accepted the challenge with that caveat and two others: that the company be indemnified by the U.S. government and that the facility be located in a remote area, "for safety's sake . . . because of unknown and unanticipated factors."[1]

When the project began, the technology was so young and poorly understood that a 1946 DuPont report concluded that there had been "no established tolerance limits for certain of the hazards which would be encountered."[2] Even though the federal government had not established limits for human radiation exposure, DuPont created its own standard of 0.1 roentgens per twenty-four-hour period for total body irradiation, with the caution, "It is better to spread this dose out over as long a period as possible. When the tolerance level is exceeded, we can make no definite estimate of the time required for recovery. In other words, if a man receives 26 roentgens in one day we cannot say that by removing him from exposure for 360 days we would eliminate all possible trouble, as with such exposure irreparable damage might have occurred."[3]

The medical supervisor's most serious warning was that averaging exposures over long periods may not prevent significant damage. This argument would recur continually in twentieth-century environmental science, and ran counter to tendencies by government agencies to average the exposures over many years, which disguised the severity of intense short-term exposures.

Groves recalled his own reservations about the safety of nearby communities, concerns that centered more on the possibility of a dramatic accident than on chronic exposures: "Reactor theory at this time did not overlook the possibility that once a chain reaction was started, it could . . . get out of control and increase . . . to the point where the reactor could explode. . . . We knew, too, that in the separation of plutonium we might release into the atmosphere radioactive and other highly toxic fumes which would constitute a distinct hazard . . . I was more than a little uneasy myself about the possible dangers to the surrounding population."[4]

Herbert Parker, a British radiologist, led the health and safety program at Hanford. Prior to 1946, Parker studied the movement of radionuclides in air, water, vegetation, and soil. He then explored how iodine-131 was absorbed by livestock. Parker knew that iodine-131 was emitted from Hanford's stacks in particular amounts; the next step was to measure how much of the isotope had been taken up. He directed soldiers and technicians to roam the countryside near the plant in jeeps and trucks, stopping to rope sheep, cattle, and other domestic animals. The workers then held Geiger counters to the animals' thyroids to test them for the presence of iodine-131, recording the detected levels in their notebooks. Most of the measurements were taken without the knowledge or consent of the livestock owners. One of Parker's assistants recalled visiting farms and pretending to work for the Department of Agriculture as an expert in animal husbandry. The deception was useful: "I was successful in placing the probe of the instrument over the thyroid at times when the owner's attention was focused on the next animal or some concocted distraction."[5]

Parker worried that he and other Hanford scientists did not understand how the iodine-131 releases made their way to the animals' thyroids. He was concerned that Hanford workers, and others living in the vicinity of the plant, were also being harmed by thyroid absorption of iodine-131. Even less was known at the time about the long-term effects of chronic exposure to radioactive isotopes of iodine-131, cesium-137, and strontium-90—all isotopes present in the immediate environment.[6]

Most of those who worked at the Hanford plant lived in Richland, a government-owned town built primarily to house the military servicemen and contractors. One could not live in Richland during World War II unless employed by the AEC or one of its affiliated corporations. The government also regularly warned residents against drinking milk from the animals that grazed nearby; milk was instead imported from Yakima or other distant communities. The town's extensive milk-testing program was described as an effort to track both "bacterial count and butterfat content. . . . And although more distant cities obtained their drinking water from the Columbia River, Richland purposely relied on well water, because the scientists at Hanford knew what contaminants they were discharging directly into the river. This intensity of monitoring, the restrictions, and the warnings collectively imply that the government knew something about environmental contamination and especially the vulnerability of the milk supply, all information kept from residents.[7]

Those who managed the Hanford Works struggled with the ethical dilemma of knowing that the hazardous materials they routinely released to the surrounding environment threatened the surrounding population—and of being unable to tell those people of the danger they faced. Residents of government-owned Richland were warned that they should be careful to grow lawns to prevent allergies that could be exacerbated by dusts. They were unaware that radionuclides released from Hanford's stacks had likely settled on Richland's soils, and were blowing in the wind. Residents were also prohibited from owning livestock such as chickens and cattle, due to concern that they might concentrate the radionuclides. And when the scientists found that sagebrush tended to absorb radionuclides, they told Richland residents to remove it from their property to reduce pollen-induced allergies. Together these policies were crude attempts to reduce the residents' exposure.[8]

GREEN RUN

During the late 1940s, scientists had monitored stack releases from Hanford, as well as from the Oak Ridge National Laboratory in Tennessee. While radionuclides were detected nearly fifteen miles from Oak Ridge, they seemed to extend only two miles from Hanford. These results might be explained by the differing sensitivity of detection equipment, or variation in the monitoring efforts between the two areas.

Hoping to track particles over longer distances, the Air Force designed an experimental release of radionuclides into the atmosphere at Hanford, code-named "Green Run," in 1949. The Green Run was only one of thirteen intentional releases in the United States, including experiments conducted at Los Alamos, New Mexico; Dugway, Utah; and Oak Ridge, Tennessee.[9] All of the tests were designed to better understand the movement and fate of atmospheric radionuclides. Normally fuel elements were cooled for ninety to a hundred days before being dissolved, which allowed for more radioactive decay and lower emissions of radiation. Elements for the Green Run test, however, were cooled only for sixteen days, and scrubbers, normally used to remove radioactive gases from stack releases, were turned off. Significantly, Healy had assumed that the fuel had cooled for twenty days, a miscommunication that was to lead to his further underestimate of the magnitude of xenon-133 that would be released.

On the day of the test, December 3, 1949, a weather front stalled near Hanford for several days, and radionuclides swirled slowly in clouds near the plant, settling in far higher concentrations than expected. In the end, nearly 8,000 curies of iodine-131 and 20,000 curies of xenon-133 vented from the stack, settling over large areas of southeastern Washington and northeastern Oregon. A radioactive cloud lingered over the plant and other communities within one hundred miles, including the city of Spokane.

The detection equipment itself became so contaminated that the Hanford staff realized their readings could not be true. Radiation levels in vegetation were found to be 400 times higher than concentrations that would injure livestock. A report of the experiment was prepared by the Atomic Energy Commission in 1950, but was classified as secret for nearly forty years. As Healy later recalled:

> At first, everything went fine. The wind was blowing from the south to the north, which was relatively uninhabited. The inversion held well. That was one of our criteria for dissolving [proceeding with the experiment]. But the problem was that at about two o'clock in the morning, we got a dead calm, and then the wind reversed direction 180 degrees. Generally, the

noble gases come out first and the iodine comes out a bit later.
About two o'clock in the morning, we had the calm; and then
the wind, instead of blowing to the north, began to blow down
the river over Richland, Pasco, and Kennewick[, Washington].
This, I believe, was the reason for the high [iodine] levels that
we got on the Green Run. . . . Because the iodine had deposited
on the ground. People would go out and back in, step on the
shoe and foot counters, and set them off. That disrupted their
operations very badly. I was very unpopular, believe me. . . .
My calculations showed that we managed to contaminate an
area about twice the size of the Windscale[, England, nuclear
plant] accident [in 1957], with about one-fifth of the amount
of iodine released. . . . When we replaced Bob Thorburn
from under the stack in the Green Run [in Washington State],
you could read him with a Geiger counter from here [in
New Mexico].[10]

Reflecting on the Hanford experiments decades later, Healy remembered:

The biggest problem, as it turned out in the long run, was that
Herb and I did not see the milk-to-man pathway. Although,
I had inklings that Herb later did begin to see the problem. . . .
The real thing was that it was a pathway from the air to the
ground. We knew that the iodine deposited very heavily on
the ground. We knew that from our own experience. There
was also the fact that A. C. Chamberlain in England had
done some experiments releasing iodine on a cricket pitch
and found that it deposited very heavily on the grasses. In fact,
when I went over to England for the Windscale review, A. C.
greeted me like a long-lost brother, because everybody else
doubted his experiments, but our experience backed his findings very closely. From the grass, the iodine went to the cow
and then to milk, which was then consumed by the most sensitive portion of the population: the children. Now, I'm sure that
if Herb had realized this pathway existed, he would have put
up a fight as only Herb Parker could. . . . I'll repeat that if
I hadn't been so ignorant, and had caught on to the air-grass-
cow pathway, I think the Green Run would never have taken

place. We had a Biology Group by that time. They were busy feeding iodine to sheep. But no one who really was in the business caught on.[11]

Questions about Green Run, as well as the degree of hazard posed by Hanford's day-to-day operations, would remain unanswered for decades. How much radiation was released from the plant's facilities? What concentrations of various radionuclides were emitted to the air, to the soil, and to the Columbia River? Was well water contaminated? Once released, where did the radionuclides go? How far could they travel? Could they accumulate in plant, fish, wildlife, and human tissues? What are the effects of exposure to different radionuclides, and other chemical compounds released to the environment as waste or pollution?

Due to the highly secret nature of the project, these questions were not publicly asked or debated, even if those Hanford scientists who knew the plant's purpose and dangers grew increasingly concerned. Still, such worries were eclipsed by the sense of urgency that grew from World War II, the Korean War, and the Cold War, conflicts that encouraged aggressive experimentation with nuclear weaponry. To national leaders facing threats to national security, the risks associated with these new technologies seemed reasonable.

THE STORY OF RADIUM

In the 1950s, the field of health physics—the study of how human illness could be induced by radiation—grew quickly. Although people were increasingly being subjected to radiation as part of medical diagnoses and treatments, and in occupational settings, little was known about the long-term effects of such exposures, which generally occurred at levels that produced no obvious and immediate symptoms.[12]

Radiologists and physicists within the AEC were well aware of the history of radium, especially the deaths of young women who had worked for the U.S. Radium Corporation in New Jersey during the 1920s. The "Radium girls" had worked from their homes, painting radium on watch dials to make them glow in the dark. The industry grew quickly during World War I to meet the military's demand for luminescent dials. But in 1926, a New Jersey dentist noticed that several of the young women were suffering from similar degeneration of their jawbones. The women, by licking their paintbrushes to make a point, had exposed themselves to disfiguring and sometimes lethal doses of radium.

During the 1920s, radium was widely touted as a therapy for conditions as diverse as nervousness, baldness, diabetes, and high blood pressure.[13] Patented medicines and therapies containing radium continued long after the watch-dial controversy was common knowledge. Radium capsules were inserted into the nose to treat children with middle ear infections, with little research on the long-term health damage that might result. The Navy and Air Force experimented with similar therapies for those suffering from inner ear disorders that worsened when air pressure changed rapidly. And in the 1940s, Johns Hopkins University researchers experimented with the capsules on several hundred children to understand the potential of radium to prevent deafness.[14]

The ambiguity of risk associated with radioactive isotopes, together with the growing impression of their diagnostic and therapeutic value, helped to lessen public anxiety over the nuclear weapons program. Moreover, health physicists were not yet able to identify safe levels of exposure to the unstable elements created by weapons testing. This uncertainty meant that those responsible for producing and releasing hazardous technologies and substances had little chance of being held accountable and responsible for adverse health effects. It was argued that if medical experts could not identify with precision a threshold between a safe and dangerous exposure, there was little need to control environmental releases, or to monitor what happened to the isotopes once they had been released into the air, soil, and water. By dodging this responsibility, too, the AEC perpetuated the cycle of inadequate information: the lack of surveillance left environmental concentrations unknown, which in turn made it impossible to estimate human doses. The absence of evidence provided cover for the AEC's aggressive and covert experimentation—a strategy rationalized by the scientists' belief that they were already pursuing the most essential goal, that of preserving national security.

EARLY EXPERIMENTS ON CHILDREN

The crude scientific experiments conducted by the AEC were extended to the realm of human testing: in particular, testing children for radiation exposures. Beginning in 1944, and continuing until 1974, the year before the commission became part of the U.S. Department of Energy, the U.S.

government sponsored pediatric research with radioactive compounds to better understand children's physiology, the pathology of the radiation in children, and the potential of newly minted diagnostic tools. The AEC conducted nearly 4,000 radiation experiments during this period, most often with the intent to use radioactive tracers to understand physiological processes, rather than the hazards to humans. The nature and extent of these studies was kept quiet until the mid-1990s, when a committee within the Department of Energy brought to light nearly eighty such studies, and further found that twenty-one of those supported by federal funds were "non-therapeutic," that is, offered no possible benefit to the participating child. The committee also found very little evidence regarding how authority was obtained to conduct these studies, leading it to conclude that authorization by parents or guardians was not likely sought.[15]

In fact, the AEC disguised these studies of radioactive compounds in the human body as "nutritional" research. In the late 1940s and early 1950s, Massachusetts Institute of Technology scientists participated in research to understand the relations between nutrient intake and radio-nuclide behavior in children. Students living at the Walter E. Fernald School, a Massachusetts institution for "mentally retarded" children, were fed breakfast food containing radioactive iron and calcium. Nine boys between the ages of ten and fifteen, and one twenty-one-year-old, were divided into two groups, with the first receiving an intravenous injection of calcium-45, and the second eating it in tainted breakfast cereals. Several letters were sent to the parents of students to encourage them to join the study; they were promised that the children would be welcomed into the "Science Club" if they agreed to participate.[16] The second letter, sent in May 1953, explained:

> Dear Parent: In previous years we have done some examinations in connection with the nutritional department of the Massachusetts Institute of Technology, with the purposes of helping to improve the nutrition of our children and to help them in general more efficiently than before. For the checking up of the children, we occasionally need to take some blood samples, which are then analyzed. The blood samples are taken after one test meal which consists of a special breakfast

meal containing a certain amount of calcium. We have asked for volunteers to give a sample of blood once a month for three months, and your son has agreed to volunteer because the boys who belong to the Science Club have many additional privileges. They get a quart of milk daily during that time, and are taken to a baseball game, to the beach and to some outside dinners and they enjoy it greatly. I hope that you have no objection that your son is voluntarily participating in this study. The first study will start on Monday, June 8th, and if you have not expressed any objections we will assume that your son may participate.

Sincerely yours,

Clemens E. Benda, M.D., Clinical Director and
Malcom J. Farrell, M.D., Superintendent[17]

The research was classified as "nutritional" with no mention of the children's exposure to radioisotopes, or the risks such compounds pose. The breakfasts involving a "certain amount of calcium" give no hint that the additive is radioactive. When some children resisted the offers, one of the MIT researchers wrote, "it seemed to [him] that the three subjects who objected to being included in the study [could] be induced to change their minds . . . [by emphasizing] the Fernald Science Club angle of our work."[18] The Massachusetts Task Force on Human Subject Research, which in the 1990s investigated the use of radioisotopes at state-owned facilities in Massachusetts, concluded that these practices violated the children's human rights.[19]

Another study was conducted in Tennessee at the University of Tennessee College of Medicine, and the John Gaston Hospital. In 1951, two African-American women, ages twenty-two and thirty-three, were administered iodine-131 while nursing their four-month-old infants. Both mothers had been diagnosed with thyroid disease. The infants continued to nurse, and the levels of iodine-131 were monitored in the mothers' breast milk and thyroids, as well as in the infants' thyroids. Observing that the infants' accumulation of iodine-131 in their thyroids was significant, the researchers warned against the administration of iodine-131 to lactating women. The study not only is appalling for its lack of concern for the mother and infant subjects, but also shows how early the AEC recognized that atmospheric fallout could contaminate mother's milk, and that this contamination concentrated dangerously in the thyroids of infants nursing from affected mothers.[20]

Researchers at Harvard Medical School, Massachusetts General Hospital, and Boston University School of Medicine conducted an experiment by administering radioactive iodine to seventy students at the Wrentham State School for "mentally retarded children" in 1961. The experiment was designed to understand whether nonradioactive iodine could block thyroid uptake of radioactive iodine-131. If it had a protective effect, health loss following nuclear war might be significantly reduced.[21]

In another study, conducted by a University of Rochester graduate student, seven subjects younger than twenty-one were placed on an iodine-restricted diet, then drank milk from a cow fed iodine-131 for a minimum of fourteen days. One of the children developed a benign thyroid nodule that was surgically removed.[22]

In 1953, pediatricians at Harper Hospital in Detroit administered iodine-131 to sixty-five premature infants, hoping to understand thyroid uptake and function.[23] Twenty-five newborns at the University of Iowa were also administered iodine-131, eight by oral administration and seventeen by injection, demonstrating more rapid thyroid uptake via injection.[24]

Researchers at the University of Iowa administered iodine-131 to pregnant women about to have therapeutic abortions in 1953. Following the procedure, levels of iodine-131 were measured in the embryos. From this experiment, the scientists learned that iodine-131 could cross the placental barrier, possibly endangering developing fetal thyroids.[25]

Between 1945 and 1949, scientists at Vanderbilt University Hospital administered the radioactive isotope iron-59 to 829 women who were between 10 and 25 weeks pregnant, discovering that the more developed the pregnancy, the more the isotope was absorbed. The children born were followed between 1964 and 1967, although no health problems were anticipated. Among the 679 children exposed in utero, one developed leukemia and two developed sarcomas. There were no malignancies in the control group, which consisted of 705 children.[26]

University of Utah researchers studied the metabolism of cesium-137 and rubidium-83 between 1965 and 1972. Subjects included five infants, five healthy children between five and ten years old, three children between the ages of four and eleven who had muscular dystrophy, six pregnant women, and twenty-three nonpregnant adults. The children with muscular dystrophy retained the lowest amounts of the two isotopes, supporting the researchers' hypothesis that cell membrane permeability was somehow associated with the illness.[27]

In yet another instance, researchers at Johns Hopkins University injected iodine-131 into thirty-four children ages two months to fifteen years who had been diagnosed with hypothyroidism. An unknown number of healthy children were also injected as a control group.[28]

Outrage over such unethical human experiments occurred only after Eileen Wellsome wrote a series of articles in the *Albuquerque Tribune* in 1993 identifying individuals who had been injected with plutonium during government-sponsored medical experiments. The subjects had been hospitalized for other ailments and had received the "treatments" without their consent, all in the name of Cold War national security. The outcome of the experiments was expected to guide the setting of standards for protecting workers who processed and disposed of nuclear materials. Hazel O'Leary, Secretary of Energy during President Clinton's administration, read Wellsome's 1993 articles and expressed shock that the agencies and officials responsible could have operated with such disregard for human rights. The Department of Energy she managed had grown from the AEC. She and others convinced President Clinton to create the Advisory Committee on Human Radiation Experiments to find and disclose all cases of government experiments with radiation. The committee's report resulted in the public release of hundreds of thousands of formerly classified documents. The cases provide insights into a secret society of government, industry, and academic scientists, and into the moral logic they relied on to justify their research. The cases also demonstrate the absence of government principles or standards to protect the rights of citizens who were exposed without their knowledge or consent.

The experiments conducted on children were especially haphazard in design, and overlooked essential information about dosages and safety. O'Leary's committee concluded in 1994: "It is evident that investigators using radioisotopes in children were not employing available information on organ weights in children to calculate tissue exposures at least until the mid-1960s. . . . [Thus] investigators may have significantly and systematically underestimated effective tissue dosages in children [that is, those dosages that caused a change in tissue]. It is notable that the highest levels of risk posed in the experiments reviewed were to infants administered iodine-131."[29]

In this context, it may seem surprising that the AEC's Isotope Division in 1949 created rules to evaluate research proposals submitted for government funding of radioisotope experiments on "normal children": "In general the use of radioisotopes in normal children is discouraged. However, the Subcommittee on Human Applications will consider proposals for such use in important researches, provided the problem cannot be studied properly by other methods and provided the radiation dosage level in any tissue is low enough to be considered harmless. It should be noted that in general the amount of radioactive material per kilogram of body weight must be smaller in children than that required for similar studies in the adult." Regardless of this recommendation, O'Leary's committee noted the absence of any requirement for informed parental consents, and that "important researches" and "harmless" were not defined by the AEC. Although all twenty-one pediatric studies examined in the early 1990s might be construed to be important, eleven of these experiments were found to pose greater than minimal risk to the subjects.[30]

ASSESSING CANCER RISK

By 1955 federal scientists understood that everyone in the nation had been exposed to the radionuclides released during nuclear tests. What remained uncertain were how much of a dose people located in various geographic regions had received, the relative potency of the most abundant radionuclides, and the differences in vulnerability of population groups, especially children, to these contaminants. Did radioactive isotopes produced by nuclear weapons testing and the experimental release of radiation cause mutations, genetic damage, and cancer? The answer requires an explanation of "dose-response" relations—that is, the relation between exposure and either cancer or genetic damage. For much of the past seventy years, generally two theories regarding cancer etiology have governed debates over acceptable levels of exposure to carcinogens. One theory suggests a "threshold effect," meaning that a specific dose must be exceeded to pose any risk of cancer, and below that dose there is little chance of health loss. The second theory, by contrast, claims that the relationship between exposure and cancer risk is linear: any exposure increases one's risk, and radiation risks accumulate through life. Of the two, the threshold theory is more comforting, since few people experience exposures that are high enough to cause immediate illness or death.

Not surprisingly, the AEC's risk assessments assumed the threshold-effect theory. It was far better for the AEC's long-term goal of nuclear-weapon development to support claims that low-level exposure was completely safe, and to allow the radionuclides to either remain undetected, or to be averaged with milk or grains that had been pooled among various farmers or regions before testing. Federal scientists also avoided the identification of "hot spots" where fallout and exposures might exceed safe thresholds.

Yet scientific evidence was building for the opposing theory. In 1956 a U.S. National Academy of Sciences (NAS) committee reported that any exposure is potentially dangerous, due to the isotopes' capacity to induce genetic mutations that are transmissible to future generations. The committee that prepared the report warned that nuclear war could make the earth uninhabitable. They also found that Americans, on average, were using up one-third of their lifetime radiation limit on medical and dental X-rays.[31]

Indeed, the debate over the safety of radionuclides increasingly focused on the concept of a "maximum permissible dose." In the mid-1950s this was defined initially for workers at energy facilities, and those who might be exposed occupationally in the military. In 1957, the AEC issued a classified report to the National Security Council that had more dire implications regarding a contaminated national food supply than was revealed to the public. "Certainly 450 roentgens received during a short period of time is lethal. Presumably, more than a few billionths of a gram of strontium, Sr90, received in the skeleton structure will also be lethal."[32]

Professor Barry Commoner testified in 1959 that understanding the health threat posed by strontium-90 would be difficult. "How harmful is the strontium-90 that comes to rest in the bones and teeth? There is no simple answer to this question. Since strontium-90 is so new we do not yet have any direct medical experience with its effects. Damage to humans might take 30 to 40 years to show up."[33] That same year, the Consumers Union collected milk from forty-eight cities in the United States and two in Canada. The group found levels of strontium-90 higher than those previously reported, but more importantly, they found considerable geographic variability.[34]

The government's failure to establish a public and national food monitoring program grew in part from the AEC's recognition that strontium-90 was discoverable in milk and bone almost whenever it was tested. The human experiments just described also reveal that private food companies had conducted their own secret tests of food contamination. While government officials feared loss of support for the weapons testing program, corporate leaders feared an international food scare and potential loss of profitability. The collective effect was widespread public ignorance and a gradual increase in tissue concentrations during the 1950s as the United States and Soviet Union tested ever more powerful weapons.

The damage to human health may seem insignificant when an individual's average dose is considered, but when the probability of cancer or loss of life is applied to the entire world's population, the numbers become more sobering. As Commoner testified,

> Dr. Edward Teller contended that the effect of fallout on health
> is trivial. To support this claim he estimated that the fallout
> from one large bomb will subtract from the life of the average
> American only three one-hundredths of a day, or about three-
> quarters of an hour. Most people would probably feel, he ar-
> gued, that three-quarters of an hour out of their life is an
> acceptably small price to pay for whatever national security is
> derived from this kind of a test. Assuming uniform distribu-
> tion of fallout, the total cost of fallout damage from one bomb
> to the entire human race would be 300,000 man-years. This
> is equivalent to 1 year off the life of 300,000 people. It is also
> equivalent to killing 10,000 people at the age of 40. The same
> calculations show that tests concluded up to 1957 will cost
> in human life more than the battle deaths in World War I.[35]

The U.S. National Cancer Institute (NCI) has developed the most definitive estimate to date of cancer risk associated with atmospheric testing. It concluded in 1997 that 5 billion people in the world were exposed to radionuclides released by nuclear weapons testing. Between 11,300 and 212,000 additional cases of thyroid cancer were likely to result in the United States alone from exposure to iodine-131 that resulted from the seventeen-year atmospheric weapons testing era.[36]

The NCI estimates, which broke down likely exposures by county

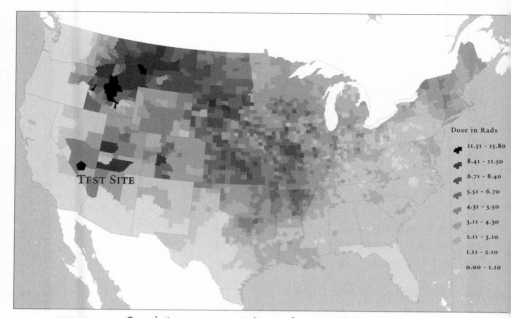

Dose in Rads

11.51 – 15.80
8.41 – 11.50
6.71 – 8.40
5.51 – 6.70
4.31 – 5.50
3.11 – 4.30
2.11 – 3.10
1.11 – 2.10
0.00 – 1.10

TEST SITE

FIGURE 3.1. Cumulative exposure to iodine-131 from Nevada tests, 1951–1962. Scientists with the National Cancer Institute and Centers for Disease Control and Prevention reconstructed spatial patterns of iodine-131 exposure attributable to the U.S. weapons testing era. Little fallout reached Pacific coastal states due to prevailing winds, and the patchiness elsewhere is explained predominantly by the fact that greater fallout occurred when the dust cloud met up with storms, and sometimes this did not occur until the debris had reached the East Coast or beyond. *Sources:* National Cancer Institute and Centers for Disease Control and Prevention.

depicted in Figure 3.1, were seriously hampered by the AEC's haphazard monitoring efforts during the testing era. For most atmospheric tests conducted in the United States, only gross beta radioactivity was measured; none of the isotopes—including iodine-131, strontium-90, or cesium-137—were monitored independently. The absence of monitoring data made it nearly impossible to understand precisely how and where the particles were moving and settling across the nation. Estimating exposure became a sophisticated guessing game, aided by computer simulations of possible radioactive releases, potential pathways across the nation, and weather patterns (Figure 3.2).[37]

The National Academy of Sciences reviewed the National Cancer Institute study in 1999 and agreed broadly with the dose and risk estimates.

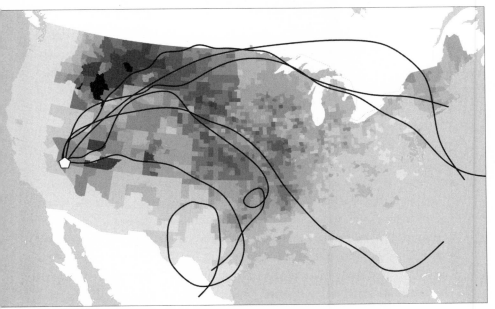

FIGURE 3.2. Overlay of Stokes and Shasta test debris pathways on U.S. exposure to iodine-131. Debris would often travel to the north and east of the Nevada Test Site. When the trajectories of debris from the Stokes and Shasta test are superimposed on the iodine-131 exposure map, there appears to be some relation between the data sets. This illustration presents a simplified image of deposition patterns that in fact resulted from many more tests, both by the United States and Soviet Union. *Sources:* U.S. Atomic Energy Commission, National Cancer Institute, and Centers for Disease Control and Prevention.

The NAS concluded that children who were very young at the time of the testing were most at risk for developing thyroid cancer from iodine-131 exposures; their thyroids are approximately ten times smaller than those of adults. Moreover, those children who routinely drank milk from cows that grazed on small farms were claimed to be at highest risk, since they were most likely to have consumed the milk shortly after milking, when iodine-131, with its short half-life, would have been its most potent. Delays in milk consumption, due to milk's being collected, pasteurized, containerized, shipped, and sold, would have made the milk in markets less contaminated with that particular isotope. In addition, milk cooperatives of that era would mix milk from areas of higher fallout with less contaminated milk, diluting the iodine-131 levels. Dilution could also occur if hay or grains stored from previous months or years were used to feed the cattle, because that would have allowed time for the iodine-131 to decay, reducing its energy when consumed. The National Cancer Institute concluded in 2000 that the vast majority of cancer risk from tests conducted at the Nevada Test Site was associated with iodine-131 that tends to build in the human thyroid. The institute scientists also concluded that most of this risk was derived from milk ingestion, and the youngest in society experienced the greatest cancer risk.[38]

The NAS cautioned, however, that even if contamination by iodine-131 had been reduced, nearly 200 radionuclides had been released by atomic fission or fusion: other compounds including carbon-14, cesium-137, and strontium-90 would have entered the human body via tainted milk and other routes, accumulating in various tissues and posing additional cancer and other risks not well understood.[39] The NAS committee recommended that total radiation exposure should be limited to 10 roentgens from human sources during the first thirty years of life, above and beyond average background levels, which were believed to be 5 roentgens. The committee estimated that average doses from fallout would not likely exceed 0.5 roentgens over a similar thirty-year period. It warned the medical community, however, to be more careful in the use of medical and dental X-rays, and condemned the practice of taking X-rays of fetuses to demonstrate their perfect form to expectant mothers, as well as their use to fit shoes.

SCRIPTING THE NARRATIVE

Not surprisingly, knowledge of the presence of radionuclides in the bones and teeth of nearly every child in the world was unsettling to most Americans, especially when it was accompanied by projected estimates of expected cancers and birth defects among the nation's young and their offspring. Public discontent led the AEC to redirect its public relations efforts to convince the nation and the world that the doses from fallout were insignificant—even as behind the scenes, the committee was coming to a different conclusion. The tone of the AEC's then-secret annual report to the National Security Council in 1955, for example, demonstrates the group's growing concern over the scale of pollution caused by thermonuclear testing the previous year. According to the report, "Debris from the CASTLE [high-energy nuclear] tests continues to fall out all over the

world, apparently because considerable amounts are still in the stratosphere. The milk from cows contains strontium-90 at about the same
levels as the soils. Human bones contain it at somewhat lower levels
than the soil, children and young people having more than older people,
presumably because of the more rapid growth."[40]

The report remained classified for nearly four more decades, but
electoral debates in 1956 led to greater transparency and disclosure. Adlai
Stevenson, the Democratic candidate for president, warned of three dangers associated with weapons testing: genetic hazards to future generations; the risk of bone cancer and other illness to the current population;
and the inability to conduct medical investigations due to background
contamination levels in human tissues.[41] Stevenson also charged that
President Eisenhower and the AEC had withheld vital information regarding strontium-90 in the nation's milk supply.

The White House "denied that the nation's milk supply had been
contaminated by radioactive strontium 90 in the fall-out from hydrogen
bomb explosions." AEC chairman Lewis Strauss responded for the president: "Mr. Stevenson has been misinformed and his public statement is
untrue. . . . The purpose of the [Libby] study was to 'determine the feasibility of purifying milk if it were ever contaminated by fall-out should we be
attacked by an enemy using atomic weapons.'"[42] But Libby, by then an AEC
commissioner, had by this time been monitoring milk for several years
demonstrating the presence of strontium-90, and Strauss knew it.

Moreover, soon after the election and within two weeks of Strauss's
denial, another AEC scientist publicly reported that a "steep rise" in the
radioactive content of metropolitan New York's milk supply had occurred
several months earlier, in September 1956. The AEC disclosure was carefully crafted, emphasizing the triviality of all fallout in comparison to the
effects of natural radiation. It also declared that, on average, exposure to
nuclear testing fallout was far less worrisome than the energy of radon
emitted by soils and cosmic rays. It compared the risks of fallout to the
effects of nuclear warfare, which would produce devastating physical,
chemical, and biological results. (As one AEC official was to report to the
president in 1959: "In case of all-out warfare, the world-wide effects of
fall-out would be infinitesimal in relation to the losses occasioned in the
target areas.") The testing program was sold as preventive medicine, with
unpleasant side effects. Finally, the government claimed that atmospheric

testing would serve to protect public health, because it offered the opportunity to conduct experiments that would guide the design and use
of evacuation routes and shelters to minimize human exposure in the
event of war.[43]

Despite these assurances, anxiety grew at the highest levels of government, perhaps in part because of further disclosures by the AEC. In
particular the commission revealed that it was disposing of radioactive
waste by burying it in soil pits, or by dumping it into rivers, deep wells, or
at sea. By 1959 President Eisenhower himself was reconsidering the wisdom of the atmospheric testing program. Notes taken at a White House
meeting in March 1959 documented Eisenhower's deep concern: "The
President said there seems to be growing evidence that testing is having
bad physiological effects. He referred to recent articles about contamination by strontium 90. He is coming to the conclusion that our position
should be that we will not test in the atmosphere. We will leave the underground and outer space tests out of any treaty."[44]

Similarly, in notes taken at a meeting the very next day, "the President
. . . turned to the question of our position on nuclear testing. He feels it
is no longer quite right for us to be rigid in the details of such matters as
inspection merely because we have been rigid in the past. All available
evidence indicates that nuclear testing is bad. The allowable dose of strontium 90 is being approached in some foods in some areas of the country.
With this development the President feels that we would no longer test
atomic weapons in the atmosphere."[45] Eisenhower later lamented "the
terrible consequences that have arisen out of the discovery of nuclear fission, endangering the whole future of civilization. . . . If the world could
be completely free of nuclear weapons, the U.S. would be better off." Also
during the meeting, Eisenhower recalled his earlier worry in 1953 that
because of atomic weapons, for the first time in the history of the nation
"we have reason to fear for the safety of our country."[46]

By 1960, the government had lost control over its nuclear narrative
and presidential support for above-ground testing slipped away. Greater
inconsistencies emerged between government reports about the dangers
of weapons testing and those of respected scientific critics, and were increasingly reported in the popular press. Those concerned about the potential global devastation of nuclear warfare, the worldwide contamination

from continued testing, and the likely proliferation of weapons to other nations began to garner broadly based political support for a prohibition against further testing.

The Limited Test Ban Treaty was an outgrowth of this political movement. Adopted in 1963, it prohibited atmospheric, oceanic, and outer-space testing of nuclear devices.[47] It was signed by 108 nations, including the United States and the Soviet Union, but neither China nor France became signatories. France was to carry out fifty atmospheric tests in the Pacific until the mid-1970s, and China conducted twenty-three tests by 1980.

Perhaps most importantly, the international treaty demonstrated broad agreement that above-ground testing placed children, infants, and fetuses at a heightened risk of genetic damage and of developing cancer. The public, once fully informed, considered these exposures unacceptable by-products of the superpowers' weapons development efforts. For the first time in U.S. environmental history, concern for children's health played an important role in reshaping domestic and foreign policy. The nation had learned a fundamental lesson in ecology about the importance of protecting the purity of the food chain for the health of all people, especially children.

CHAPTER FOUR

Nuclear Accidents

ON OCTOBER 10, 1957, farmers and townspeople in the rural village of Yottenfews, England, were confused and alarmed when nearly 3,000 workers from the nearby Windscale power plant suddenly appeared, commandeered their trucks and cars, and formed a caravan that steadily but quickly evacuated the area. Nearly ten years earlier the British government had built the Windscale facility to produce plutonium for nuclear weapons, modeled after the Hanford, Washington, enrichment site. The air-cooled Windscale reactor had overheated and was venting radioactive iodine and uranium into the atmosphere from its 440-foot stack. J. Robert Oppenheimer, who had cautioned the U.S. Army to locate Hanford in a remote location, had predicted such an accident several years earlier.

Yottenfews residents were left confused, angry, and nervous about the exodus. Not until almost twenty-four hours later did the British Atomic Energy Authority announce the accident and reassure local residents that it posed no hazard to human health. By 2 A.M. the following Sunday, however, a full fifty-eight hours after the release of the radionuclides, farmers in the region were awakened by police knocking at their doors, warning them against selling their milk. They were instructed to wait for public officials to collect the milk for testing and disposal. The government initially decided to ban sales of milk from cows living within a 14-square-mile area surrounding the plant, but scientists soon recognized they had underestimated the extent of the release and contamination of pastureland. By the following Tuesday, officials had extended the milk-sales ban to a 200-square-mile area. Soon, milk from nearly six hundred dairy farms within the 200-mile zone had been collected and dumped into the Irish Sea—although most of the milk produced in the contaminated area during the more than two days between the time of the accident and the ban had already been shipped to markets.[1] In all, after the Windscale accident nearly a half million gallons of contaminated milk from the region were dumped into the Irish Sea. A study by the Radiological Protection Board of Great Britain concluded in 1983 that 260 cases of thyroid cancer could reasonably be attributed to the incident, predominantly from exposure to iodine-131.[2]

Within a week of the accident, radioactivity was detected in the air in London, nearly three hundred miles to the south, at a level twenty times greater than normal.[3] Workers who tested film at Kodak's plant in London, too, were confused about what could have exposed their film. Soon they heard about the Windscale release and realized it was the cause. The company had some experience with contamination from radiation fallout. Following the 1945 Trinity atomic bomb test in New Mexico, Kodak had received complaints from customers who had purchased foggy X-ray film. A company physicist discovered that the film had been packed in material made from corn husks at an Indiana plant, and it seems that fallout from the Trinity test had rained on the corn fields that provided those husks. The company became a key contractor to the AEC and eventually the nuclear power industry due to the sensitivity of its films for detecting nuclear radiation.[4]

The Windscale accident quickly undermined public trust in the British government, as those living near the plant were left to cope with declining cattle prices, lost sales of milk, and problems securing mortgages for the purchase or improvement of contaminated lands.[5] The government, anxious to restore public confidence in the British nuclear weapons and atomic energy program, within a month produced a white paper concluding that the accident had resulted from "faulty and inadequate instruments . . . and faulty judgment on the part of the operators . . . attributed to organizational weaknesses."[6] Much like the Bhopal, India, explosion at a Union Carbide India Ltd. pesticide plant that killed several thousand and injured hundreds of thousands in 1984, and the space shuttle *Columbia*'s disintegration in 1986, the Windscale accident eventually was explained by poor managerial judgment and organizational defects, rather than technological or mechanical failures. In fact, each of these disasters illustrates humans' incapacity to manage highly hazardous materials, a failure to warn exposed populations, and the absence of any effective emergency response or restoration plans.

Shortly after the accident, Windscale's name was quietly changed to Sellafield in an effort to improve the image of the facility, the BAEA, and the region. Public and private investments in the facility soared over the following decades, as Sellafield became a repository for spent nuclear fuel rods from other European nations. Throughout the facility's existence, operators disposed of radioactive wastes in the sea. The amounts of radiation released intentionally were compounded by accidents, including one in 1986 that released half a ton of uranium into the Irish Sea.[7]

The unfortunate history of Windscale offers many important lessons in ecology, health, and economics. The movement of radionuclides from the facility into the atmosphere, the contamination of pastures, and the excretion of these toxins in cow's milk, together explain pathways leading to human exposure. Winds and rains carried and deposited the isotopes far from their site of release, providing yet another example of how climate shapes patterns of contamination. Windscale also demonstrated the potential for nuclear facilities to inflict long-term damage on the economies of agricultural communities that often surround them.

Those responsible for handling and processing nuclear materials face a public relations dilemma. If they warn the public about site dangers and radiation releases, public support would quickly erode for their programs. Yet if they fail to provide adequate warning and the public becomes aware of contaminating releases, trust in the managers (public and private) declines quickly. Windscale in particular is a case study in the gross insufficiency of disaster management planning. For even when dangerous facilities are built in remote areas, population growth and development will eventually spread as close as allowed by zoning laws—which are often administered by local authorities eager to generate higher property tax revenues. And as this growth occurs, the rapid evacuation needed to prevent dangerous exposures becomes more complex and less feasible.

Fortunately, environmental monitoring of the Windscale accident was precise enough to allow government scientists to calculate an area of the west Cumbrian coastal strip, roughly ten kilometers to the north and twenty kilometers to the south of the site, that had experienced the highest levels of iodine-131 deposition. Scientists and public officials fortunately employed precautionary judgment, deciding to ban the consumption of milk produced within this zone—a strategy that reduced exposures to infants and children, and undoubtedly protected their thyroids from later disease. The year, 1957, marked after all one of the most intense periods of atmospheric weapons testing in history, and British scientists by then understood many of the dangers posed by nuclear materials, due in part to lessons learned from the U.S. weapons testing program.[8]

CHERNOBYL

Twenty-nine years following the Windscale accident, early in the morning on April 26, 1986, nearly all were asleep in Pripyat, a small Ukrainian village lying 2.5 miles from the Chernobyl nuclear power facility, when plant operators began an experiment. The first step was to turn off several safety and emergency warning devices, a process that reduced the cooling capacity of the core. Workers planned to reduce power to 20 percent of capacity, a safe minimum level, but neglected to press a switch that would have prevented a further reduction, one known to produce a rapid and dangerously unstable buildup of gases such as xenon. Power fell quickly to 1 percent of capacity, then surged 1,500 times higher within five seconds. The operators recognized their error and tried to drop the control rods into the core in a desperate attempt to reverse the runaway reaction. But the rods took ten seconds to put in place, too late given the explosive power surge. The Soviets later admitted that the graphite control rods were shorter than they should have been, because of an effort to reduce operating costs.[9]

The reactor exploded, blowing a 6-million-pound steel plate off its roof and propelling radioactive gases and debris 3,600 feet into the sky. Graphite that surrounded the fuel rods caught fire and exploded, landing near the core. As the fires continued, radioactive particles continued to vent from the site for at least two weeks. Soviet and U.S. physicists estimated that 25 percent of the radiation released from the accident was emitted within twenty-four hours of the initial explosion, well before the public was notified and an evacuation began.

Two engineers were sent to check on the reactor core several hundred yards from the command room, and if necessary, to lower the control rods manually. When they entered the core building, they found a mass of twisted wreckage. The control rods they had hoped to force into place no longer existed; instead, their remnants were ablaze, giving off an eerie bluish-red glow and overwhelming heat. The two men struggled back to the control room to report the disaster to the operators, who reacted with disbelief. They informed their superiors in Moscow that the reactor was still "intact" but there had been an accident. A third engineer was sent to inspect the core and to physically implant the rods, but he too found only the blazing radioactive debris. All three engineers sent to assess the damage were immediately hospitalized with severe radiation burns and all suffered agonizing deaths within ten weeks as their organs dissolved.

Firefighters arrived ninety minutes after the explosion and confronted flames that shot nearly one hundred feet into the air from what was left of the roof. Their boots sank into molten asphalt. Some were overcome by the dizziness, fatigue, and vomiting that generally accompanies intense radiation poisoning. All those who collapsed were hospitalized and eventually moved to Moscow. Other than potassium iodide pills available in their first aid kits, the firemen had no special protective equipment, just normal respirators and heat-resistant clothing. Six of the first-responder firefighters also experienced painful deaths from radiation burns and poisoning.

Chernobyl was designed with no protective dome to prevent the release of radioactive gases into the atmosphere in the event of a damaged core. But the Soviets were not the only ones to forgo this safety feature. That very year, 1986, the U.S. Department of Energy was operating eight plants without full containment structures, having concluded that domes were "not cost effective" given the low risk of an accident. The Hanford reactor is most similar in design to Chernobyl, being light water cooled and reliant on a graphite shield, but it too has no protective structure covering the core area. By 1986, Hanford had experienced ten times the number of shutdowns experienced by other U.S. reactors.[10]

For nearly forty hours after the catastrophic Chernobyl accident, no public warning was issued. Although the incident happened in the early hours of Saturday, April 26, later in the day children went to the school in Pripyat, an outdoor wedding was held, and several townspeople went fishing at a favored site, the cooling pond to the reactor. During this period, officials in Moscow were aware that a reactor fire was raging.[11]

Worldwide Alarm

On the morning of April 28, two days following the Chernobyl explosion, a worker arrived at a nuclear plant in Sweden. As he entered the door, radioactive particles on his shoes triggered an alarm. The plant was immediately evacuated and other emergency procedures instituted. Within hours, other plants in Sweden reported similar alarms. Upon determining that all operations seemed normal, Swedish scientists began to suspect that their detection equipment was sensing radionuclides in the air, perhaps as a result of atmospheric weapons testing or a nuclear accident. They checked the weather patterns over the past several days, which suggested that the particles had most likely originated within the Soviet Union, someplace in the Ukraine. Swedish officials immediately demanded to know if the Soviets had experienced an accident. Following several days of denial, late on April 29 the Kremlin disclosed "An accident has taken place at the Chernobyl power station, and one of the reactors was damaged. Measures are being taken to eliminate the consequences of the accident. Those affected by it are being given assistance. A government commission has been set up."[12] The next day, a Soviet bulletin claimed that although nearly two hundred people had been hospitalized near the Chernobyl nuclear power facility, the water, food, and air in Kiev were safe.

Several days later, radiation in Sweden reached a thousand times normal background levels. Winds then shifted toward the south, blowing radioactive particles toward Romania and Yugoslavia. Within five days, both the Polish and Swedish governments had issued warnings about radionuclides in their milk supplies, and farmers had been cautioned against grazing their cattle on pastures. By the end of the first week, the fallout was detected in Great Britain and most other Western European nations. Within two weeks, the EPA had announced that the fallout, traveling at 100 to 150 miles an hour across the Pacific with the jet stream, had reached the United States.[13] The U.S. government scientists reported that Chernobyl's fallout was measurable in surface waters, including some public water supplies, but on average if one liter of rainwater were ingested containing levels detected in Santa Fe, it would be equivalent to receiving approximately one-quarter of a chest X-ray.[14]

Nearly thirty-six hours after the accident, the Soviet government issued a terse 250-word statement that gave no hint of the accident's devastating health effects or officials' inability to contain the ongoing disaster. The absence of information was in part legitimate; the Soviets did not fully grasp the enormity of radioactive material released to the atmosphere or where it had traveled. Yet much was known in the early days that could have been relayed to other countries but wasn't—and the government's delay in releasing available data raised international suspicion that the Soviets would cover up information necessary to protect citizens of other nations. Within days, governments around the world were condemning the Soviet failure both to warn other nations of the accident and to track the contamination with precision. This international criticism would soon target the nuclear industry at large.

The Evacuation

After thirty-six crucial hours had passed, during which the most intense levels of radiation were released into the surrounding environment, civil defense officials were finally able to convince government and plant authorities to evacuate the Pripyat and Chernobyl communities. More than a thousand buses lined up to move residents from the highly contaminated regions.

Families first received word of planned evacuations from radio announcements instructing everyone to take "documents, money, food and clothing for three days." *Pravda*, the Communist Party daily, reported that nearly 60,000 children were taken by bus from areas thought to be most contaminated. Children were separated from parents and sent ahead to Pioneer camps, the Soviet version of Boy and Girl Scout camps, dispersed between the suburbs of Moscow to the Crimea. The buses were stopped en route while the children were checked for radiation. If a child's radiation level was too high, his or her clothes or belongings were burned. Parents and children were reunited days or sometimes weeks later.[15] After their evacuation, families were prohibited from returning to the areas of highest contamination to live, although by September officials had decided to allow each family to send one member to visit their home once—only once—to retrieve belongings.

Ten days following the release, the area requiring evacuation was defined as an eighteen-mile radius around the plant known as "the zone." The zone was intended to include the most radioactive lands. But politics also played a part in identifying the threshold between safe and dangerous.

Moreover, the intensity and movement of continuing releases from the still burning reactor made the boundary unreliable.

Damage Estimates
Haphazard radiation monitoring after the accident failed to detect hot spots of contamination, but later, more careful monitoring demonstrated that many lived in highly contaminated areas beyond the eighteen-mile zone. In 1990, three and a half years following the accident, the government adopted a more protective standard of acceptable radiation exposure —35 rems in a seventy-year lifetime—which caused the danger zone to expand. (A "rem" is a unit of damage to human tissues caused by radiation exposure, now replaced by the sievert.) Even so, the Soviets calculated exposure by estimating the environmental contamination detected long after the explosions; the intense radiation released shortly after the disaster began was not accounted for in their estimates.[16]

Despite these early underestimates of contamination and its hazard, the scale of the problem was and remains staggering. By 1990, 4 million residents of Byelorussia, the Ukraine, and western Russia were living on seriously contaminated soils, including 5 million agricultural acres, roughly the size of Massachusetts, in the Ukraine alone. Many of these people faced the reality that they had spent the previous three to four years living in areas that were no longer considered safe. They now worried about relocation, as well as the health consequences of their unsuspected radiation exposures. By 1989, ten thousand new homes had been constructed for those originally displaced, but due to the new estimates thirty-seven additional villages would be required.

In 1991, a commission appointed by the Ukrainian Supreme Soviet estimated that 5 million people in Ukraine lived on lands so contaminated that evacuation should be unconditional. Although the Soviet law set a standard of allowable exposure at 35 rems, the Ukrainian commission recommended a maximum lifetime exposure of 7 rems. They also argued that the standard for allowable soil contamination should be lowered from 15 to 5 curies per square kilometer.[17] (A curie is a unit of measure to identify the amount of radioactive material present.) As the quality of radiation surveillance improved, and as radiation standards were tightened, more land and people fell into the "at risk" category. The national Soviet government, however, resisted these revised standards; it was concerned about legal liability and costs related to the size of the evacuation zone, responsibility for medical care and monitoring, compensation for property losses, site restoration, and the building of new communities for evacuees.

Both dry fallout from the Chernobyl accident, along with rain that had absorbed particles from the air, had quickly reached surface waters in the area. Within the first week following the accident, the Soviet government warned its citizens against drinking this water. Warnings were often delayed for the consumption of contaminated fish and other aquatic life, however, due to an absence of environmental monitoring. The Kiev Reservoir, which lies nearly seventy miles from the Chernobyl plant site, then provided water for nearly 46 million people. Despite efforts in the days immediately after the accident to seed clouds and cause rainfall away from the reservoir, its water was contaminated as the radioactive cloud moved south. The reservoir's enormous size diluted the concentration of fallout in drinking water, and it was not declared off-limits for fish consumption until 1989, three years following the accident.[18]

As the radiation in the atmosphere swept quickly across Europe, economic disaster quickly followed. No one wanted to risk exposure to potentially contaminated products from the area. Soon after the accident, nearly five hundred tons of beef from cattle that had grazed in Byelorussian pastures was shipped to markets in Georgia. Upon discovering the source of the shipment, however, the Georgians refused to accept it. The radioactive beef then embarked on an odd, multiyear journey to find a home. It was eventually returned to Chernobyl and as of 1990 was still there in forty refrigerated cars.[19]

Other countries were also alarmed. The Austrian government banned the import of milk, fruit, and vegetables from Eastern bloc nations. The Dutch agriculture ministry prohibited cattle grazing. The Swiss warned its citizens against providing fresh milk to young children, and against drinking rainwater from cisterns. Italy banned the sale of its own leafy vegetables—resulting in the waste of nearly 10 million pounds of produce. (Carrots were exempted, with the logic that they had grown underground.) Children under the age of ten and pregnant women were cautioned against drinking milk in Italy. Poland, too, the nation closest to the damaged reactor, was devastated. On May 3, the United States warned women of child-bearing age and all children not to travel to Poland due to the health hazards inherent in radiation levels five hundred times greater than normal. But often the directives were less clear; instead, confusion reigned as government officials issued restrictions that implied danger but did not specify its level or particular health significance. One Italian television announcer summarized the frustration: "If it's not dangerous, why are they banning it? If it is dangerous, why aren't they telling us?"[20]

Nearly 3 million sheep were banned from markets in Scotland, Wales, and Cumbria, a county in northwest England, due to extensive contamination of northern Great Britain, and the government warned it would maintain the ban until the "last traces of fallout from the disaster are gone." Britain eventually provided $10 million worth of compensation to the sheep farmers.[21] Nearly 400,000 claims for compensation for lost milk and vegetable revenue were filed with the West German government from farmers forced to destroy their products.[22]

Standards for the acceptability of radionuclides in different foods varied among the European nations, which led to a chaotic set of national bans on specific products such as milk and fresh produce. West Germany reported radiation levels that were sixty times normal, causing widespread fear that its food and water supplies were unsafe. The public was warned against allowing children to play on playgrounds due to fallout in sandboxes, play equipment, and fields. Early in 1987 the German government decided to destroy 3,000 tons of powdered milk containing nearly ten times the allowable level of radiation allowable for human consumption. Bavarian officials created an international uproar when they suggested that the contaminated milk be diluted by adding it to animal feed and then sold to Angola and Egypt. They argued that the European Community standard could be met by mixing milk from different sources. Officials from these developing nations, however, were less than pleased at the thought that food found too dangerous for Europeans might be sold and consumed in their countries.[23]

Finally, the Europeans decided to ban meat, live animals, and produce from all areas within 1,000 kilometers (630 miles) of the site, about the distance between New York City and Chicago. The ban excluded East Germany, due to the insistence of West Germany. The countries were unable to agree on a uniform standard of acceptable contamination, but decided that exporting nations must respect the standards of the importing country. The delay in reaching the agreement became an embarrassment to all nations, because it meant consumers had no way of knowing the source of foods in their markets, and thus the contamination levels of these foods. Instead, consumers were left with vaguely worded reassurances of safety by food companies and ambiguous warnings by governments.[24]

Entire cultures were made vulnerable by the Chernobyl accident. Lapland is home to the Same, an indigenous people living in northern Sweden. Originally, they were nomadic fishermen and hunters, and their culture is closely tied to nature, as demonstrated by their language, which has eighty words describing different kinds of snow. But as isolated as they seemed, they were still at risk from Soviet radionuclides that emanated from an accident nearly a thousand miles away. By mid-September 1986, the Swedish government had found that 97 percent of Lapland reindeer meat—a Same food staple—exceeded the national acceptable standard for radiation. Due to the Chernobyl tragedy, the Same food chain will be contaminated for hundreds of years. The detection in the reindeer of cesium-137 with a half-life of more than thirty years was especially worrisome; it means the problem is virtually irreversible in a single lifetime.

In some highly radioactive areas, the reindeer were initially ferried into double-decker trucks across hot spots to feed lots, in an effort to prevent additional exposure. Later, the Same community decided to feed all of the reindeer uncontaminated grains, not a minor challenge given a herd of nearly 500,000 animals. The Swedish government eventually decided to purchase the deer meat and allow its consumption by mink and fox in farms. The contaminant was thus removed from the human food chain, while saving the Lapps from economic ruin, but there is no satisfying conclusion to the Sames' predicament.[25]

Prospects for Cleanup

What should be done with the landscape contaminated by Chernobyl—the soil, rock, trees and other plant life, fish, and wildlife, as well as the homes and other structures left behind by former residents? The United States has faced this problem near many of its nuclear testing facilities, and following some accidents. During the 1950s, a U.S. military aircraft crashed in Spain, spilling the plutonium from its nuclear weapons across hundreds of acres of wheat. The United States responded by stripping the radioactive topsoil with bulldozers and shipping it to the United States for disposal. And several decades following the nuclear testing in the

Marshall Islands, the government scraped off the upper layers of soil and buried it so that crops might again be planted, and human exposure could potentially be limited.[26] But the area near Chernobyl believed contaminated with 5 or more curies per square mile is nearly 13,000 square miles in size—nearly 25,000 times the size of the Spanish wheat field. In addition, those who participated in the Chernobyl cleanup effort report that as soon as contaminated cars, trucks, and ambulances were buried, thieves would dig them up to remove their parts for sale on black markets, even though many parts were radioactive.

To restore contaminated farmland poses a special challenge. Nearly a thousand square miles were designated to be most dangerous, and Soviet workers excavated surface topsoil, burying it in pits, to reduce health hazards and restore the land's agricultural potential. Grazing animals were an immediate concern. If cesium-137 is ingested by dairy cattle, it will contaminate milk and dairy products made from it. Wheat futures on the Chicago Board of Trade soared 13 percent during a single day shortly after the fire as investors speculated that the Soviet's need for uncontaminated North American grains would lift prices. Similarly, meat, livestock, and cotton futures reached their daily trading limits, as traders speculated that European foods would be avoided or prohibited in international markets.[27]

Long-Term Health Effects
The U.S. Department of Energy estimated in 1994 that the Chernobyl explosion exposed 2.9 billion people (nearly half the world's population) to radionuclides, which would eventually result in an estimated 17,400 fatal cancers.[28] During the early 1990s, an increased incidence of several diseases among those living closest to the plant was gradually recognized. At a scientific conference held in Kiev during 1991, Yuriy Spizhenko, the Ukrainian minister of health, noted a rapid decline in the health of children living in contaminated areas. The Ukrainian Institute of Gynecology, too, reported a decrease in fertility, as well as increases in deformities among children born after the accident. Incidence of uterine bleeding increased by 300 percent. Dmitri Grodzinsky, a biologist at the Ukrainian Academy of Sciences, reported in 1990 that 150,000 Ukrainians suffered from thyroid disease—of these, 60,000 were children and 13,000 required chronic treatment.[29] These statistics may not reflect the effects of the accident

on human health with any precision, in part because the Soviets did not monitor fallout carefully at the time and the quality of surveillance has improved since the 1980s. Moreover, the most intense and dangerous exposures, which clearly occurred during and immediately after the April fires, were poorly monitored and recorded—despite the obvious signifi- cance of the event. And no comprehensive health surveillance system was or is in place in the affected area to monitor human illness.

More than twenty years after the event, the health effects of the ac- cident remain difficult to estimate, although most medical experts fa- miliar with disease trends believe that illness caused by both short-term exposure to intense radiation, and long-term exposure to lower levels, are contributing to numerous immunological, neurological, cardiovascular, reproductive, and endocrine disorders, in addition to an elevated incidence of cancer. Soon after the crisis began, thirty-one people died, and nearly five hundred were hospitalized. The death estimate is extremely controver- sial, due to the absence of effective health surveillance. By August 1986, a Soviet government commission had issued a sobering report on the events leading to and following the explosions. Compared with the vague and deceptive comments released just after the accident, the commission was startling in its criticism of the government's handling of the accident. The officials anticipated 4,750 deaths from radiation exposure far from the plant, and 280 deaths from exposures nearest the accident. An ad- ditional 1,500 deaths were predicted from thyroid cancer from exposure to iodine-131 derived from contaminated food, especially milk.[30]

Nearly 24,000 people living near the site received "very substan- tial doses of radiation," but the actual levels remain unknown due to the poor quality of environmental and health surveillance.[31] Anticipated longer-term effects from lower-level exposures include mental retarda- tion, genetic defects, and cancer. All of these effects are delayed, and thus more difficult to attribute to radiation exposure. The threat of damage at lower levels of exposure depends in part on age, with fetuses and very young children being the most susceptible. Estimates of future cancers attributable to the radiation vary between several thousand and a million. As recently as 2007, scientists concerned that the nation's forests have accumulated levels of cesium-137, strontium-90, and plutonium worried that droughts in the years 2004 to 2007 could increase the risk of exten- sive forest fires—blazes that would release these radionuclides into the atmosphere and create new patterns of fallout, contamination, economic damage, and health loss.

Only months after the Chernobyl accident, in September 1986, forty-nine member nations of the International Atomic Energy Agency, including the Soviet Union, approved two international conventions. One requires immediate notification of accidents, and the other pledges international aid to contain damage in the nation where an accident occurs. The cooperative agreements were sorely needed: at the time nearly 270 nuclear power plants existed in the world, with a hundred others under construction. But they could have been more comprehensive. The United States and Soviet Union signed both agreements, but only after successfully excluding leaks from weapons and underground weapons tests from notification requirements. The Soviets agreed that they had a moral but not a legal responsibility for the damages caused by the Chernobyl disaster. An international ban on attacks against civilian nuclear facilities, suggested by developing nations deeply concerned over the Israeli attack against a nuclear power plant in Iraq, was not approved.[32]

Environmental and health surveillance are crucial to identify the location of hazards and those with associated illnesses. Identifying the source of contamination, as well as those harmed by it, creates a moral responsibility to warn, to reduce exposures, to provide health care, and to restore property. But these costs, estimated to be hundreds of billions of dollars in the case of Chernobyl, would bankrupt many nations; consequently, governments are eager to avoid rigorous surveillance. An international insurance fund supported by the nuclear power industry would encourage adequate cooperation and information gathering.

Officials in the U.S. nuclear power industry and government predict that the risk of a Chernobyl-like disaster is exceptionally low in the United States due to more protective plant engineering designs than existed at Chernobyl, increased vigilance against terrorism following the World Trade Center attacks, better worker training, and evacuation planning. But when one carefully considers the proximity of most nuclear facilities to major population centers and the extent of the contaminated landscape in the Ukraine and Byelorussia, assurances that millions of people could be evacuated in time to prevent serious exposure seem suspect. In 1986, those living in Chernobyl and its surrounding areas had little understanding of their vulnerability; the limited potential of evacuation plans; the economic collapse that would be associated with loss of farmland, forests, homes, and businesses; and the near impossibility of restoring a landscape to an uncontaminated condition. Those hoping to manage threats from the rise of terrorism during the twenty-first century would be wise to consider the lessons of Chernobyl and other disasters when determining how to manage the highly radioactive materials that are now dispersed among nuclear facilities used for defense and power in the United States.

PART TWO TRAINING FOR WAR,
WHILE WASTING NATURE

CHAPTER FIVE

Sowing Seeds of Protest

IN 1941, RADAMES TIRADO was eight years old and living on the island of Vieques, off the east coast of Puerto Rico. His father built and repaired ox-drawn carts used by the island's sugar companies to haul bundles of cane to local mills, where they were crushed for their juice. Radames, along with his seven brothers and four sisters, lived in the small village of Resolución, on the northern side of the island. The children often walked or rode horses to the beaches, where they would swim and dive for conch or lobster in the waters near shore, or search for land crabs.[1]

The northern side of the island borders the Atlantic, while the Caribbean Sea lies to the south, just past beaches of fine white sand. On sunny days, the sea is a palette of shimmering blues and greens: turquoise over sands, aquamarine over shore seagrass beds, and emerald green to cobalt blue as waters deepen. Coral reefs protect many of the beaches from storm waves, and provide shelter to grouper, yellowtail, parrot fish, conch, and lobster. Tarpin, barracuda, mackerel, shark, dolphin, tuna, and manta rays swim in deeper waters, while manatees loll among the seagrass beds on the northwestern side of the island. Groves of palm trees anchor some of the beaches, standing guard over dense mangrove lagoons that lie behind dunes of sand. Great blue herons, frigate birds, pelicans, and hawks soar overhead in the steady trade winds. Several of these lagoons—those with narrow openings to the sea—are filled with magical dinoflagellates, single-celled algae that glow when disturbed.

The Tirado family, like most others living on the island in the 1940s, worked for one of four sugarcane companies, earning only several dollars per week. Poverty, poor nutrition, an absence of health care facilities, poor quality water, crude sanitation, and an absence of electricity in some bar- rios made life difficult for many.[2] Despite the poverty, most Viequenses had the advantages of homes, jobs, rich land, productive seas, and a stun- ningly beautiful island on which to live and raise their families. Yet nearly 95 percent of the rural population owned no land.[3] Nearly two thousand families held informal tenancy rights to build houses and farm plots— often several acres in size—large enough for fruit trees and vegetable gardens, and sufficient to graze goats and sometimes sheep and cattle. Some managed to buy small tracts of land to run shops and other small businesses that serviced the communities' needs. But the *centrales* never yielded land title to their workers.[4]

During the summer of 1941, Radames's father returned from work to tell the family that they were being forced from their home. The U.S. government had decided to construct a large military base that would include Vieques and the nearby island of Culebra, and in a dramatic first step, the U.S. Navy had purchased the sugar company lands, nearly 70 percent of the island. The two companies had little choice in the matter. If they had refused, the United States would have exercised its right to condemn private property for national security purposes.[5] But they also were probably receptive to the proposal. The Great Depression had dev- astated the sugar industry in the Caribbean basin. One of the Vieques sugar processing companies had closed in 1927, and by 1935 nearly 64 percent of Vieques males were unemployed. The islanders' dependence on a single export crop made them especially vulnerable to fluctuations in international sugar markets, and falling sugar prices in the 1920s and 1930s had led to declining land values.

The remote location of Vieques—an island off of an island in the northeastern corner of the Caribbean—had discouraged more intense development, while increasing its appeal as a potential military base. The completion of the Panama Canal early in 1914 further increased the strategic importance of the Caribbean basin and especially Puerto Rico. Indeed, secure access to the canal immediately became a U.S. military and economic priority, one that endures to this day. There were also very immediate national security issues at stake during the 1940s, when the Tirado family first learned it would be displaced. By 1940, the United States was facing the very likely scenario of involvement in the growing European conflict, and Puerto Rico could offer Great Britain a safe haven for its naval fleet if British bases were further damaged or captured. The Japanese attack on Pearl Harbor in 1941 underlined the dangers of concentrating troops, vessels, and other facilities in a single harbor. Vieques and Culebra would provide a remote theater in which to train troops for invasions and test new weapons systems.

The U.S. Navy's task of removing seven hundred families from their homes to make way for weapons storage bunkers, radar facilities, docks, an airport, and housing facilities was not as difficult as it might have been. Only four sugarcane company owners controlled 19,000 of the 21,000 acres needed for the base, and these were acquired for an average $50 per acre. Many of the tenant farmers considered legal action to prevent the appropriation, or to gain increased compensation, but their poverty, ambiguous tenancy rights, and common illiteracy all reduced their capacity to resist.[6] Indeed, the cane companies had acted as both employers and landlords, providing the workers with few rights and the company with few obligations. Sugarcane was grown on the richest agricultural soils, preventing its use for growing locally consumed foods, even if the workers and their families had access to less desirable company lands to grow subsistence crops and graze their pigs, chickens, cattle, and goats. There was little that thousands of landless, impoverished, and powerless peasants could do to change the decision, or control the future of their island—even though the shift in landlord from the private sugar companies to the U.S. military was to be catastrophic to the island workers, most of whom lost both their jobs and their homes.[7]

The government chose to resettle the 4,600 displaced Viequense in the island's midsection, Santa Maria, on some purchased company land. The western end of the island would be used for weapons storage, docking, and radar facilities, while the eastern end would be transformed from grazing and palm plantations into a live-fire bombing and artillery range.

Radames Tirado's father told his family that they must move nearly seven miles to the east, and that a truck would arrive soon to carry their belongings. They, like the other families, were given a piece of paper with a four-digit number on it corresponding to a rectangular plot of land fifty feet by ninety feet, laid out in a simple rectangular pattern that neglected the hilly terrain, seasonal streams, and low-lying swales that would flood with rushing runoff during tropical storms. Like other U.S. government settlement schemes such as the Homestead Act that encouraged settlement of the American West in the late nineteenth century, the plan completely ignored variations in terrain and ecology, which meant that some of the plotted area was uninhabitable. The U.S. Navy's plans also neglected to include any roads, water, electricity, sewerage, or storm drainage for the islanders. In addition, before their move the families had been free to plant or graze on open lands not under cane cultivation, but the tightly clustered settlement and hills of the new location made doing so much more difficult. Continuing to use the original areas, even surreptitiously, was also not an option. Soon after the resettlement, government workers constructed fences that prevented residents from returning to their homelands and even their cemeteries.

At the appointed time, the Tirados climbed on the back of a flatbed truck with whatever belongings they could carry, and were driven to Santa Maria. When they neared their assigned plot, they were told to get off the truck. They were left standing in the brushy field with no shelter, not even a tarp to protect them from the heat of the sun and tropical downpours; their modest piles of belongings lay beside them. They used palm fronds and poles from the forest to fashion a makeshift shelter, where the family lived for months until they were able to build a more permanent home on the site. The boys helped their father salvage wood from their demolished hacienda and carry it to their assigned plot. Each family dug pits for human waste and wells for water. The small plots assigned to each family were connected by crude pathways more suitable for horses than vehicles.

Thousands of other displaced islanders were similarly deposited at the new settlements and left outdoors to fend for themselves for months until they could construct modest shelters. Living outdoors, they faced additional threats. Human and animal wastes often polluted the water, and much of the island had neither electricity nor running water. In addition, Vieques, like most other Caribbean islands, has long harbored diseases such as malaria, yellow fever, and dengue fever, which because they are carried by mosquitoes especially endangered refugees, military personnel, or others without shelter. No statistics were kept on malaria incidence on Vieques, but many still recall the recurrent fevers, a telltale sign of the illness. The nearest medical care for serious cases was on mainland Puerto Rico, easily a full day's journey.

Those who ignored the Navy's warnings to move were eventually given twenty-four hours to leave before bulldozers arrived. Ramón Rucci and his wife were among those forced to vacate their house in this way. The night they were trucked over to their new plot of land, his wife gave birth while lying on the ground in a makeshift shelter that Ramón had fashioned from sugarcane stalks and palm fronds. Radames recalled many other women who also gave birth in crude shelters during the rainy season following their moves. Several thousand Viequense eventually moved to nearby islands, including the main island of Puerto Rico and St. Croix.

The U.S. Constitution requires that "just compensation" be paid to U.S. private landowners if the government takes their property for public purposes such as a military base, highway, school, or park. But although the Navy compensated the plantation owners for their property, the Tirados and thousands of other tenant families were provided between $5 and $25 for their loss, while others received nothing. The governor of Puerto Rico wanted the Navy to pay for relocation expenses, but the military officials would only provide some of those who had been dislocated with wood and modest construction materials.[8]

Radames and his family lived outdoors for several months while his father found materials to build a new shelter and to feed the young family. Mangos, papayas, passion fruit, and coconut were plentiful, and many islanders were proficient at catching fish, shellfish, and land crabs.

Severina Guadalupe tells a similar story of her family's history. She recalled that before the U.S. military purchased the island, her father had grown sugarcane, plaintain, and yucca. Her extended family lived in a wooden farmhouse on twenty-seven acres. When she was thirteen, the family received notice that they must vacate within twenty-four hours. To the family, such a move was unthinkable. "We refused to leave our land, our farm, so when that machine came it destroyed everything—our cooking pots, our clothes—they left us with nothing. . . . They came up with this huge mechanical thing and came up to the porch and whup!

They just drove over our house." Her father was compensated $350, but lost his job as a farmer along with the house. He moved to the small village of Isabelle Segunda and sustained his family by taking on odd jobs. Severina eventually married, and remained on the island along with four brothers, each of whom contracted cancer. Nearly sixty years following their eviction, Severina still seethed with resentment and anger.[9]

Today, looking down from the air, almost no trace of the former villages remains. Once the Navy successfully emptied the islanders from their villages, the homes, stores, and offices were bulldozed into pits, burned, and covered with earth. Plant life has overgrown the streets, foundations, and even the former village plaza. The cane fields have since grown into a dense semitropical forest that shadows the brick ruins of the cane processing mills. Only the once graceful residence of a former island judge remains visible high on a ridge.

After the family's displacement, Radames's father used his carpentry skills to become an accomplished builder of fishing boats. Radames himself became an English teacher, and eventually served as mayor of Vieques between 1976 and 1980, when he worked to convince Governor Carlos Romero Barceló to sue the Navy for the environmental damage inflicted on the island. The federal district court found that the Navy's release of weapons into coastal waters did constitute pollution, and so was a violation of the Federal Water Pollution Control Act, but it did not prohibit future training.[10] It found that "appreciable harm" had not been caused by the Navy's "technical violations" and that demanding a cessation in bombing "would cause grievous, and perhaps irreparable harm, not only to Defendant Navy, but to the general welfare of this Nation."[11] The Court of Appeals for the First Circuit did not agree, ordering the Navy to obtain a permit before continuing to bomb.[12] But the U.S. Supreme Court overturned the appellate court's reversal, so the bombing continued.

The lawsuit forced the Navy to disclose its activities to the court and the public, and to meet the requirements of the National Environmental Policy Act. To do so, the Navy prepared an environmental impact statement that explained its operations, weapons use, and history of polluting the island. In fact, under the provisions of the act, the Navy is required to prepare these statements for each of its sites in both draft and final form, although it never completed the final Vieques analysis. Governor Romero Barceló agreed to settle the lawsuit in 1983, if the Navy would provide greater economic support for community development and compensate fishermen for the income lost during training exercises.

The governor of Puerto Rico created a Commission for the Island of Vieques to study the naval presence and make recommendations that would hopefully resolve conflicts between the islanders and the Navy. Radames was appointed as the assistant commissioner, in recognition of the community's trust in his leadership. In 1999, the commission recommended that the Navy immediately "cease and desist" their military activities. Finally, in 2002, President Bush decided that the weapons training range was no longer necessary in an age of electronic warfare, and that the Navy should leave the island.

More recently, Radames reflects that the Navy's legacy is not simply one of human rights abuse, environmental destruction, and the loss of health for the Vieques villagers, but also deliberate economic stagnation, despite the potential to grow a vibrant economy. As he contends, "The fundamental problem here is that the Navy never has had the intention of helping the people of Vieques." He believes that the Navy actively sought to block the municipality's search for development funds, and to discourage island tourism that would only intensify protest against the Navy's training and bombing. His claim is supported by the comments of Howard Hunt, a Navy captain who described his reaction to a proposed $100 million federal investment in the Roosevelt Roads Naval Complex that was to include Vieques: "We're not going to throw away such an investment so that Vieques should be converted to a Mecca for tourism."[13]

CARLOS ZENÓN

Carlos Zenón was four years old when a U.S. Navy representative knocked on his family's door. Their house and land had been acquired by the Navy, and the representative was there to order them to pack and leave. In later years, Carlos's mother was often to repeat her protest: "But I didn't sell it to anyone. How could they take our homes against our will and without paying us?" Her resistance was met with a demand to leave within twenty-four hours. The next day, a truck and bulldozer arrived and waited outside the house while his mother ran to collect whatever belongings she could quickly throw into bags. Carlos recalls his fascination with the large and powerful machine as he watched their home crumble, even as his mother stood weeping. The Zenóns, like the Tirados, then piled what belongings they could on the back of the government truck. They too were given a piece of paper with a number that designated what was to be their small assigned plot in the new area.[14]

A local barber heard about the Zenóns' plight and offered to give the family his one-room barbershop for shelter. The shop was dismantled, placed on the truck, and then rebuilt on their lot in Santa Maria. The windows and door of the shed were broken during the move, however, so the family was forced to use fertilizer bags to keep out the rain. The Zenóns, like most of the other displaced islanders, had no electricity, water, sanitary facilities, or roads when they arrived, nor did they have access to lands for grazing animals or planting crops. Their homes and jobs disappeared overnight, leaving them with little shelter or means to make a living—as well as deep resentment.

The relocated families, which never had held title to the land they were forced to leave, were also tenants with no legal property rights in the new settlement. The government wanted the option to displace the islanders again should it decide to expand the base at a later time. To a community disoriented by one move, the absence of "land security" generated even more distrust of the government's intentions. The lack of legal property rights had reinforced the power of sugar companies, and now the Navy.

The government wielded its power of land appropriation in 1941 and again in 1947–48. In 1961 the Defense Department secretly devised a plan to abolish the municipality of Vieques and remove all residents, so as to prevent any conflict between training activities and local residents. Puerto Rican Governor Muñoz Marin, however, appealed to President Kennedy, who eventually decided against the plan. In 1964, too, the Navy proposed that the military acquire all property along the southern Caribbean shore of Vieques. This plan was also met with overwhelming local opposition, and was abandoned.

Several years after moving to Santa Maria, Carlos Zenón's mother cut her leg. When the wound would not heal, and her condition worsened, she was unable to work. Carlos begged for sugar and rice from neighbors. He had no shoes or clothes for school and knew he needed to find a job. He spent many hours watching the fishermen, who would row their small, fourteen- to twenty-foot skiffs out to sea to set their traps. At one point, Carlos approached a fisherman he had long admired with a proposal. If

Carlos rowed the boat while its owner fished, could he receive some payment, even if only a small portion of the catch? The fisherman laughed at the twelve-year-old's youth and slight build, and asked why he thought he could row on the rough seas surrounding the island. Carlos thought for a moment and offered another idea. He commented that he had watched the fisherman return to the docks after a long day, only to spend several hours cleaning their catch in the hot sun before being able to sell it. What if Carlos cleaned the fish on the skiff as soon as they were caught, and in return received perhaps twenty-five cents a day and several fish? The fisherman could then deliver his catch to market before the others, and avoid the least pleasant chores of the trade. Carlos started the next day.

The 1950s were difficult years for Viequenses, both economically and socially. Most servicemen were stationed on the main island of Puerto Rico at the Roosevelt Roads base headquarters, where hundreds of millions of dollars were invested. By contrast, only several hundred servicemen were stationed on Vieques to maintain security at the island's modest facilities: Camp Garcia, an airfield, Mosquito Pier, weapons storage magazines, and several observation posts. Hardly any of the anticipated services and jobs on Vieques came about. Instead, dozens of bars and brothels opened to attract the tens of thousands of troops that temporarily trained on the island. Few of the islanders welcomed these enterprises. Elderly islanders today recall drunken sailors and marines drifting from bar to bar, sometimes bursting into private homes or relieving themselves late at night on the sides of buildings. Young girls would run and hide in ditches when they saw or heard groups of servicemen approaching, fearing abuse and sexual assault. Islanders' resentment, which had first flared from the 1940s evictions, only intensified with continued military presence.

Island teenagers, including Carlos, organized to protect the residents. When sailors wandered through his resettlement community of Santa Maria, the teens would attack, kicking and beating them with sticks, then sending them back to their base as a warning to others. Several years later, Carlos joined the Army, in part to better understand the foreign culture that had such an influence in shaping his life and island. The Army sent him to school to study English, building his capacity to understand, question, and confront those who had abused his community. The Army also taught Carlos the military strategy and tactics that he would use in later acts of civil disobedience.

From the Viequense perspective, the military presence was a cultural invasion. The residents were increasingly dispirited and impoverished. Nearly three thousand had left the island between the Navy's arrival in 1942 and 1960. Many had fled when the sugar companies closed their operations, having lost both their homes and jobs. But others had stayed, hoping to find work in construction, facility maintenance, or services, jobs promised by the Navy but normally filled by nonislanders. In fact, the Navy's activities on Vieques required little service-sector support, especially when compared to the facilities needed to house, feed, entertain, and educate the thousands of sailors and family members stationed at Roosevelt Roads only six miles across the strait.

The primary purpose of Vieques for the United States was to provide a physical theater for mock invasions and for weapons training. Given the scale of the exercises, it is hardly surprising that the islanders felt exploited, wasted, and contaminated by the troops—and during some seasons as though they were under near constant bombardment. Nearly two hundred different types of weapons were used or tested on the island during the last half of the twentieth century. And military training exercises on Vieques and nearby Culebra often involved tens of thousands of troops, predominantly those enlisted in the Navy and Marines, but sometimes including the Army and Air Force. Eventually other NATO nations, especially British, French, and German forces, were asked to practice and coordinate invasion tactics with the U.S. forces. Aircraft carriers, destroyers, and cruisers would launch attacks on the island, deploying thousands of troops in landing craft that would assault the powdered shell beaches, all with live fire. Once ashore, the troops would bury land mines on the beaches and move inland to join or engage those landing on the opposite side of the island. Simultaneous bombing and shelling of the island would occur in the designated live impact area.[15]

Since this was a "training" ground, errors in aim were common. Bombs, torpedoes, missiles, and rockets often missed their mark; some were undershot and sank in waters near the shore, while others landed miles outside the live impact area, closer to the residential community of Isabel Segunda that lies in the island's midsection. Hundreds of millions of pounds of ordnance have been dropped or exploded on Vieques and the surrounding waters since the military occupation began in 1941.[16]

Not only were the Viequense under siege; their livelihoods were often carelessly put at risk. Naval captains and pilots paid little attention to the small bobbing buoys set by local fishermen, hardly noticing when their propellers sliced fishing and trap lines during maneuvers. The fishermen's loss in such cases was substantial and included not only the trapped fish, but more importantly, the cost and time required to replace the traps, lines, and buoys. Fishermen were rarely compensated for these losses, despite repeated protests over military exercises that were conducted 180–250 days per year.[17]

Beginning in 1975, following the Navy's decision to pull out of Culebra due to the intensity of local protest, military maneuvers on Vieques intensified—as did conflicts with local fishermen. In 1978, Carlos Zenón lost 131 traps due to severed lines. After repeated attempts to recover his costs, he sued the Navy in U.S. district court in San Juan. The Navy countered that it wanted evidence of the lost traps, including receipts for line, traps, traps, and buoys. Despite his inability to produce a full record of his losses, and the obvious difficulty he faced in producing evidence of the lost traps, the Puerto Rican judge awarded Zenón a reasonable estimate of the value of his loss. The Navy's lawyers decided to appeal, and were successful in relocating the case to a court in Norfolk, Virginia, where they believed the judge would react more favorably to the uncertain evidence. They also believed that the Zenóns would not have the resources to travel to Norfolk and testify.

Carlos chose to press ahead and fight the appeal. Relatives and friends sold baked goods for several months to help finance Carlos's travel costs to Virginia, as well as those of a few close supporters. At the beginning of the hearing, the judge summoned lawyers from both sides to the bench. The judge prefaced his statements with the comment: "I've never seen a fisherman who wasn't poor," then chastised Navy attorneys for forcing the fisherman to spend nearly $1,000 for the trip to be present for his case, on the assumption that a court nearer Washington, D.C., would react more favorably to the Navy. He warned that the Navy should settle the Zenóns' claims, including travel expenses. An agreement quickly followed.[18]

The judge's decision surprised Carlos, and he left Virginia with a growing sense of confidence. For the first time in his life, he understood that the Navy could be held accountable to a set of principles other than its own. He returned to Vieques ready to convey to his fisherman's association that the Navy might be more vulnerable than they had assumed.

That same year, the Navy announced its intention to hold maneuvers for a twenty-eight-day period and issued a notice that fishing would be prohibited during the exercises. The fishermen were enraged, and Carlos as leader of their association made the short trip across Vieques Strait to the office of the Roosevelt Roads commander. Carlos asked, "Why can't we be allowed to fish from dawn to 10 A.M.? This would allow us time to bring our catch back to the island without significant loss, and you would have the remainder of the day to hold your exercises." But the commander held firm, arguing that the complex exercises involving several NATO nations and dozens of ships and planes had already scheduled their war games from dawn until dusk each day. Nothing would be changed.[19]

Carlos returned to Isabelle Segunda and quickly gathered those in his fishing association. After the group had debated the issue for hours, Zenón concluded that negotiations were futile, and their only recourse was civil disobedience. The fishermen were afraid that they might be shot or imprisoned, and that their boats might be destroyed. Still, nearly forty of the fifty members agreed that the time had come to make their stand. They stayed up all night planning nonviolent tactics to disrupt the maneuvers scheduled to begin the next morning. At dawn their modest armada of brightly colored fishing skiffs left port.

As the war games began, the small fishing boats swerved into the paths of landing craft filled with troops. Instead of ramming the small boats, the commander ordered that the fishermen be captured unharmed. The protesters had been given an unexpected opportunity to capitalize on their detailed knowledge of the shoals and reefs surrounding the island. A chaotic game of cat and mouse followed, with the fishermen luring the naval vessels into shallow and often invisible reefs from which only the fishermen knew escape routes. The fishermen also attached lengths of rope to heavy chains as a way of stopping the fleet. When the Naval craft approached a fleeing skiff, the fishermen would throw a rope beneath the Navy hull. As the rope became entangled in the propeller, the chain would also be pulled into the craft's propellers, causing in some cases serious damage. Although several fishing boats were harmed, so too were several naval craft. No gunfire was exchanged, no one was seriously injured, and the Navy chose to stop its operations. The shallow coastal waters had become a play within a play, a mock battle between unarmed fishermen and the most powerful Navy in the world.[20]

During the next several years, similar encounters occurred regularly between the fishermen and the Navy. As the fishermen became more experienced in civil disobedience, their tactics grew increasingly sophisticated. Advance notice was normally provided to islanders by the Navy before holding maneuvers that closed their fishing grounds. These warnings provided Zenón and others with ample time to alert news networks, which increasingly came to document the mismatch in power between the antagonists. Film soon became an unexpected ally of the protesters, giving them a powerful counternarrative to the U.S. military's projected calm. Each confrontation became a public relations nightmare for the Navy, as high-powered landing craft with machine gun turrets chased yellow and pink fishing skiffs through turquoise and royal blue tropical reefs, all while destroyers, cruisers, and aircraft carriers stayed nearby. The fishermen would stand at the bow of their skiffs, hurling rocks from slingshots at the pursuing craft. The islanders' bravado and the Navy's frustration were instantly telecast around the world via satellite on international networks including CNN and BBC.

Local activists were often sons or grandsons of those who had been forced to leave their homes and resettle. For some their anger and resentment grew from their parents' or grandparents' forced relocation; for others the closure of fishing grounds and ensuing economic hardship were the most compelling issues. And many of the islanders were frustrated by the Navy's relentless bombing and shelling, which hindered the development of a tourist economy. Consequently the protesters had one target—the U.S. military—but various objectives. Some groups fought for fishing rights or compensation. Others sought property rights to Navy lands. And still others favored complete political independence, meaning they argued for closure of the base, as well as severance of Puerto Rico from U.S. territorial claims.

The protests intensified when in 1999 David Sanes Rodriguez was killed by a Navy bomb that was accidentally dropped on the Navy's observation post adjacent to the live impact area.

MILIVY ADAMS

José Adams is a baker and a fisherman who lived on Vieques when his daughter Milivy, a two-year-old, developed a lump on her skull. Milivy's condition worsened and was eventually diagnosed as neuroblastoma, a form of cancer that ravages very young children by sending malignant tumors to different parts of the body. If diagnosed early, and proper medical treatment is received, there is a 90 percent survival rate when identified in one-year-olds. Survival rates decline rapidly with later diagnoses, however. In Milivy's case, the cancer spread quickly to her lung, kidney, leg, and hand. She fought bravely, as did her parents José and Zulayka, and the family moved to New Jersey so that Milivy could receive treatment at the nearby Children's Hospital of Philadelphia. She received two bone marrow transplants before her doctors concluded that her frail body could not sustain additional treatment. She passed away on November 17, 2002.

Tears welled in José's eyes as he recalled his daughter's brief, traumatic life and his family's desperate search for a cure. He resents the difficulty they faced in obtaining quality medical care on Vieques, especially their inability to receive an early and accurate diagnosis, and afterward more specialized advice and treatment. Vieques, with its population of nearly 10,000, has never had a pediatrician, let alone an oncologist. José's very first trip off the island was to obtain specialized medical advice and treatment for his daughter. The early misdiagnosis, search for specialized care beyond Vieques, and insurance problems delayed the prompt treatment that Milivy needed to survive. Her courage and hope inspired all who heard her story, and she quickly became a symbol of the island's future. Her face is painted on the wall of the island's school, in a colorful mural facing the village square.

José has the sense that somehow the environment of Vieques caused the death of his daughter, and he worried especially about the family's habit of eating fish that he regularly caught and brought home. Fearful that his one-year-old son might experience the same fate as Milivy, he clipped a lock of the little boy's hair to be sent to a medical laboratory and examined for heavy metals and explosives. José grew to blame himself for his daughter's cancer.[21] But many other friends, family, and community members are quick to blame the Navy, given its chemical and physical abuse of the island's environment. It is obvious to all that the Navy created an enormously hazardous environment in Vieques, dropping or firing bombs, artillery, and nearly two hundred different types of ordnance on the island for nearly half a century. Some of the chemicals that make up munitions are lead, mercury, cadmium, and explosives, which are recognized to be persistent carcinogens, neurotoxins, and endocrine disruptors.

The Navy readily acknowledges that it has fired weapons and released chemicals into the area, but it consistently argues that these compounds are stable, and that their distance from the community has prevented significant hazard to the islanders. The government's strategy—admitting contamination, but denying human exposure—has been effective given the constraints of tort law, which requires, as a first step toward any award of compensatory damage, strong evidence that the activity is responsible for the defendants' loss of health.

One such compelling case may involve neuroblastoma, a kind of cancer that produces within immature nerve cells malignant tumors that tend to metastasize rapidly. Children are most prone to the disease; human susceptibility to nervous system tumors, including neuroblastomas, appears to decline with age. The only known environmental cause of pediatric nervous system tumors is ionizing radiation. Perhaps other environmental carcinogens, or mixtures of these carcinogens, are successful in inducing the disease, but these have not been tested experimentally in animal studies.[22]

Neuroblastoma is rare among infants: in the United States, three cases are normally discovered among 100,000 children younger than one year, and this incidence drops to one case per million children by the time the children are ten years old.[23] Since Milivy's diagnosis, two additional cases of childhood neuroblastoma have been diagnosed among Vieques's 3,000 children. Three cases among a population of fewer than 3,000, roughly a rate of one in 1,000, is a risk significantly higher than the national average of 3.5 among 100,000. The rarity of the cancer, however, makes the statistical significance of this rate difficult to judge. Should this be considered a cancer cluster? The extremely small number of cases makes it possible that the group of cases arose by chance, rather than resulting from environmental exposures that occurred on the island. Many believe that the Navy is responsible for Milivy's death and the high prevalence rates of many other illnesses among the Viequense, but proof remains elusive. For now, Viequense residents have the burden of producing a preponderance of evidence to demonstrate that their illnesses are related to exposures caused by the Navy. Until the burden of proof is switched to the Navy to demonstrate that their chemicals played no role in the prevalence of illness, the islanders face an uphill scientific and legal battle.

ZAIDA TORRES

Zaida Torres raised a daughter and two sons on Vieques. She practiced nursing for nearly eighteen years, eventually becoming the emergency room supervisor at the local hospital, where she often was the first health professional to have contact with seriously ill islanders. The absence of physicians, and generally poor recordkeeping, meant that for nearly two decades she probably had one of the best perspectives on Viequense health trends.

By 1975, Torres believed that the islanders were suffering from a higher than normal rate of cancer. Within the next decade, she had come to believe that her community also had an abnormally high incidence of neurological, endocrine, cardiovascular, and respiratory diseases—although no disease registry for these conditions existed on Vieques or in Puerto Rico, so comparisons of Viequense health to that of other communities were not possible. Her own daughter was diagnosed with cancer in 1985, leading to a tragic and desperate struggle that ended with death at the age of fifteen. When her daughter lost all of her hair during chemotherapy, Torres shaved her own head to make her daughter feel less self-conscious. In 1993, Torres herself developed cancer, a battle that continues today. This collection of experiences has led her to believe that somehow the Vieques environment, which had been steeped in hazard-ous substances during the Navy's half-century of weapons testing and training, must be playing a role in the islanders' illnesses. Although she did not have proof, the severity of both contamination and disease were sufficient for her to believe that a causal relation existed. To her the con-clusion is simply common sense.

In 1998 Torres created the Alliance for Women and Children, or Allianca. Its purpose is to define and promote women's and children's rights, to establish registries for illnesses that afflict them, to pursue research that explores relations between environmental contamination and illness, and to educate women—especially young teenagers—about birth control and abortion rights.

The statistics compiled by Allianca are grim. Nearly 54 percent of women on the island are unemployed, and most are single parents. Vieques now has the highest teenage pregnancy rate among the seventy-five municipalities in Puerto Rico (34 percent). These pregnancies often result in the preterm birth of infants with low birth weights, a condition that increases the risk of other serious health problems later in life. Teen pregnancies also elevate the school dropout rate, and increase reliance on government welfare and Medicaid.

Torres has a plan to break Vieques women and children from this cycle of poverty, poor health care, minimal education, poor nutrition, and associated health problems. The Viequense face so many challenges, she argues, that they should not have to worry about the metals, pesticides, solvents, explosives, and other chemicals that she believes pose special threats to island women and children. The high incidence of numerous health problems, together with the islanders' limited capacity to manage both economic and health risks, have led her to conclude that Viequenses should not have to bear additional hazards posed by the Navy's chemical contamination. By tracking and publicizing disease prevalence and the sluggish pace of government cleanup efforts, she hopes to attract public and private funds to the island.

When asked about the future of Vieques, Torres reflected for a moment before articulating her vision. Her ideal society would be transparent and democratic, conditions absent for most Viequenses during the twentieth century, when they were dominated first by sugar farm owners and then the Navy. The islanders would become economically self-reliant, breaking their dependence on welfare or other forms of government support. They would be well-educated, and have access to excellent health care at a reasonable cost. Their environment would be restored to its pre-Navy condition, and the people would be free from fear of environmentally induced disease. She knows that the community has a long way to go.

In August 2004, Johnny Rullan, former secretary of public health in Puerto Rico, published a comparison of the incidence of various illnesses on Vieques to that in the entire Puerto Rican population. Rullan's staff had conducted a door-to-door investigation of Vieques and found a cancer rate 27 percent higher than among those living on the main island. The prevalence of other illnesses was also higher: hypertension rates were elevated 34 percent; asthma, 16 percent; diabetes, 28 percent; epilepsy, 116 percent; and cardiovascular disease, 130 percent. The same study found no significant differences between the two groups in health-related behaviors such as smoking habits, levels of physical activity, numerous demographic characteristics, and vitamin intake.[24]

Other indicators suggest that the Vieques islanders are living under extremely stressful conditions. In 2000, a full 81 percent of Vieques children were living beneath the U.S. poverty line, with most being raised by a single female. In 2002, 18 percent of island babies were born with "low birth weight," meaning they face a greater risk of asthma, diabetes, and abnormal neurological development. In 2003, the pregnancy rate among Vieques girls between the ages of twelve and eighteen was 34 percent. The risk of infant death during the first year of life was 25 percent higher for those babies born to Vieques women than for those born in Puerto Rico at large.[25]

Many have challenged islanders' claims that the Navy is responsible for these health conditions. They argue that there is little evidence of human exposure to toxic chemicals, and that even if exposures were demonstrated, their association with disease would remain unproven. Yet there can be little debate that the Navy is responsible for Vieques's depressed economic conditions, and that the poorest in society normally suffer from the highest burden of disease. The islanders' poverty and remoteness also prevent them from obtaining quality medical care and insurance. Underlying this community under stress is an environment bearing an enormously complex mixture of toxic substances released by the Navy during its tenure. What were these chemicals, and how and where were they introduced to the island's environment?

CHAPTER SIX

Ravaging Landscapes and Seacoasts

THE NAVY FULLY INTENDED to build an Atlantic version of Pearl Harbor on the triangle defined by Ceiba on the Puerto Rican mainland and the islands of Vieques and Culebra. The two islands offered the opportunity for staging mock military campaigns that demanded coordination among the Navy, Air Force, Marines, and Army, often with participation of other nations in the NATO alliance. The Defense Department used Vieques as a practice and training site prior to actions in Guatemala in 1954, Cuba in 1961, Vietnam between 1965 and 1972, Santo Domingo in 1965, Chile in 1973, Granada in 1983, Panama in 1989, and Iraq in 2002 and 2003.

During the exercises, aircraft carriers would lie offshore and launch fighter planes while marines, riding hovercraft that were launched from enormous sea-level doors in transport ships, stormed the beaches of crushed coral. Destroyers, cruisers, and battleships would fire artillery shells, rockets, and missiles from offshore locations. Underwater mines were suspended in harbors and land mines were buried on beaches. Once ashore, tanks and artillery fired live ammunition at advancing troops, and after the exercises had concluded, other units would practice sweeping the area—often unsuccessfully—for live munitions.

Fighters and bombers, many launched from nearby aircraft carriers or the Roosevelt Roads Naval Air Station in Ceiba, strafed and bombed targets on the eastern end of Vieques and Culebra. Several airstrips were cut into the Vieques landscape; one, nearly a mile long, provided access for large jets. Another was lined with dysfunctional tanks, aircraft, armored vehicles, and some structures—all of which became target practice for pilots as they fired machine guns and rockets while dropping bombs.

A network of roads rises from the shoreline into the densely forested hillsides on the western end of the island, where soon after the land was acquired from the sugar companies, almost two hundred Walmart-sized concrete bunkers were built into the hillsides, then covered with soil and planted with sod for camouflage. After closing in 1948, they were reopened in 1962 following the Cuban Missile Crisis and housed weapons until 2001. They stored missiles, rockets, bombs, and chemical, biological, and nuclear weapons. The island eventually became one of the largest munitions containment facilities in the Atlantic Basin.

In the 1960s, the Defense Department opened the island to other nations for training troops and deploying weapons. It soon became the largest multinational training and weapons testing site for NATO nations and other allies. Germany, France, England, and some Latin American nations sent their air forces to Roosevelt Roads. Raytheon and General Electric supplied and sold weapons stored on the island. By the middle of the 1990s, the United States was receiving nearly $85 million a year from its allies. As the Navy's Vieques website advertised, the island offered several advantages for friendly nations seeking to try out a variety of conventional and unconventional weapons: "The complex can support air-to-ground, mine delivery, naval gunfire, artillery exercise, and subsurface assaults. Real-time critique data is available for all exercises and a visual scoring system is capable of measuring bomb drops within one foot. The complex allows a full amphibious assault to be conducted in the most realistic training environment in the world."[1]

CHEMICAL AND BIOLOGICAL AGENTS

Vice Admiral John Shanahan has estimated that between 1980 and 2000 the Navy dropped nearly 3 million pounds of ordnance on Vieques every year. Records were not well kept before this time, but a similar intensity of munitions release during the entire fifty years of training would have contributed 150 million pounds of ordnance to the island's environment. In fact, the Navy disclosed in 1979 that during peak training periods,

7,600 bombs were dropped on the island per month. The intensity increased during the 1970s in part because in 1973 Congress, responding to resident protests, had directed that the nearby air-to-ground training range on Culebra be closed. More recently, government inventories have detailed that an average of 14 million pounds of ammunition were dispatched by the Naval Ammunition Storage Depot each year in 1993 and 1994, suggesting a scale of testing greater than other estimates.[2]

In the shallow waters near the island's shores, bombs of various sizes and shapes lie surrounded by mangled and rusting iron. Some may have missed their land target altogether, while others probably first bounced off the land like a flat rock skipping across the water. Many came to rest at the bottom of the sea, providing homes for coral, crabs, and damselfish that ironically have found them to be a safe haven.

The Navy has attempted to minimize the problem of unexploded ordnance in coastal waters by claiming that only 5 percent of bombs were likely to have fallen in near shore waters. Yet even this rate of error would leave 45,000 bombs lying beneath the sea. Similarly, the Navy has estimated that 72,000 pounds of explosives were dropped into coastal waters each year, meaning that over the past three decades, several million pounds of ordnance has likely come to rest under Vieques coastal waters.[3]

Not only were bombs dropped from aircraft; millions of artillery shells were lobbed at the island from offshore ships. The Navy in 1979 disclosed that each year between 120 and 130 U.S. and foreign ships took part in bombing and naval gunfire support exercises, which normally spanned 100 to 115 days per year. The government estimated that 40 percent of the shelling—which involved primarily two- to five-inch artillery shells—missed their land targets and instead sank in coastal waters. With millions of rounds fired, that means that hundreds of thousands of rounds likely lie offshore.[4]

The closing of the nearby Culebra naval bombing range increased military training on Vieques. Between 1983 and 1998, Vieques was bombed, strafed, or targeted with naval gunfire an average of two hundred days a year.[5] The U.S. Agency for Toxic Substances and Disease Registry has estimated that thirty-nine tons of high explosives were likely dropped on the live impact area on a single high-use day, and that 90 percent of these bombs probably did not detonate. The agency also estimated that

hundreds of thousands of rounds of automatic weapon rounds were fired
in nearby areas. In 2000, the Navy reported that it had dropped between
3 and 4 million pounds of ordnance on Vieques each year between 1983
and 1998. If one assumes this to be an average figure for the larger span
of 1950 to 2000, this could amount to 200 million pounds of weaponry
deposited on or near Vieques during the U.S. occupation.[6]

Chemical weapons were also released to the Vieques environment.
Agent Orange, the Vietnam-era defoliant containing the potent combination of the herbicides 2,4,5-T and 2,4-D, was sprayed at various locations
on the island. Napalm, gasoline in gel form, was dropped from aircraft.[7]
Many pesticides—including DDT, which was discovered recently in a
former government landfill—were released on the island to control the
thriving insect population. And solvents used to clean engine parts were
dumped into pits, contaminating soils and nearby groundwater.

Another source of chemical contamination was fuel. During the
Navy's tenure on Vieques, fuel supplies, including diesel, gasoline, and
jet fuels, were spilled or deliberately released from vehicles. Fuels were
also intentionally sprayed on the landscape, set on fire, and then treated
with flame-retardants during training exercises. The firm Baker Environmental Consulting has documented no fewer than 7,000 pounds of
rocket fuel—amines and nitric acids—dumped at three locations on the
island.[8]

Diesel fuels were delivered to the Camp Garcia Base via tanker. A
pipeline reached from the rocky shore into deeper ocean waters, where it
lay submerged. Each time the fuel was replenished, the line was purged,
sending nearly a thousand gallons of diesel fuel into the Caribbean Sea.
The tanks were filled four times a year, which means that over the pipeline's twenty-five years of operation, nearly 100,000 gallons of fuel may
have been released into waters that Vieques fishermen have long considered a prime fishing habitat.

Chemical and biological weapons testing is also a part of the island's
troubled legacy. Robert McNamara, secretary of defense under President
John Kennedy, ordered a series of tests to assess troop readiness to with-
stand assault by chemical and biological weapons. The Pentagon declassified documents in 2003 that report the military had previously tested
various chemical and biological agents on nearly 5,500 troops in Vieques,
Alaska, Hawaii, Florida, Canada, the Marshall Islands, and Britain. In

1969, for example, Marines conducting a mock assault of the island were sprayed with trioctyl phosphate from A-4 Marine jets. The Pentagon acknowledged that trioctyl phosphate causes cancer in some test animals, but asserted that its effects on humans were unclear. Because trioctyl phosphate was classified as a "chemical simulant" rather than a "harmful warfare agent," "no efforts were made to protect personnel." Other chemicals tested during the program included the nerve agents sarin, soman, tabun, and VX; bacteria; viruses; fungi; phthalates; flame retardants; riot control agents; and *Aedes aegypti* mosquitoes, presumably to determine their potential to infect troops with parasites that cause illnesses such as malaria.[9]

President Nixon renounced the development and use of biological weapons on November 25, 1969, restricting research to "techniques of immunization and measures of controlling and preventing the spread of disease."[10] This ended the testing of biological weapons on the island, but not chemical or more conventional explosives.

The U.S. Institute of Medicine within the National Academy of Sciences is now charged with the responsibility to explore any illness that may be associated with the U.S. Department of Defense research effort. There is no evidence that the military examined the potential of trioctyl phosphate to accumulate in marine food webs, although it was sprayed directly over some of the island's most productive fisheries habitat, and directly over the nesting site of endangered sea turtles.

TOXIC METALS

During its exercises, the Navy also routinely released "chaff" from aircraft flying over or near the island. Chaff was first introduced by the United States during a bombing raid on Bremen, Germany, in 1943 as a way of hiding planes from German radar. It is composed of strips of aluminum, which conduct electrical signals, wrapped around silicon strands. Bundles and rolls of chaff are generally dumped from airplanes at high altitude and the material slowly descends, drifting with the winds. When the area is scanned by radar, the aluminum in the chaff distorts the signal. But this confusion isn't always desirable. Sometimes when chaff was used in bombing runs, San Juan weather reporters would mistakenly predict stormy conditions even when skies were clear. In addition, the material can confuse air traffic control regulators, thereby endangering civilian flight safety. It has also been blamed for power outages, and some have raised health concerns. Lead was a known component of chaff until at least 1987: more than a decade later, in 1998, the Air Force still maintained stocks of leaded chaff.[11]

The Centers for Disease Control has estimated that as much as 133 tons of chaff were released each year near Vieques. Contamination levels are more difficult to estimate. The Navy has never measured ground-level concentrations of lead, aluminum, silicon, or fiberglass, even though it leased much of its acreage in the vicinity of the bombing range to land-less farmers for grazing. Significantly, the Navy frequently touted the contribution of the low-cost leasing program to the local economy, but it never sought to determine whether the pollutants it was releasing into the area had been absorbed by plants or cattle—or by the people who used them for food.[12]

RADIATION

In 1994 the Navy admitted to firing 267 rounds (88 pounds) of depleted uranium ammunition from a fighter jet machine gun. Depleted uranium is used to harden the casings of some ordnance so it can pierce armor or other fortified surfaces. Uranium is a natural element mined from the earth and enriched for use in nuclear reactors. Highly enriched uranium may be used to construct nuclear weapons, and depleted uranium is a by-product of the enrichment process. It is nearly 40 percent less radioactive than natural uranium, but its use in 1994 required a special permit from the U.S. Nuclear Regulatory Commission—a kind of permission never requested by the Navy. Intense controversy surrounds the use of depleted uranium weapons, primarily because it is difficult to ascertain how much contamination is released and whether human exposure to the casings and ordnance may lead to health problems. Depleted uranium ordnance emits alpha, beta, and gamma radiation, with gamma being the most dangerous. Moreover, when depleted uranium ammunition is used, a significant percentage of the uranium vaporizes upon impact, making it difficult to detect and clean up.

Once the Navy realized that it had fired the rounds illegally, it reported the incident to the Nuclear Regulatory Commission, while attempting to recover any remaining shell casings. Fewer than half of the depleted uranium rounds fired were found; most were pulverized or had vaporized

upon impact. Uranium, however, does not disappear, and some islanders worry that fine particles and vapor have spread out over the island,
contaminating the land, plants, and marine life.

Another source of nuclear contamination came about from the testing of nuclear weapons. The battleship USS *Killen* was commissioned in
1944 and was on active duty during the U.S. invasion of Borneo and other
Philippine battles, when it survived a direct bomb hit that killed fifteen
of its crew members. After supporting the U.S. postwar occupation of
Japan, the ship was decommissioned. But at the height of the atmospheric
nuclear weapons testing program, the Navy, anxious to know the resilience of its fleet to a nuclear attack, brought the ship out of mothballs and
anchored it, with dozens of other vessels near or at ground zero in the
Marshall Islands, for the Wahoo and Umbrella blasts during the spring
of 1958.

The Baker Shot had taught the Navy and the Atomic Energy Commission that the force unleashed by underwater nuclear explosions was
enormously destructive. Baker sent nearly 2 million tons of radioactive
water a mile into the air. As this mass fell from the sky into the lagoon,
waves radiating from the center reached a height of ninety-four feet. All
ships in the vicinity were drenched by radioactive water, foam, salt spray,
particles, and vapors. Radiation levels on these ships were so high that
monitors reached their limits within minutes.

Robert Oppenheimer, the physicist who led the development of the
first atomic bomb at Los Alamos, had predicted these effects and in 1946
had objected to the Navy's testing of nuclear weapons. He argued that
the use of nuclear weapons against naval vessels would be unlikely, because enemies would favor targets of concentrated populations, such as
those in Hiroshima and Nagasaki. Oppenheimer was further concerned
that underwater blasts would produce an exceptionally toxic marine environment, perhaps destroying coral reefs and associated marine life for
decades.[13]

His intuition proved to be accurate; the radiological contamination of
ships lingered far longer than the Navy had anticipated. Crews sprayed
and scrubbed the surfaces of those ships not sunk by the blasts in order
to remove the radiation, but the area was so toxic that they could stay on
the vessels for only a few minutes at a time. The radioactive particles were
very difficult to remove after they had worked their way into chipped paint

and rusted surfaces. The sailors experimented with different methods to pressure-wash them away, including blasting crushed coconut shells, rice, and ground coffee at the ships' surface, but none of these techniques significantly improved the cleanup efforts.

Moreover, soon the contaminants had spread throughout the lagoon, and because active-service ships used seawater for cooling purposes, the irradiated lagoon water soon contaminated the pipes and engines of manned vessels. Barnacles and algae that had attached to the ships' hulls also absorbed radiation, increasing the crews' exposure. Once the commanders realized they were living in a contaminated soup like that which Oppenheimer had predicted, they moved the manned vessels beyond the lagoon to uncontaminated waters. Thirteen of the less-contaminated target vessels were remanned and sailed back to the United States. The more contaminated ships, however, had to be abandoned near the Bikini or Kwajalein atolls, or towed to the Hawaiian Islands and then sunk.[14]

The Killen, however, was to have a different fate. After the damaged ship was towed through the Panama Canal and Caribbean Sea to Vieques, it was anchored near the live impact area, where it served as a target for aircraft and naval artillery. For nearly twelve years beginning in 1963, the Killen was barraged by rockets, mortars, machine-gun fire, artillery rounds, and bombs. Much of the ordnance missed its mark only to sink unexploded to the ocean floor. The Killen eventually came to rest in seas only thirty-three feet deep, just 500 meters from shore at the head of an exquisite bay, Bahia Salina del Sur. The bay also happens to be one of the most productive fishing sites in the island's coastal waters. Today the rusting hulk acts as an artificial reef covered with algae, coral, and barnacles, and is home to many creatures including damselfish, which dart in and out of its five-inch gun barrels.

Although the Killen has sunk out of view, many questions remain about its history and what it might mean for the peoples and environment of Vieques. How radioactive was the vessel following the nuclear blasts? Have radioactive particles contaminated and damaged coral, fish, and other marine life in the area? Could those who have eaten fish caught near the Killen have been exposed to radionuclides, and if so, what might be the health implications? The same questions might also be asked of those vessels towed from ground zero to other locations, including the Hunters Point Naval Shipyard in San Francisco, where radiation was sandblasted from the surfaces of many ships that participated in the nation's Pacific nuclear experiments.

A HISTORY OF TRIAL AND ERROR

The term "training ground" implies a location where error and imprecision are expected. The purpose of training is to increase the accuracy of real military operations, and to avoid unintended injury and damage. Television images of electronically guided bombs and missiles during the air assault on Baghdad leave an impression of technological precision and the capacity to avoid unintended damage. But the situation on Vieques over the fifty years of its use as a military training area was vastly different than that in Iraq. Most of the munitions released on the island were guided instead by human judgment, and a significant proportion of the ordnance missed its target, landing in nearby coastal waters.

On a balmy April evening in 1999, David Sanes Rodrigues was working his shift as a civilian security guard at the observation post overlooking the bombing range. His job was to prevent unauthorized personnel from entering the area, and he was often kept busy turning away those protesting the Navy's presence on Vieques. As he manned his observation post that evening, two Hornet fighters left the USS *John F. Kennedy* aircraft carrier a dozen miles offshore, preparing to bomb the airstrip nearby. The pilots, traveling at four hundred miles an hour, aborted their first two passes over the target due to cloud cover and poor visibility. On the third pass, the flight leader ordered the wingman to drop two 500-pound bombs on a target lying on the air strip. But the pilot locked his guidance system to the wrong end of the runway, that nearest the observation post. When the bombs exploded, Sanes was knocked unconscious and soon bled to death. Four others at the post, including observers and security guards, were also injured by the explosions.[15]

The Navy was quick to point out that this event marked the first civilian death on Vieques related to military activity during nearly sixty years of training and bombing exercises. Protesters, however, countered that pilots had made similar errors in the past, dropping ordnance on the same observation post in 1995 and 1998, causing extensive structural damage even if no injuries. Other near-tragedies had also occurred. In 1993, five live bombs had been mistakenly dropped near a local landfill, nearly five miles from their targets. Another had been dropped in a residential neighborhood though it fortunately caused no injuries. In 1995, bombs had been dropped near the boat *Coki Ayala* with no damage or injury, and in 1996 an underwater explosion had injured a local fisherman.

THE VIEQUES PARADOX

Some of Vieques's best fishing sites lie just off the shores of the bombing range. Within Bahia Salina del Sur, schools of damselfish, yellowtail snapper, grouper, angelfish, and parrotfish are often visible in a glance, and sea turtles, barracuda, and sharks are not uncommon. Seagrass beds within the bay are also rich with a favorite seafood of the Viequense: corrucho or conch, some the length of a forearm and with shells ranging from deep red to coral and orange.

Which pose the greatest ecological and human health threat, the offshore unexploded munitions, or the chemical soup leaking from contaminated land into the lagoons and bays? The offshore weapons most likely create the greatest hazard for those species and individuals that spend time near them, including people who catch them in fishing nets or are foolish enough to retrieve them as souvenirs. In 1999 James Porter at the University of Georgia discovered that rusting bombs were indeed leaking explosives into surrounding seawater.[16] But the large volume of seawater, and its movement with tides, waves, and currents, probably dilute the hazardous chemicals quickly.

In September 2004, Tropical Storm Jean stalled over Vieques, drenching the island with more than two feet of rain in just two days. As streets turned into rushing streams and rivers, many Viequense watched from their porches as uprooted plants, construction debris, barrels, and trash raced down the streets toward the sea. Human sewage burst from manhole covers, and animal waste flowed freely across lands grazed by cattle, horses, sheep, and chickens. Pipes carrying water from the main island of Puerto Rico were shut down for five days. Large sections of asphalt disappeared, and water running over unpaved roads carved ruts that could easily swallow a car. Rain of this intensity may have a cleansing effect on the landscape, but the brown, sediment-laden water that surrounded the island for weeks following the massive storm suggested that the coastal waters had become seriously contaminated.

Tropical Storm Jean had a different effect on the low-lying bombing range. During an ordinary rainy season, the bomb craters often fill with water, but Jean turned the entire area into a toxic lagoon in which various weaponry, some exploded and some intact, rusted, leaking their chemical components. As the lagoon waters rose with rainwater from Storm Jean, they gradually breached the dune (part of a barrier of dunes that the Navy had for decades believed would effectively contain the chemical mixture), spilling into Bahia Salina del Sur on two sides of the bay. The waters of Bahia are normally a brilliant turquoise color where they lie over sand, and teal blue above the seagrass beds. For several weeks after the storm, however, murky lagoon waters painted the nearby Caribbean Sea clay-brown.

The waters in Vieques bays are somewhat sheltered, meaning that both currents and waves have less of a diluting effect than occurs in the open sea. The bays and lagoons are also nurseries for many species of commercially valuable fish and shellfish. But as stormwater rushes around and through the cracked and rusting remnant munitions, what chemicals are released to coastal waters? Do they concentrate in bottom sediments and marine organisms, then move up a marine food chain that ends at the islanders' dinner tables? Are exposures higher than would be found in the diets of other Caribbean islands? Is the mixture of chemicals in Vieques fish similar to the mixture of chemicals released from military weapons? And do Viequense people carry a similar chemical burden, in a pattern that ties them uniquely to the Navy's weapons?

During a recent visit to the island, I came to realize that these same types of questions could be asked of all seriously contaminated sites in coastal areas—and that for most communities, they will never be answered. It simply costs too much to investigate and understand the history of contamination, human exposure, and any potential health effects. As I swam along the shore of the bombing range back to the boat anchored above the *Killen,* I paused and turned toward the shore. I sensed what I now call the Vieques paradox. The water's clarity and shimmering blue-green hue, the lush backdrop of reemerging plant life on the hillsides, and the brilliant colors of tropical fish beneath the surface were all nearly impossible to reconcile with my understanding that this is one of the most contaminated sites in the world.

CHAPTER SEVEN

Mercury

IN 2003, CARMIN ORTIZ-ROQUE, an obstetrician in San Juan, Puerto Rico, led a study of the concentration of mercury among Puerto Rican women. She made a troubling discovery: women of childbearing age living on Vieques had mercury levels nearly nine times higher than their counterparts living in northeastern Puerto Rico, and nearly seven times higher than levels found in the United States. Ortiz-Roque reported that eleven of the forty-one Viequense women tested had levels of mercury "sufficient to cause neurological damage in their future children." She then noted a correlation between mercury levels and fish intake, and that Viequense women reported consuming fish at nearly twice the frequency of their counterparts in Puerto Rico, all of which suggested that the mercury levels might be explained by the fish the women were eating.[1] Consumer response on Vieques was immediate. Many residents stopped buying fish of all types, fearing for the future health of their children. Fish prices dropped, and commercial fishermen lost considerable revenue.

Studies such as Ortiz-Roque's may be valuable first steps in answering questions about environmental contamination on the island, but they also raise many more questions than they answer. The implication is that Vieques fish are contaminated, but what species of fish were consumed? Where were they caught? What other sources of mercury exposure are plausible? And did the women involved in the study respond accurately to the study's questions?

ABOUT MERCURY

Mercury is a naturally occurring element, a component of the earth's crust that is released to the environment from volcanic eruptions, forest fires, the weathering of rocks, and soil emissions. Nearly 158 tons of mercury are emitted annually into the air from the United States. The EPA recently estimated that four human activities account for 80 percent of emissions: coal fired boilers (33 percent), municipal waste combustion (19 percent), commercial or industrial boilers (18 percent), and waste incinerators (10 percent).[2] (Incinerators are included in this list because many people simply throw away mercury-containing batteries, thermometers, blood-pressure meters, fluorescent lamps, compact fluorescent bulbs, and electronic switches.) Throughout much of the twentieth century, organic mercury was also widely used as a pesticide and fungicide. By 2000, however, many of these uses had been restricted in wealthier nations due to concern over mercury's persistence in marine and aquatic ecosystems, and its toxicity.[3]

Ice cores collected from the Upper Fremont Glacier in Wyoming contain a record of atmospheric mercury deposition that geologists have studied to estimate natural versus anthropogenic mercury sources over the past 270 years. Overall, human activity contributed 52 percent, volcanic emissions 6 percent, and preindustrial background levels 42 percent. Over the last hundred years, however, human activities have accounted for nearly 70 percent of the total mercury measured in the cores.[4]

Once mercury is released to the atmosphere, it can move long distances attached to fine dust particles or water vapor before settling to earth. Its airborne movement and eventual descent to earth is similar in many respects to nuclear fallout. The length of time it spends airborne depends upon its "species" or chemical form.[5]

Mercury tends to exist in one of its inorganic forms, such as $Hg(II)$, and once on the earth's surface tends to be quite mobile, traveling with surface water. It can bind to organic compounds—those containing carbon—through a process called methylation that occurs in the presence of microorganisms. The product, methylmercury (MeHg), is easily absorbed by aquatic species and makes up nearly 95 percent of the mercury found in fish. As larger organisms eat contaminated smaller ones, the methylmercury concentrates, so that very large predatory fish may have mercury levels 1 to 10 million times higher than surrounding waters.[6] Humans are exposed to methylmercury predominantly by consuming fish, and the levels of mercury in fish vary significantly by species, age, and location. In general, larger (and older) predatory fish such as sword-fish, shark, and tuna have higher levels than smaller (and younger) fish and shellfish like shrimp, salmon, scallops, and oysters. The low levels in salmon are explained in part by their young age when caught, but also by the fact that many are raised in hatcheries, where they are fed fish meal to promote rapid growth.

Studies of methylmercury levels in other foods including meats, poultry, vegetables, fruits, and cereals have found levels to be 1,000 to 10,000 times lower than those found in fish and shellfish. Variation in diet, then, explains most of the differences noted in individuals' methylmercury concentrations.

Mercury has been known to be poisonous since the late nineteenth century, although today most people have little understanding of its danger or how they are exposed. Once inside the human body, methylmercury is readily absorbed into the bloodstream through the gastrointestinal tract. It is moderately lipophilic, meaning that it binds to fat, and it moves easily through cell walls where it attaches to proteins. It migrates to most human tissues, but concentrates in the brain, liver, and kidneys. Mercury moves easily across the placenta of pregnant women to circulate in the vascular system of fetuses, and it crosses the blood-brain barrier.[7]

The half-life of mercury (that is, the time it takes the human body to reduce its concentration by 50 percent) ranges between 44 and 180 days. Mercury ingested from a single meal could remain in your body at some level for more than a year (Figure 7.1). If a pregnant woman weighing 50 kilograms ate a tuna fish sandwich with mercury at the high end of the range detected in 2001 (0.75 parts per million in canned tuna), and if the longer half-life of 180 days is assumed, her blood levels would not return to the maximum safe level recommended by the National Academy of Sciences for nearly three years. The average level of mercury detected in tuna by the FDA in 2001 was 0.17 parts per million. Assuming the more rapid excretion rate (half-life of 44 days), and this more average level of mercury, blood levels would not return to the recommended safe level for

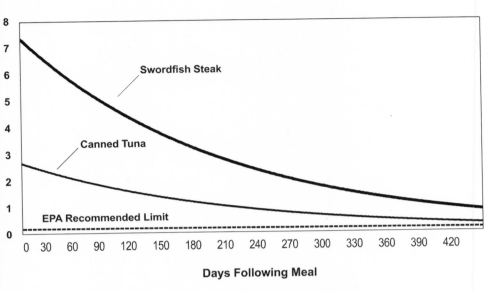

FIGURE 7.1. Mercury in human tissues. Eating a single serving of some species of fish can result in mercury concentrations that exceed the EPA's recommended limit. This example assumes that a 22-kilogram child consumes six ounces of fish at one serving. Mercury levels are derived from the U.S. FDA fish testing program for canned tuna fish and swordfish steaks. Mercury levels are normally far lower in smaller and less predatory fish such as shrimp, scallops, and salmon. *Source:* U.S. Food and Drug Administration.

nearly eight months. So mercury from a single sandwich might remain in your body between eight and thirty-six months.

Since mercury in a pregnant woman will cross the placenta, how long will it take for the fetus's blood to return to a safe level? Because of the similar blood concentration in mother and fetus, but the fetus's much smaller size, fetal exposure can be expected to be between 15 and 500 times higher than the mother's, depending on its age and weight. Given the earlier assumptions about half-life, then, some mercury from a single exposure could thus remain within the fetus (or newborn) for more than 400 days. Since the dominant routes of methylmercury excretion include lactation, the baby will also likely receive some additional exposure via breast milk after birth. Clearly, mercury exemplifies all of the traits of a very dangerous element: it is persistent, mobile, and toxic.

Intense accidental exposures to methylmercury have demonstrated its ability to cause mental retardation, cerebral palsy, deafness, blindness, and sensory and motor impairment in adults.[8] Exposure to mercury at lower levels has been associated with adverse effects on the cardiovascular system and a recent study has found that high blood mercury levels, possibly caused by eating seafood, are associated with infertility in both men and women.[9]

Methylmercury especially threatens normal growth and development of the human nervous system. At higher doses, the effects on fetuses in particular can be severe and include mental retardation, cerebral palsy, as well as visual and auditory deficits. More recently, subtle neurological effects including deficits in cognition, memory, and learning capacity are associated with early-in-life exposures.[10]

Several studies of island populations have contributed to an understanding that prenatal exposure to methylmercury can adversely affect a child's normal growth and development. The two largest have been controversial. One study conducted in the Faroe Islands in the North Atlantic Ocean found adverse neuropsychological effects (deficits in memory, language, and attention) associated with maternal consumption of pilot whale meat contaminated with mercury. The effects on brain function associated with prenatal methylmercury exposure were detectable at exposure levels commonly considered safe.[11] A second study, conducted in the Seychelles Islands in the Indian Ocean, reported no adverse developmental outcomes in children following both prenatal and postnatal exposure from fish.[12] In an effort to resolve differences in the results between the Faroe and Seychelles studies, the U.S. National Academy of Sciences convened a panel to review the evidence. The panel's experts concluded that on balance the evidence supported the more worrisome Faroe study findings.[13]

In a study of 149 children on the island of Madeira, off the coast of Morocco, children were exposed to methylmercury due to their mothers' high intake of black scabbard fish while pregnant. Maternal hair mercury concentrations on the island ranged from 1.1 to 54.1 micrograms per gram. Children of mothers with hair mercury concentrations higher than 10 micrograms per gram experienced auditory and visual deficits.[14]

Additional research has found that elevated levels of mercury in hair are associated with neurodevelopmental abnormalities among populations in Morocco and Iraq, along the Amazon River, and in Greenland.[15] An Iraqi study that analyzed eighty-four mother-child pairs poisoned by consuming mercury-contaminated grain in the winter of 1971–72 confirmed that mercury in mothers' hair at levels over 10 parts per million appeared to be related to neurodevelopmental abnormalities in their children.[16]

A 1999 study of three villages along the Tapajós River, a tributary of the Amazon, found that more than 80 percent of 246 children had hair-mercury concentrations of about 10 micrograms per gram, and were experiencing associated neuropsychological declines in motor function, attention, and visual-spatial performance.[17] Similarly, a 2002 study of Inuit children supports evidence that "prenatal or early postnatal exposures to methylmercury may cause subtle neurobehavioral deficits."[18]

Babies exposed to methylmercury in utero may also experience changes to their blood pressure. A study published in 1999 found in a birth cohort of a thousand children from the Faroe Islands that their prenatal methylmercury concentrations were correlated with higher blood pressure at age seven. Among those children who had lower birth weights, the effect was even greater.[19]

Elevated levels of mercury in humans have been found throughout the world. In addition to the studies already cited:

- A 2001 study of the Wayana population from French Guiana (where fish are known to have high levels of methymercury) found that 57 percent of the Amerindians had mercury levels higher than the World Health Organization's recommended maximum safe concentration (10 micrograms/gram); all those tested over one year of age exceeded this level.[20]
- In March 2001, the U.S. Centers for Disease Control and Prevention found that one in ten women of childbearing age in the United States is exposed to mercury at levels that could be harmful.[21]
- In 2002, researchers at the Chinese University of Hong Kong found that high blood levels of mercury are associated with infertility in men and women.[22] These findings are worrisome for any population that depends on fish as a major source of protein. Often poor coastal and island populations consume more fish than others. A recent study in California too, found that older

Americans with cardiovascular disease may purposely increase their intake of fish to obtain the vascular benefits of omega-3 oils, and so may inadvertently augment their mercury intake, and perhaps their blood pressure.

• A Finnish study of 2,005 men with coronary heart disease found that those 25 percent with the highest hair mercury levels faced a 60 percent increased risk of death from cardiovascular disease, and a 70 percent higher risk of having coronary heart disease.[23] A separate study in the United States, however, found that although mercury levels were highest among those eating the most fish, after controlling for smoking, age, and other risk factors, mercury was not associated with an increased risk of coronary heart disease.[24]

One of the most important problems in the field of environmental health is how best to understand the dose that damages human health. Identifying a threshold between safe and harmful doses has been elusive for mercury. Over the past several decades, scientists and government officials have lowered the "acceptable daily intake" of mercury in response to increasingly sensitive studies demonstrating more subtle effects at lower concentrations than previously tested. This process of discovery and regulation follows a pattern much like that which occurred with many pesticides—including some containing mercury and licensed by the government for much of the past century.

On the basis of the advances in research in this area, in 2001 the EPA recalculated the "benchmark dose" for prenatal exposure to mercury to 58 micrograms per liter of umbilical cord blood. This benchmark dose is also the maximum level recommended by the U.S. National Research Council panel in 2000. The benchmark was selected because it was the dose at which the number of children with subnormal scores on neurobehavioral tests doubled.[25]

The EPA also has developed a "reference dose" for mercury, which is defined as that level of exposure that is "likely to be without risk of deleterious effects during a lifetime." The EPA recommended that this level be set at 0.1 micrograms per kilogram of body weight per day, based on the agency's concern that methylmercury in human tissues has a long half-life and threatens children's neurological development. This means that a 50-kilogram woman should not consume more than 5 micrograms of methylmercury per day. The World Health Organization set its exposure limit recommendation five times higher than this EPA health goal, but its recommendation was designed to protect adults from such extreme effects as developing paresthesia, numbness, and tingling of the extremities and lips.[26]

Some researchers worry about variation in susceptibility among children. There is some evidence to suggest that umbilical cord blood levels are 70 percent higher than maternal levels, and differences in body weight mean that tissue concentrations are likely to be far higher in fetuses than mothers. After analyzing data from an Iraqi mercury poisoning incident, for example, several scientists estimated that susceptibility may vary by a factor of 10,000 across the population of children and adults who were exposed.[27]

Uncertainty in scientific evidence and subjectivity in government interpretation of evidence also are common—and seem to become more of a factor in setting limits for exposure to hazardous chemicals as the economic consequences for polluters, or for those responsible for hazardous site restoration, increases. If one wished to be cautious given the uncertain evidence, the acceptable daily intake level would be set very low. When the costs of precaution are high, and the evidence highly uncertain, however, a common government tactic is to set the acceptable exposure levels higher, at least until the evidence is stronger. Uncertainty such as this is often managed by applying "safety factors" when setting acceptable levels of exposure. In practice, this is accomplished by looking at the most sensitive population and discovering the lowest concentration that induces any adverse health effect. This level would then be divided by the safety factor to identify a "safe" level of exposure. The effect of this precautionary method may be to remove products from the marketplace due to "excessive contamination," depending on the magnitude of safety factor chosen. Within the United States the longest history of experience with this regulatory approach has occurred in the area of pesticide management. EPA, however, chose to employ only a tenfold safety factor for methylmercury and this level corresponds to an allowable intake of 0.1 microgram per kilogram of body weight per day, a level the U.S. National Academy of Sciences agreed with.

MERCURY LEVELS IN ADULTS AND CHILDREN

If mercury exposure can retard normal growth and development of fetuses and small children, why has the public not been warned about the threat? The Food and Drug Administration is responsible for setting allowable limits for mercury in fish, and these have been set at 1 part per million. The National Academy of Science in 2000 examined all available studies of the health effects of methylmercury exposure and concluded that neurodevelopmental abnormalities occurred at doses lower than any other adverse effects. They identified the lowest level of exposure that was associated with measurable neurodevelopmental problems (58 parts per billion of methylmercury in maternal cord blood) and agreed that a tenfold safety factor should be used to set the acceptable daily intake at a level that would assure that cord blood levels would not exceed 5.8 parts per billion. Based on these evaluations, the NAS found that the EPA's recommended limit of 0.1 microgram per kilogram of body weight per day would protect public health. The NAS also called for harmonization of policies of the EPA, FDA, and other regulatory agencies to adhere to this standard. The FDA continued to employ a limit of 0.4 micrograms per kilogram of body weight per day, thereby allowing exposures from fish four times higher than the EPA or NAS suggested, at levels corresponding to 1 part per million in fish.[28]

Assume, for example, that you purchase a swordfish steak that contains mercury at the 1 part per million government recommended limit for pregnant women. (The mean detected level reported by the FDA in 2006 was 0.97 parts per million.) If the steak is half a pound (226 grams of fish), and 1 microgram is allowed for every gram (1 microgram of mercury per gram of fish is 1 part per million), this meal would result in the consumption of 226 micrograms of mercury. For a 50-kilogram (120 pound) pregnant woman, this would result in an exposure of about 4.5 micrograms per kilogram of body weight for every day the legally allowed limit is consumed, an exposure forty-five times higher than the level that the National Academy of Sciences considers to be safe.

Given these data and the government's analysis, what should you do to keep your blood levels of mercury beneath the EPA recommended limit? Consume at most one normal meal of swordfish or tuna every forty-five days. This suggested guideline assumes an average level of fish contamination; it also takes for granted that forty-five days after the meal, your mercury level will have dropped to zero, even though it may take far longer to fully clear a single dose of methylmercury from the body (Figure 7.1).

Body weight is crucially important for determining exposure. If a child ate the same fish steak, but weighed only 25 kilograms (55 pounds), the concentration per kilogram would be double or triple that of the mother. For a fetus weighing only 1 kilogram, the concentration per kilogram of fetal body weight would generally be at least fifty times higher than the mother.

How much mercury has been found in human tissues? The levels are surprisingly high. One recent study of eighty-nine patients visiting a private internal medicine practice in California found an average level of mercury of 15 micrograms per liter of whole blood, ranging between a low of 2.0 to high of 89.5 micrograms per liter. On average, blood levels were three times higher than the NAS- and EPA-recommended levels of 5 micrograms per liter; surprisingly 89 percent of all those tested had levels above this ceiling. The patients reported consuming nearly thirty different species of fish; swordfish consumption, however, was most strongly correlated with elevated mercury levels.[29]

For several decades the fishing industry and food processors have convinced Congress and the FDA to manage mercury levels in fish by advising the public of the problem. Instead of banning sales of fish contaminated with mercury and regulating air emissions from power plants to reduce mercury emissions, the government relies on an end-of-the-food-chain warning system known as "fish advisories." The FDA's system relies on individual states to monitor their water and fisheries; states also have the responsibility for issuing public advisories. By 2008, forty-one states had issued advisories warning the public about the presence of mercury in fish. But the absence of advisories within the other ten states simply indicates the absence of testing, not that mercury was undetected. Advisories may be issued for entire lakes or rivers, or specific areas or segments. Statewide advisories have been issued for mercury in freshwater fish in only thirteen states; eleven others have issued statewide advisories for coastal fisheries.[30]

Again, fish are not prevented from being sold in the marketplace, regardless of their mercury content. Instead, the FDA assumes that the public has the ability to read signs posted near lakes and rivers, and limit consumption to levels that would prevent dangerous levels of mercury from entering the human body. Realistically, however, how many people read these signs and make the calculations to understand whether—based on their weight, gender, age, and if they are pregnant—they should eat the fish they just caught?

One way the FDA could ban the sale of contaminated fish is by using its authority to set and enforce "action levels" for food contaminants that threaten public health. When action levels are exceeded, the FDA can remove these "adulterated" foods from the marketplace, although they rarely do so. In practice, action levels have been nonbinding informal guidelines. Furthermore, action levels for mercury have undergone dramatic changes over the years. The first action level set for mercury in fish was 0.5 parts per million in 1969. Ten years later, the agency doubled the allowable contamination level to 1 part per million, and in 1984, it changed the way that it measured mercury, reporting only methylmercury, rather than total mercury. This change brought more fish into compliance with the 1 part per million standard, even though other forms of mercury can be transformed into methylmercury once inside the human body. The FDA fish sampling program was further restricted in the late 1990s. In 1998 and 1999, after testing only eighteen samples of swordfish and shark and finding that 50 percent contained methylmercury at or above the 1 part per million action level, the FDA stopped its mercury testing program for fish, leaving consumers to rely on fisheries and fish-processing firms to help them avoid what is now well-recognized in the scientific literature to be one the most important chemical hazards in our food supply.

This reliance on the private sector to monitor itself, and to report violations to the government, has long been standard procedure among environmental and health protection programs in the United States. The seafood inspection program is no different. The FDA's regulations require that private seafood processing firms sufficiently test their products to identify important health hazards, and to develop a plan to manage these risks. But the regulations are generally not followed. By 1999, 54 percent of fish processing firms did not have a plan in place to identify and manage human health threats from their products or suppliers. The FDA is even unable to estimate how many large fishing ships should also be subject to its regulations because the vessels process the fish on board.[31]

The lack of information about the scale of the problem is startling. The

U.S. Government Accountability Office reported that it took the under-staffed FDA two years to visit all registered processing plants once. This was an improvement from previous years, when a processing firm could expect a visit from FDA inspectors only once every four years. And when the FDA inspectors finally do arrive for an unannounced visit, 48 percent of the time they cannot conduct their review because seafood is not being processed at that time. When the inspectors do directly inspect products, they have found one or more serious violations of their hazard control plans in 55 percent of the cases.[32]

In March 2001, the FDA issued an advisory warning pregnant women and women of childbearing age to stop eating shark, tilefish, swordfish, and king mackerel. Tuna was noticeably absent from the FDA warning, although oddly enough, FDA officials did draft a separate health advisory warning consumers against eating more than three meals of tuna fish per month, to limit mercury exposure. In fact the March 2001 advisory encourages continued fish intake by women, stating: "FDA advises these women to select a variety of other kinds of fish—including shellfish, canned fish, smaller ocean fish or farm-raised fish—and that these women can safely eat 12 ounces per week of cooked fish." But an average 50-kilogram woman, consuming 12 ounces of canned tuna per week, containing mercury at the average detected level of 0.17 ppm, would exceed by 65 percent the maximum average daily intake recommended by the NAS and EPA. A 25-kilogram child consuming this amount of tuna in a week would consume three times the agencies' recommended safe limit. Perhaps significantly, between September and November 2000, only four months before releasing the final version of the health advisory that did not mention tuna, FDA regulators had met privately with representatives of the Bumble Bee, Starkist, and Chicken of the Sea corporations.[33]

Given this history, it is perhaps not surprising that FDA inspectors have routinely advised industry managers and FDA inspectors "not to identify methylmercury as a hazard reasonably likely to occur."[34] With this loophole, the seafood industry truly has no incentive or requirement to self-test for mercury.

The problem may be underreported, but it has not been solved. The EPA in 2004 doubled its estimate of children at risk, noting that "about 630,000 children are born each year at risk for lowered intelligence and learning problems caused by exposure to high levels of mercury in the womb."[35]

A 2003 study of fish intake by women and children by Susan Schober and her colleagues estimated that 7.8 percent of U.S. women consume mercury-tainted fish above the recommended daily intake level.[36] Mercury levels were significantly higher among those women who consumed three or more meals of fish per week. Nearly 4 million children are born each year in the United States and this same study found little difference between the blood mercury levels of pregnant women when compared to women who were not pregnant. If 7.8 percent of pregnant women have levels of mercury above the recommended blood level, and if mercury concentrations in cord blood are 70 percent higher than maternal levels, then more than 300,000 fetuses are receiving exposures that are likely to be approaching the dose where neurobehavioral deficits have been reported.

Children are not the only ones in danger of health loss from mercury. The average U.S. resident consumes about fifteen pounds of seafood per year (18 grams per day), 80 percent more than the EPA and NAS recommend. And the NAS committee in 2000 estimated that 5 percent of the U.S. population (15 million people) consumed enough fish to place their blood levels of mercury above the NAS recommended maximum concentration.[37]

The United States imports nearly 3.9 billion pounds of seafood from 160 different nations, and monitoring and surveillance pose a formidable problem. Consequently, it pays to ask where your fish come from. The answer is likely to be Canada, China, Chile, Equador, or Thailand, which account for more than 50 percent of all fish imports. If the U.S. FDA is lax in regulating fish contaminants, what is likely happening with food safety in more polluted nations that have weaker laws and more limited resources to ensure food safety?

THE LATEST WARNINGS

The EPA and FDA in 2004 jointly issued a notice advising pregnant and nursing women, women who may become pregnant, and young children to consume no more than twelve ounces of fish per week, but albacore tuna should be limited to six ounces per week, approximately one modest meal. The agencies suggested that intake of canned light tuna should not exceed two average meals per week.[38] Tuna companies vigorously fought the new advisory, and their influence resulted in the following addition to the press advisory: "FDA and EPA want to ensure that women and young children continue to eat fish and shellfish because of the nutritional benefits and encourage them to follow the advisory so they can be confident in reducing their mercury exposure as well." The California attorney general filed suit against major tuna processing companies in 2004 based on their failure to warn consumers of the presence of mercury, recognized by the state to be both a carcinogen and reproductive toxin. The attorney general also sued sixteen supermarket chains for failing to warn consumers in markets about the presence of mercury in fresh and processed fish.[39]

The EPA regulates pesticides very differently than the FDA regulates mercury. For example, human exposure to pesticides found to be developmental neurotoxins (including some mercury compounds) must be restricted to 1,000 times lower than the dose found to induce adverse health effects, under provisions of the Food Quality Protection Act of 1996. Under pesticide law, the government must also assume additional sources of exposure to the active ingredient—perhaps from indoor, lawn, and garden use or contaminated water supplies—when setting maximum daily limits for food. Further, all exposures to pesticides allowed by law must provide health protection for children, infants, and fetuses.[40]

By contrast, the government only recommends that the public limit their fish intake to control against the health threat of fish-derived exposure to mercury. Why should one standard be used for methylmercury in meat, poultry, vegetables, and fruits, while another less protective standard is used for the same toxic substance in fish?

During the waning days of the George W. Bush administration, in December 2008, the FDA issued a draft report proposing to recommend increased fish intake by women; for example, it proposes overturning the EPA-FDA warning to limit albacore tuna intake to six ounces per week with a recommendation to consume at least twelve ounces a week. The EPA, however, is challenging the claim by the FDA that the health benefits from fish-derived omega-3 fatty acids outweigh the dangers to fetuses from methylmercury.[41]

METHYLMERCURY TESTING AROUND VIEQUES

Standing on the stern of the *Pumba*, about to dive into the turquoise waters of Bahia Salina del Sur, Carlos Ventura made a fist and pounded his chest as he described his family's diet. "Fish give you a strong heart. We will live a long time, unless our fish absorbed too many of the Navy's chemicals." Given the study by Carmin Ortiz-Roque that found elevated levels of mercury among Vieques women, who eat large amounts of fish, and the consequent increase in concern among Vieques residents, scientists from the U.S. Agency for Toxic Substances and Disease Registry, within the CDC, sought to understand whether contaminated fish may be responsible for the abnormally high disease incidence among islanders.

Researchers from the ATSDR converged on the island for several days in 2001 to collect and test area fish for contamination. They chose fish from several different locations near the island's shore, and collected some samples directly from one of the island's fish markets. After testing the fish for numerous chemical elements and explosives residues, the agency concluded: "There is no apparent health hazard from fish consumption. . . . It is safe to eat a variety of fish and shellfish every day. . . . It is safe to eat fish and shellfish from any of the locations sampled, including from around the LIA and the two sunken Navy target vessels. . . . It is safe to eat the most commonly consumed species, snapper, every day."[42]

These are clear and confident statements meant to reassure Vieques residents about their health, and the future of the island's productive fishing industry. A closer look at the government study demonstrates that the ATSDR collected fish samples at six different sites on the island: two off the east end of the island in close proximity to the live impact area; one near the village of Esperanza on the south coast, nearly ten miles from the live impact area; one northeast of the village of Isabel Segunda, twelve miles from the bombing range; Johnny's Pescadoria, the local fish market; and one off the largely undeveloped western end of the island, the site of weapons storage bunkers.

Yellowtail snapper is the fish most consumed by Viequense, but only six were caught and tested. None were collected from the contaminated east end sites, one was collected off of Esperanza, two were taken north of Isabel, and three were found off the west end, farthest from the range. Five were then collected at the local fish market, with no understanding of where they were caught. Lobsters are also favored among Viequense fishermen, but in this case, too, only one was collected near the live impact area, while five were collected at a local fish market, again with no knowledge of where they were collected. The choice of six different sampling locations demonstrates the government's interest in understanding spatial variability in contamination levels. But the scientists undermined any potential for comparison by collecting too few fish of the same species at each site. For example, twenty species of fish and shellfish were sought at five different sites, creating one hundred distinctive interpretative possibilities. No fish were collected for 54 percent of the combinations, and two or fewer fish of the same species were collected 75 percent of the time, far too small a number to generalize about safety or hazard.

And why would the government test fish collected from a fish market, with no knowledge of where they were caught? This might make sense if fish markets were the primary source of fish consumed on the island. If so, samples taken from that source could provide a representative sample of average fish intake, and be a legitimate basis to estimate human exposure to contaminants. But many Viequense consume fish they catch themselves, or that they purchase directly from professional fishermen.

What would be the best sampling design to explore the hypothesis that Navy chemicals contaminated fish consumed by the Viequense? One would need to know which fish are most consumed by the Viequense, and these fish should be tested, if they are known to inhabit coastal waters near Vieques. (Large predatory fish such as swordfish, tuna, shark, and barracuda migrate long distances, and have little chance of absorbing chemicals originating on or near Vieques, so should not be tested.) Fish should be collected as close as possible to the live impact area, with additional fish of the same species collected both two miles away and ten miles away. This strategy would provide information on whether concentrations decline with distance from the bombing range. (The ATSDR sampling design followed this pattern, with the exclusion of intermediate sites.) The sampling sites should also be known favored fishing grounds for Vieques fisherman. Ideally, too, fish would be collected during different times of the year, because the species of fish caught often vary with the seasons. Finally, a similar sample size should be collected for each species at each location, and a control population of fish should be collected around a distant island, one with no chance of being contaminated with chemicals similar to those released on Vieques.

A new study would also benefit from including the extensive knowledge of the local fishermen. Most Vieques fishermen know the life histories of at least several dozen species of commercial fish, when and where they are most likely to be found. Many have a mental map of the island's reefs, Spanish shipwrecks, fallen Navy fighter planes, sunken Navy vessels, and areas of the sea floor littered with bombs and piled with artillery shells. They also recognize that some species like the whelks, snails that grow to the size of a softball and are prized as a local delicacy, tend to remain close to shore and move slowly across short distances. Barracuda, in contrast, will spend much of their early life near shore, but then cruise through much of their adulthood in open ocean waters. Shorter-lived fish or those that move over larger distances are less likely to pick up local contaminants than longer-lived fish that spend more time closer to shore. Working with twenty species of marine life daily, the island fishermen are ecologists, with deep knowledge of those species they depend on for economic survival; this knowledge should be put to use for the community's own benefit.

Are levels of mercury in Vieques fish dangerous? Do they result in exposures that exceed the maximum levels recommended by the NAS? As we learned earlier, in 2000 the NAS concluded that the most scientifically defensible limit for human intake of methylmercury is 0.1 microgram per kilogram of body weight per day.[43] ATSDR for some unexplained reason assumed that a level three times higher than the NAS recommendation should be the appropriate safety threshold. Using average concentrations of mercury detected in fish collected at all six locations, all exceed the NAS recommended limit by six- to elevenfold for children, and by three- to fivefold for adults.[44]

In response to criticism that it had failed to test yellowtail snapper, the islanders' favorite fish, the ATSDR did test eleven of the fish for mercury concentrations and concluded, "It is safe to eat snapper every day." Yet in its own tables, it has highlighted its findings with the caution that some detected concentrations result in exposures that exceed the NAS recommended intake. Had ATSDR employed the lower, more health-protective limit, the exposures may well have been deemed to constitute a health threat to children and others depending on their age, body weight, and fish consumption, as well as fish contamination levels.[45]

The government eventually did address the implications of its study for the health of children, but here too the finding is more reassuring and optimistic than cautionary. "ATSDR recognized that infants and children can be more sensitive to contamination of their food than adults because children are smaller; therefore childhood exposure results in higher doses of chemical exposure per body weight. Because children can sustain permanent damage if these factors lead to toxic exposure during critical growth stages, ATSDR as part of its public health assessment process is committed to evaluation of their special interests at sites such as Vieques. ATSDR specifically evaluated the exposure to children and determined that they can safely eat fish and shellfish from Vieques."[46]

The ATSDR's findings were not credible to many in the Vieques population. The islanders find it unthinkable that hundreds of millions of pounds of bombs and other weapons could be exploded on the island's land or within its coastal waters and be undetectable in its marine food webs. Some islanders believe the ATSDR is an apologist for the military's pollution history, and given the lack of a scientifically defensible study showing otherwise, are perhaps even more fearful and receptive to claims that mercury levels in fish are to blame for the high rates of illness on the island—claims that have frightened some Viequense from eating this high-protein, low-fat, and inexpensive food source. Even if they just ignore the ATSDR study, they are still left without essential advice regarding which species and locations yield clean fish.

The islanders' risk of exposure is almost certainly high. During the summer of 2003 Marina Spitkovskya and Laura Hess, two Yale College undergraduates, developed a food intake survey under my direction and administered it to nearly fifty Vieques fishermen. The results were surprising: the islanders ate fish for lunch, fish for dinner, fried fish, baked fish, grilled fish, fish stew, fish broth, fish paste, fish salad, fish tacos, fish pastries . . . endless variations on the theme. In fact, the Vieques fishermen sometimes consumed twelve pounds per week, thirty-two times more than the FDA and EPA had predicted. Consequently, even low levels of chemical contamination in these fish could result in significant chemical exposures for the fishermen.

Stock depletion of the most valuable species in the best fishing grounds haunts the fishing industry worldwide. Sophisticated technology such as dragnets, radar, sonar, and refrigeration all contribute to the increased rate of harvesting. Generally, however, no one knows the reproduction rates of the seafood until well after local or regional populations have crashed. Viequense fishermen, by contrast, rarely have any electronic equipment, even a VHF radio. They use small boats, often with single outboard engines. Many fishermen pull traps by hand, and those who are younger and can tolerate the water pressure spearfish for the largest and most valuable catch. On a recent research trip, I passed a fifteen-foot skiff with two island fishermen hauling traps several miles offshore. It is a risky business: steady twenty-five-knot trade winds had pushed up eight-foot swells. But their ability to fish successfully points to the remarkable abundance of fish near the island due to modest fishing pressure, low technology, and the absence of capital.

The Viequense fishing culture may be sustainable now in terms of fish abundance, but it faces an insidious chemical threat. Although little is known about the chemical content of Caribbean fish—they rarely have been tested—all fish on the planet contain pollutants produced by human activity. Tests of fish taken from remote Canadian lakes hundreds of miles from the nearest village demonstrate elevated levels of methylmercury. Whales taken from the far reaches of the Arctic Sea contain even higher levels of mercury, PCBs, and numerous pesticides such as DDT, which was banned in wealthier nations during the last quarter of the twentieth century.

Some Viequense fishermen do not wish to explore the issue. After all, the absence of evidence of hazardous chemicals is normally interpreted as proof of safety, especially among those who normally buy and consume fish. Other fishermen, however, want answers, in the hope of putting to rest the worries of those who purchase and consume Vieques fish. What chemicals do Vieques fish contain? How do they compare with levels found in fish elsewhere? What threat could they pose to human health? If some fish contain hazardous levels of chemicals, could the continued harvest of others that tend not to accumulate the toxins sustain the fishing economy? Still others understand a strategy long used by those who sell products vulnerable to contamination: market mixing. Finding contaminants in fish does not necessarily foretell the demise of the Vieques fishing industry. Fish, like coffee, vegetable oils, hamburger, and many other foods, are often mixed in the marketplace. Hormone-treated beef from the United States may be mixed with hormone-free beef produced in the European Union. Coffee grown using pesticides banned in the United States may be mixed with coffee grown using pesticides that are neither toxic nor persistent. And Viequense fish may be mixed with fish caught in nearby St. John or St. Thomas, where fish have not been tested. Consumers rarely ask where foods come from, and this form of product dilution in a very international marketplace provides an opportunity for hiding contaminated goods. Although an unsettling practice at best, it is a common survival tactic for businesses faced with the increasing chemical complexity of the environment, and it is often condoned by federal laws that demand testing of the marketed products, not their component ingredients.

Now that the Navy has left Vieques, the fishermen are free again to catch what they can. If government agencies can provide them with accurate, comprehensive information about what is in the fish they catch, the government could encourage the consumption of fish that are least contaminated, while strictly testing and regulating the sale of those species that are most risky. This strategy could go a long way to protect human health, and given the many delicious fish species in the Caribbean, it should not harm the fishing industry. If the Food and Drug Administration adopted this model, the safety of all seafood would be far greater, while the public would be assured access to the significant health benefits that toxin-free fish provide.

CHAPTER EIGHT

Wasteland or Wilderness?

IN 2001, PRESIDENT BUSH decided that the Navy should leave Vieques. His decision had grown in part from the islanders' persistent and embarrassing protests, but he was also responding to a financially starved Congress and the expense of the impending war in Iraq. The federal Base Realignment and Closure Commission supported the recommendation for closing the entire Puerto Rican base known as Roosevelt Roads, which included Vieques. The base closure would reduce defense expenditures by nearly $300 million a year.

Congress directed the Navy to transfer most of the base's lands, assessed at $1.7 billion, to the U.S. Fish and Wildlife Service (FWS) by 2003, although 4,250 acres were given to the municipality of Vieques for future uses yet to be decided.[1] The shift in federal responsibility recognizes the island's extraordinary biological diversity. Seven species of sea turtle nest or routinely visit Vieques, with five of these considered either endangered or threatened. Leatherbacks, Greens, and Hawksbills swim to shore to bury their eggs on the island's remote, sandy beaches. The young turtles hatch beneath the sands, scratch their way to the surface, and then use their fins to slide across the sand to the ocean.[2]

VIEQUES: A LIVE IMPACT WILDERNESS AREA

Also in 2001, with a quiet Congressional vote, Vieques's 900-acre live impact area became a federal wilderness area.[3] The Wilderness Act of 1964, however, specifies criteria for the wilderness designation that appear to preclude its use on Vieques:

A wilderness, in contrast with those areas where man and his own works dominate the landscape, is hereby recognized as an area where the earth and its community of life are untrammeled by man, where man himself is a visitor who does not remain. An area of wilderness is further defined to mean in this chapter an area of undeveloped Federal land retaining its primeval character and influence, without permanent improvements or human habitation, which is protected and managed so as to preserve its natural conditions and which (1) generally appears to have been affected primarily by the forces of nature, with the imprint of man's work substantially unnoticeable; (2) has outstanding opportunities for solitude or a primitive and unconfined type of recreation; (3) has at least five thousand acres of land or is of sufficient size as to make practicable its preservation and use in an unimpaired condition; and (4) may also contain ecological, geological, or other features of scientific, educational, scenic, or historical value.[4]

The live impact area meets none of these criteria, perhaps with the exception of recovering wildlife habitat (Figures 8.1, 8.2). Why then, would they give the area wilderness status? One possible answer may have to do with money. With the wilderness designation, Congress gave the FWS the authority to exclude public access. Yet the designation also provided the Navy with an opening to argue that exposure to munitions and chemical hazards could be prevented by denying access, an approach that might obviate the need to expend hundreds of millions of federal dollars to restore the site to its natural condition.

The wilderness logic became more tortured during the summer of 2004, when Sila Calderón, then governor of Puerto Rico, designated the same live impact area a federal Superfund site, a status reserved for the most seriously hazardous areas in the nation.[5] Each state and U.S. territory

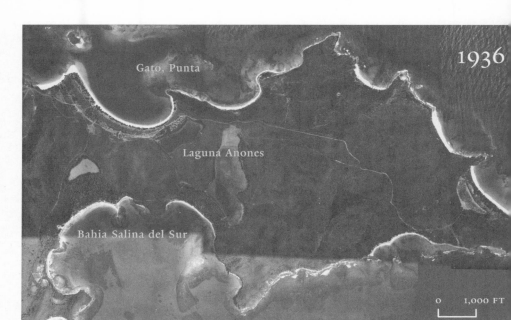

FIGURE 8.1. The east end of Vieques before the U.S. Navy's tenure, 1936. The east end of Vieques in 1936 was used predominantly for grazing cattle. Note Laguna Anones, which lies between the Atlantic Ocean on the north and the Caribbean on the south, separated from the sea only by low dunes and palm plantations. Bahia Salina del Sur long has been a favorite site for local fishermen, especially given the darker seabed grasses that provide habitat for Queen Conch. *Source:* Puerto Rico Department of Transportation and Public Works.

can unilaterally add one—just one—site to the Superfund's national priority list. Puerto Rico chose to add Vieques.[6]

The Vieques National Wildlife Refuge, which spans more than 18,000 acres, is now the largest reserve in the Caribbean. Upon opening this new refuge, Oscar Diaz, its first supervisor, immediately faced many difficult scientific and political questions. Where are the contaminated areas? How dangerous are they to wildlife, the public, and his staff? How should the lands and coastal waters be restored? How should he control the demands of the Viequenses to visit the eastern beaches, which have been closed to the public for fifty years?

Diaz arrived on the island with two staff members, several vehicles, no boat, and no available experts on munitions or toxic substances. By 2006, the refuge had employed only one full-time biologist and one

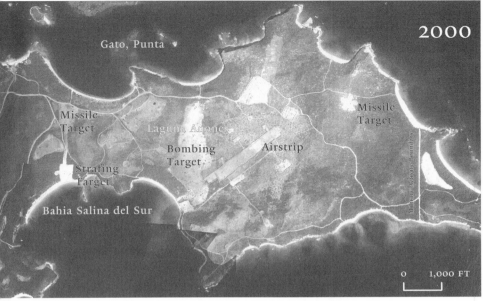

FIGURE 8.2. The east end of Vieques, 2000. The U.S. Navy's fifty-year history of bombing and shelling Laguna Anones has nearly severed the island into two pieces. Palm plantations were bulldozed so beaches could be invaded in mock assault by NATO forces. Aircraft and tanks were scattered as targets, and now the land is pockmarked by craters, the bays hide many unexploded bombs, and still-undetonated artillery shells remain piled underwater. The Military Activity Boundary on the right side of the photo did protect habitat for threatened and endangered species. *Source:* NASA.

ranger, but no one had the expertise necessary to identify hazardous areas within the reserve, with the exception of unexploded weapons lying on the land's surface.

The future roles of the FWS and Navy in managing the refuge remain unclear because their legal jurisdictions overlap. The FWS is responsible for protecting the welfare and future of fish and wildlife, and it is the agency with the clear legal authority to prevent physical damage to the habitat and to regulate public access. Yet it lacks the expertise to judge the chemical hazards to human visitors posed by Navy wastes. In addition, the effects of these chemicals on wildlife and fish, and in particular the native rare sea turtles, have never been studied, leaving Diaz with a quandary. How can he know what level of cleanup is necessary to assure their protection under the Endangered Species Act? Given the uncertainty, where should the burden of proof lie: with the Navy to demonstrate sufficient safety, or with the FWS to demonstrate unreasonable risk?

The EPA's task of identifying hazards and restoring lands and coastal waters also requires the involvement of the military. Although the EPA is legally responsible to identify chemical hazards, and to plan for their effective control, the Defense Department and its contractors are clearly controlling the process of site access, evaluation, and cleanup. Federal funding for analysis and cleanup flows from Congress to the Defense Department, not to the EPA. This places the agency in a reactionary role, with insufficient staff and expertise to challenge Defense Department plans. Given the enormous resources of the Defense Department relative to the FWS and EPA, the entire process of evidence gathering, scientific analysis, interpretation of uncertainty, and restoration planning is controlled by the Navy. Even the pace of cleanup is determined by the Navy's budget; perhaps not surprisingly, then, the military has chosen to extend the process for several decades to reduce annual expenditures.

Because the Navy is the only agency involved with the expertise to identify, disarm, and dispose of munitions, it has retained control over access to the live impact area and other parts of the island's contaminated eastern end. That means FWS staff are not permitted by the Navy to enter parts of the refuge due to the threat of unexploded munitions. FWS staffers, however, are not the only ones seeking access. Although the Navy has posted warning signs to discourage public access to the eastern end of the island, yachts normally anchor near the isolated and seemingly idyllic beaches that have not yet been cleared of munitions. They set their anchors in Bahia Salina del Sur, unwittingly raking a sea floor littered by munitions that missed their land targets. Visitors to the area presume a right of access to the federal refuge, yet the hazards are not easily apparent. Given that only one ranger is monitoring the refuge's nearly 18,000 acres and 100 miles of coastline, they are basically on their own.

In a related cautionary tale, Supervisor Diaz had planned to open several beaches to the public on the eve of the land transfer. Worried about unexploded ordnance given that the areas had been closed for decades by the military, he demanded that the Navy use its most sensitive equipment to scan the beaches and access roads for any remaining munitions. Several land mines and shells were discovered beneath the sands.[7]

THE MMR STORY

In 1940, the Defense Department took ownership of another training ground and military base, Camp Edwards, which was later renamed the Massachusetts Military Reservation (MMR). The base, which comprises thirty-four square miles on the western side of Cape Cod, includes runways, barracks, fuel and ammunition storage areas, and live-fire training grounds. The military also used the area to practice fire suppression, meaning that fuels were dumped or sprayed on the sands, ignited, and sprayed again with fire retardant chemicals. Most of these activities occurred with little understanding of the vulnerability of underground water supplies due to sandy soils. The MMR was most heavily used during World War II, the Vietnam conflict, and the first Gulf War in 1990. On September 11, 2001, National Guard fighters scrambled from the base in a failed attempt to intercept hijacked airliners aiming for U.S. targets.

Detergents were discovered in a nearby community well in 1978, triggering intensive testing of chemical hazards on the base. Delays in identifying the severity of the problem, and failure to inform the public, led to years of additional exposure to hazardous chemicals in surrounding communities. Over the next decade, soil and water were found to be contaminated by exploded and unexploded ordnance (explosives and metals), spilled jet fuels, chemical weapons, solvents used to clean engine parts, and paints. Unlike Vieques, the base sits on sands that cover the only drinking water source for four towns and for thousands who visit the Cape each year. The MMR was added to the Superfund national priority list in 1989, beginning one of the largest groundwater restoration efforts in U.S. history.

Nearly eight thousand wells have been drilled on the reservation to monitor the changing concentrations and location of hazardous chemicals in the vicinity of seventy-eight known pollution sources. Dozens of computer monitors line the walls of the control room, demonstrating rates of flow to and from the filtration plants. Contaminated water is pumped from the ground through three-story-high carbon filters that lie within warehouse-sized structures. Billions of gallons have been pumped through the carbon granules and then returned to the ground in aquifer recharge areas, some of which lie in cranberry bogs.

By 2004, the Defense Department had spent nearly $750 million on cleanup efforts, and it is expected to take twenty more years for the

filtration efforts to lower the concentration of contaminants to target limits. Whether these levels indicate safety, however, seems unclear. The
goal is not complete removal, but to diminish the concentration to levels
that carry an acceptable risk.[8] For these reasons, the EPA and Navy are
always alert to creative ways to allow contaminants to remain, while assuring that public exposure will be minimized. Perhaps this is the logic
behind the cleanup goal for RDX, an explosive chemical found in soils or
water on most training bases. Set at 2 parts per billion, this target level,
if it occurred in other communities, would trigger a health advisory by
the EPA.

Walking across the Cape Cod base during the spring of 2006 with
several scientists working with the Air Force Center for Environmental
Excellence, I was struck by similarities and differences between the MMR
and Vieques. Both sites had experienced a sixty-year history of intense
military training, weapons testing, and intense contamination. At MMR
plumes of pollution flow from the bases into surrounding areas, and
the Defense Department now measures their rate of success by the diminishing concentrations of hazardous chemicals, measured by testing
samples drawn from their complex network of wells. On Vieques the
toxic chemicals are less a threat to drinking water, but flow freely into
coastal waters.

Yet the differences are stark. Vieques is among the poorest of Puerto
Rican communities lying within a commonwealth that has no voting authority in Congress. The MMR, by contrast, lies surrounded by a relatively
wealthy population in 2008 represented by Senators Edward Kennedy and
John Kerry of Massachusetts, both consistent defenders of environmental
protection and human health. The extent of contamination is now well
documented and publicly known, and the MMR is on a clear schedule
for eventual restoration, although the pace of cleanup will take decades.
Vieques's more recent designation as a Superfund site means that its
"restoration" will probably take much longer. The Defense Department
has little incentive to restore the live impact area to its pre-Navy condition, and if the area is not fully restored, residents concerned about their
health, safety, and depressed economy and property values will have little
satisfaction. Aaron Wildavsky, a twentieth-century political scientist, once
wrote that being "richer is safer."[9] These thoughts help to understand the
past and perhaps the future of Vieques. The wealthy are advantaged when

it comes to identifying and managing serious threats to their health and environment.

Vieques is not alone in having contradictory status as both a hazardous waste site and a refuge or wilderness. Other former Defense Department sites have been designated as wilderness areas, including Adak Island, a former naval air station in the Alaskan Aleutian chain, which was added to the Alaska Maritime National Wildlife Refuge in 1980.[10] Other Aleutian Islands lying within the same refuge include Amchitka, the site of three underground nuclear tests during the 1960s, and Cannikan, the location of the largest underground nuclear test in U.S. history.[11] All are now managed by the FWS.

The Nevada Testing and Training Range, managed by the Air Force, includes much of the Desert National Wildlife Refuge, which was established by Congress in 1936. The refuge also lies adjacent to and northeast of Frenchman Flats, the former Nevada nuclear weapons testing site sixty-five miles northwest of Las Vegas where numerous nuclear bombs were detonated. Mushroom clouds created during the nuclear tests generally blew toward the refuge.

The Pacific Islands National Wildlife Refuge includes Johnston Atoll, the site of several high-altitude nuclear bomb tests in the 1950s and 1960s, including one test in 1962 during which two rockets misfired, spewing plutonium across the island. Johnston was declared to be a wildlife refuge in 1926, was then taken over by the Navy, and in 1948 the Air Force assumed control. Until 2000 it served as a storage and disposal site for military stockpiles of chemical weapons, including nerve and blister gases. The FWS and Defense Department, after much debate, agreed that 40 picocuries per gram of soil would be an appropriate cleanup standard for the atoll. (By contrast, the cleanup targets for Rocky Flats Reservation in Colorado range between 35 and 600 picocuries per gram of soil, depending upon the probability of human or wildlife exposure.) In 2005, the Defense Department declared that Johnston Atoll had been "restored" given that hazardous chemicals have been reduced in concentration to "acceptable risk" levels, and munitions have been removed or destroyed.[12]

The U.S. Department of Defense owns nearly 500,000 sites scattered throughout the world, ranging in size from single radar towers to full military bases. Vieques is one of at least 50,000 federal sites—predominantly Defense, Energy, and Interior Department properties—with hazardous conditions in need of restoration. Since 1990, Congress has encouraged the Defense Department to save federal funds by closing military bases that no longer are essential for national security. Most sites are being converted to other uses, including parks, wildlife refuges, schools, and other community facilities. And most are known to be contaminated by munitions or other hazardous chemicals, even if few have been tested to determine whether those living in nearby communities face significant danger.[13]

Of those Defense Department facilities that have been tested, so far nearly 2,000 have been found to be seriously contaminated—including 750 that contain unexploded ordnance or explosive wastes. In 2001, the contaminated sites were estimated to require $20 billion to restore sufficiently to permit their conversion to other public and private uses. Cleanup efforts are not expected to be complete for seventy years or more.[14]

The top five most contaminated and expensive Defense Department sites include Rocky Mountain Arsenal in Colorado, Waikoloa Maneuvering Area in Hawaii, McClellan Air Force Base in California, Nellis Air Force Base in Nevada, and the Massachusetts Military Reservation. Each is projected to cost more than $700 million to restore to levels of acceptable risk, an ambiguous concept yet to be determined jointly by the Defense Department and EPA.[15]

If we demand complete restoration of contaminated sites, we face the "problem of the last 10 percent" as described by U.S. Supreme Court Justice Stephen Breyer.[16] Breyer is a proponent of comparative risk analysis to judge the relative severity of diverse threats to human health and environmental quality. His ideal decision-making process would combine quantitative risk assessments for various hazards—that is, data about risk obtained from testing and analyzing these hazards—with economic analyses so as to understand how the greatest societal danger may be reduced at the least possible cost.

This reasoning has guided the EPA and Defense Department in their choice of cleanup priorities, and helps to explain why some sites receive more immediate attention and funding than others. Critics of these methods argue that quantitative rankings mask the subjectivity inherent in estimates of both risk and cost. They also claim that the technical complexity of the analyses means that most of those affected are unable to participate fully in decisions about issues as important as the choice of cleanup targets and what levels of contamination pose an acceptable risk to health.

The underlying reasons for the tens of thousands of toxic military sites in the United States and its territories also explain what caused universal exposure to strontium-90 and iodine-131 during the weapons testing era. The military has controlled the technologies deployed, information about its operations, financial resources, and legal authority, making public understanding and democratic challenges to weapons testing and military training impossible. National and economic security have long dominated U.S. politics, often at the expense of other ideals such as a literate society, environmental quality, and human health. Those who prepare for and conduct wars have rarely considered how these values might be balanced. Any civil society must be environmentally sustainable, but such a goal rarely has entered the minds of military leaders.

The Viequense found this pattern to be unacceptable. And to those who understand this history, the government's attempts to hide their legacy of toxins behind labels of "wilderness" and "acceptable risk" are deceitful and shameful. With no money, little technical understanding, and much courage, the Vieques islanders' protests somehow brought the Navy's chemical, biological, and physical assault to an end. This is a remarkable history of community empowerment and success against enormous odds. Yet Viequense aspirations to restore their environment and to build a sustainable economy will be no easier than their fight to evict the Navy. Earlier chapters describe how the United States has created and confronted similar problems. Congress chose to restore those Marshall Islands contaminated by nuclear weapons testing, and to compensate their residents by providing funds for basic community services. A similar plan was followed for Nevada Test Site "downwinders" and others exposed during nuclear weapons development and testing. Such a model might be applied to Vieques and other communities whose health, environments, and economic conditions suffered while U.S. officials blindly pursued military objectives.

PART THREE THE BUSINESS OF PESTICIDES

CHAPTER NINE

Capitalizing on Innocence

IN FEBRUARY 1994, Cindy and Todd Ebling moved to the Prestwick Apartments complex in New Albany, Indiana, along with their two young children. At the time, Christina, age three, and Alex, only six months old, seemed healthy and active. Within several months of the move, however, both children began to experience a variety of health problems, including seizures. Their conditions worsened, leading to the hospitalization of Alex by October, and Christina by December.[1] Cindy recalled Alex's early seizures: "First one I remember, AJ was in his walker and I thought he was choking; his teeth were gritted, his eyes watered and he wasn't making any noise."[2] She later reflected on one of the more dangerous consequences of seizures, the hundreds of falls Christina has experienced: "You have to be guarded at all times; we've had . . . a broken nose [when she fell out of] the chair at breakfast."[3]

The similarity in symptoms experienced by the two children, and the near coincidence of their onset, suggest that something about their environment, perhaps the new apartment, might have contributed to their worsening illnesses. "The fact that this occurred in two children simultaneously is pretty hard to ignore," reported Roger Pardieck, representing the Eblings in a lawsuit against Dow AgroSciences. "Their doctors have looked for every other possible explanation, and there is none."[4] When a child becomes ill, most parents struggle to identify a cause. Could the

children just be anxious about their new surroundings? Were they react-
ing to something different in their environment? Might they be responding to something in the new apartment, perhaps the paint, furnishings,
carpets, or heating system? Before the Eblings' arrival in the apartment,
its managers had contracted with a pesticide applicator who periodically
sprayed the complex for insects to prevent infestations, perhaps from
outdoor insects or even new tenants. This firm applied an insecticide
marketed as Dursban 2E and Dursban L.O., but available only to licensed
pesticide applicators due to the chemical's potential toxicity. The EPA
had registered both products in 1982 and 1984, respectively. Cindy and
Todd Ebling, who had become concerned that there might be a connection between the pesticides and their children's health problems, sought
advice from experts at the Office of Indiana State Chemist. Samples taken
in 1995, after they had moved from the complex, contained residues of
chlorpyrifos from Dursban.

CHLORPYRIFOS

Chlorpyrifos, one of the insecticides sprayed in the apartment, is an organophosphate insecticide. During the 1960s and 1970s, organophosphate insecticides increasingly replaced organochlorine insecticides such
as DDT and dieldrin. The EPA has registered forty organophosphate insecticides, permitting indoor applications, underground injection, and
use on nearly a hundred different crops as well as lawns and gardens.
Federal regulation governing pesticides such as chlorpyrifos moves at a
glacial pace, especially when the chemical in question is widely used and
highly profitable. New evidence that hazards are greater than originally
estimated must be compelling to overcome an EPA presumption that the
chemical can be used safely. Past standards were based on few or absent
scientific studies, and were guided by government regulators' former
authority to balance risks to human health against benefits to farmers,
food processors, and others.[5]

Today chlorpyrifos is among 1,200 separate pesticides licensed by
the EPA, and these are commonly mixed with various other ingredients
to manufacture more than 20,000 pesticide products—each of which
must be individually reviewed and licensed by federal authorities. Given
these numbers, it is not surprising that there is a long queue of chemicals
waiting to be reviewed to see if new evidence of toxicity or exposure might

justify tightening regulations or even banning the compounds from markets. Over the last two decades, the queue has grown longer, and the delay
for each chemical has increased. Many individual pesticides have been
"under special review" for fifteen to twenty years despite strengthening
evidence of danger. Agency scientists object that political appointees put
the chemicals in the "parking lot," waiting for sufficient political will
and scientific evidence of significant risk to force the government to act.
Activist administrations struggle to move chemicals out of the parking
lot into public attention, while anti-regulatory administrations jam more
chemicals into it while sales and use continue. During the 1990s the
parking lot was filled with organophosphate insecticides that all faced
uncertain regulatory futures given the more stringent criteria for approval
contained in the Food Quality Protection Act of 1996.

The EPA originally licensed organophosphates for use indoors, on
lawns and gardens, in recreational areas, and on food crops, forests, and
highway rights-of-way, and the chemicals are impregnated into wood
products, plastics, paints, and cleansers. Children and others may have
multiple exposures to their residues as they go through their daily lives,
encountering them in food, water, homes, schools, workplaces, recreational settings, and while using transportation. Yet when the Eblings
moved into the Prestwick Apartment complex, neither the EPA nor any
other regulatory body had considered how exposures could accumulate
across compounds, how the risks might be additive or synergistic, or how
children might be especially vulnerable.

Nearly eight hundred pesticide products contain chlorpyrifos, and
each of these was reviewed and licensed individually by the EPA and
required a separate label before it could be sold. By the time chlorpyrifos
was sprayed in the Ebling's apartment, the chemical had been registered
and used in the United States for nearly thirty years. Beginning in 1965,
the year it was introduced, chlorpyrifos had crept into the lives and tissues of most Americans. Quickly it became the primary alternative to
DDT, which had been banned in 1972; and by the end of the century,
nearly 21 million pounds of chlorpyrifos were being sold annually in the
United States, with roughly 5 million pounds being used in residential
settings. A 1997 California survey found that nearly half of state school
districts routinely applied chlorpyrifos. A national survey in 1994, the
year the Eblings moved into Prestwick Apartments, found residues of

chlorpyrifos in the urine of 82 percent of 1,000 individuals tested. Later human-tissue surveys demonstrated similar or higher percentages of detectable residues.[6]

Residues on Food

To protect against dangerous exposures in food, the EPA sets maximum contamination limits or tolerances for each crop. If chlorpyrifos residues exceed the legal limits, the food may be branded as "adulterated" and removed from the marketplace. The list of allowable residues in foods is lengthy and curious. Why are concentrations of chlorpyrifos in citrus oil allowed to be 2,500 times higher than in bananas? Similarly, residues in corn oil are permitted to be thirty times higher than in fresh corn, and the pesticide's residues in beet sugar may legally be 1,500 times higher than in eggs. These differences between the "parent" crop residues and those found in processed forms of the same plant have little to do with safe exposure limits, but they do demonstrate an unavoidable concentration during processing as the plant material is converted to oils, sugars, or dried forms. No two chemicals behave exactly the same way, making it very difficult and expensive to track residues, especially through a global food-supply mechanism that continues to grow more complex. The variations in allowable levels also illustrate the effect of the EPA's authority to balance risks against benefits, and the power of those chemical and agricultural lobbies that pressure the agency to neglect concentrated sources of exposure.

When scientists with the U.S. Department of Agriculture looked carefully, they found that the chlorpyrifos persisted on apples, tomatoes, and grapes, foods consumed commonly by many small children. During the period when the Eblings lived in the Prestwick Apartments, nearly 25 percent of the apples tested by the agency contained chlorpyrifos residues. With so many different allowable uses, and residues appearing on different foods, in water, and both indoor and outdoor environments, the agency faced a considerable challenge. Its liberal licensing system had created a scientific nightmare for anyone wishing to understand how a child might be exposed to chlorpyrifos residues while moving through daily life.[7]

Lingering Residues in Water

If tens of millions of pounds of chlorpyrifos are sprayed outdoors in the United States, what is the potential that U.S. drinking water supplies are contaminated? When the government finally looked more carefully for chlorpyrifos residues in the outdoor environment, it found the chemical in streams near agricultural and urban residential areas of New York, Georgia, Alabama, and Florida.[8] Along nearly forty-three miles of the San Joaquin River in California, for example, 50 percent of the water samples tested found chlorpyrifos doses lethal to water fleas (*Ceriodaphnia dubia*), which are often used as a barometer of both acute and chronic aquatic toxicity.[9] The U.S. Geological Survey also found chlorpyrifos in the Tuolumne River and Newport Bay in San Diego at concentrations lethal to the fleas. In particular, it found that residues were highest following storms that can send a pulse of water flowing over treated surfaces, such as farmlands or lawns, into the rivers. Because the agency had not bothered to search for residues before giving its approval for product after product, it had no evidence to suggest that persistence could be a significant worry.

In 1998, thirty-three years after tens of millions of pounds of chlorpyrifos had been released into the U.S. environment, the EPA offered its own understanding of the chemical's threat to water quality, which was really a description of its lack of understanding: "We currently have no significant data on chlorpyrifos residues in reservoirs and lakes."[10] This statement meant that the agency had not looked for chlorpyrifos, not that waters had been tested and found to be free of it. In the absence of monitoring, data regulators had relied, as is typical in such cases, on crude computer models to predict surface runoff into streams, rivers, lakes, and ponds.

Nearly 5 million pounds of the chemical were used by professional applicators each year during the 1990s to protect buildings from termites—3.6 million pounds outdoors and 1.4 million pounds indoors. Almost ten pounds was applied on average to each site, meaning that each year of that decade nearly 500,000 structures in the United States received the termiticide treatments. Often the product is injected into the soil through a hollow rod, which in areas with high groundwater levels can contaminate water nearby. This may mean a significant hazard for the more than 40 million people in the United States who derive their drinking water from wells.[11] Wells are often drilled or dug close to homes for convenience, and most of them are shallow, often less than a hundred feet deep. Many believe that shallow wells are not as safe as deeper ones because surface contaminants may quickly travel long distances, and may be drawn into the well as the well is used. Damaged well casings may also allow hazardous chemicals to enter the well directly. The federal Safe Drinking Water Act provides no protection or monitoring of water systems serving fewer than twenty-five individuals or fifteen households or service connections. Even larger systems are not required to check for chlorpyrifos since as of 2009 the EPA did not regulate the chemical in drinking water.[12]

How long does chlorpyrifos linger in the environment? The answer depends on the condition of the soil—and whom you ask. The longevity of residues depends on many factors, primarily the chemicals' stability under oxygen-deprived conditions. In aerobic soils, the EPA assumes that the chemical persists longer than years. In one soil study, under anaerobic conditions, the chemical persisted longer than four years.[13] Research conducted between 1988 and 1994 by the scientists R. B. Leidy, C. G. Wright, and H. E. Dupree, however, demonstrated that a house treated for termites by injecting the soils near the foundation with chlorpyrifos produced detectable air residues indoors nearly eight years later.[14] By 1999 Dow AgroSciences, one of the manufacturers of chlorpyrifos, reportedly had spent nearly $100 million producing almost 3,600 studies to test the toxicity and environmental behavior of chlorpyrifos. These studies were designed to promote the chemical's approval by regulatory agencies in ninety-eight nations, and to help protect the company from lawsuits filed by those who believe they have been injured by chlorpyrifos-containing products. Manufacturers claim that the chemical has the potential to control nearly 250 nonagricultural pests, and have recommended it for use in and around many different types of structures. Dow also reported earning nearly $100 million per year from sales of chlorpyrifos-containing products alone. The company maintains 350 different chlorpyrifos product labels in the world, with 142 of these for agricultural purposes. The chemical is so important to the company that it has funded a named Dow professorship at the University of Michigan.[15]

The EPA reviewed data submitted by Dow that demonstrated that some wells contained chlorpyrifos residues after structures had been treated to kill termites. Buried deep within an EPA report that reviews drinking water hazards associated with the insecticide, the agency presented its assessment: "We conclude that if chlorpyrifos is used for termite control within 100 feet of a drinking water well . . . residues can persist at detectable levels for at least 6 months."[16] Few homeowners or workers would ever think to ask a builder or former owner whether their home or workplace had been treated with chemicals that might seep into subsurface water. Even so, the EPA received nearly 250 "incident reports" between 1992 and 1997, complaining of well-water contamination. The agency acknowledged that applications to control termites could produce very high concentrations, but repeated its conclusion that the problem would only occur if wells were located within one hundred feet of the application site, and would be "localized." With wells commonly dug within one hundred feet of homes to reduce piping and trenching costs, this conclusion should be unsettling. Given the environmental persistence and movement of chlorpyrifos following outdoor termite applications, EPA in 2004 prohibited its use during the construction of new buildings.[17]

During the 1990s, Theodore Slotkin, a professor at Duke University, wondered whether chlorpyrifos interfered with the normal growth and development of brain cells in young rats. Slotkin found that chlorpyrifos did indeed disrupt DNA replication in brain cells, resulting in fewer cells than normal. Moreover, these important effects on the brain were seen at doses too small to produce any symptoms or signs of toxicity, including the cholinesterase inhibition that is used by EPA to judge health risk.[18] Consequently the standard test to determine organophosphate insecticide danger, a measurement of the inhibition of cholinesterase, would not be sensitive enough to detect these effects on brain cell development—effects that have possible implications for cognition, memory, and learning.[19]

These results were embarrassing to the EPA staff; their toxicologists had never before asked whether pesticides affected humans' developing nervous systems, nor had they demanded tests to explore these effects. This seemed to be an egregious oversight given the agency's knowledge of both organophosphate insecticides' war-gas ancestry, and the painful but well-known history of the government's delayed recognition that children's nervous systems are vulnerable to lead. In 2000, the agency eventually requested that neurotoxicity tests in very young animals be conducted for forty pesticides.

Local and private water companies resist proposals that would require them to pay for water quality tests, and if pesticides are found, to reduce their concentration to safe levels. Their resistance has been effective. Not one of the forty organophosphate chemicals in use during the twentieth century—nearly 60 million pounds per year—are regulated as drinking water contaminants, so there is no government requirement to test for their presence in public or private water supplies. Few Americans have the scientific or legal background necessary to recognize this oversight; instead most believe that the water they use and drink is known to be pure.[20]

Do Exposures from Different Sources Compound Risk?
Are children being exposed to dangerous levels of chlorpyrifos when exposures from foods, water, and indoor environments are considered together? And given the hundreds of uses allowed for chlorpyrifos by the EPA, what is the likelihood that a child could experience a dangerous or even life-threatening exposure? As of 1994, when the Eblings moved into the Prestwick Apartments, the EPA had not considered these questions about chlorpyrifos, or other pesticides. By this time, however, the agency did know full well the extent of chlorpyrifos use in the United States. In 1992, the EPA had estimated that Americans held nearly 176 million containers of pesticides in their homes, and that nearly 16 million of these contained chlorpyrifos. The manufacturer, DowElanco, estimated that chlorpyrifos accounted for roughly 25 percent of the national market share for residential insecticide use.[21]

When in 2000 the EPA finally added the different likely sources of exposure, the risks faced by infants and children younger than six years were found to be nearly ten times those experienced by adults. This was true in part because small children, whose body weights are low, consumed more residue-containing fruit products than adults (when intake is adjusted by body weight). And if indoor areas had been treated, the young, if present, would be more exposed than adults because small children spend much of their time closer to the ground or floor, where residues settle.[22]

The EPA developed the concept of a "risk cup" in the late 1990s to define an acceptable level of chemical threat to human health. That is, if after the agency adds together the risks from different sources—food and water residues, consumer products, indoor air—the risks overflow the cup, they are unacceptable. By contrast, if the EPA were being more cautious, it would not only prevent overflow, but include an extra margin to protect against errors in estimates due to scientific uncertainty or less than cautious assumptions.

In 2000, the agency found that the chlorpyrifos risk cup was overflowing, and that children were most in danger. Apples and tomatoes were believed to be responsible for most of the exposure from foods, and indoor applications were capable of producing potentially dangerous exposures via inhalation, skin contact, and ingestion. All postbloom apple applications were prohibited, use on tomatoes was canceled, and the residue limits for grapes were lowered. All indoor and outdoor residential applications were also prohibited by 2004.[23]

DIAZINON

The owners of the Prestwick Apartment complex had initially contracted with a pest control company to manage insect problems by applying chlorpyrifos. In April 1994, however, the apartment managers decided to use their own maintenance workers to take over the routine pest control by periodically applying "Creal-O," a mixture of mineral spirits and three pesticides—diazinon, pyrethrins, and piperonyl butoxide—first registered in 1974 and manufactured by the Louisville Chemical Company.[24] Diazinon, an organophosphate insecticide like chlorpyrifos, was first registered by the federal government in 1956 and manufactured by the Ciba Geigy Corporation. Today, product licenses or registrations for diazinon are held by six multinational corporations.[25] By 1999, nearly 13 million pounds were being used in the United States each year on corn, hops, beets, apples, pears, bananas, grapes, peaches, cherries, and nuts, including almonds and walnuts. When the EPA concluded its most recent review of the chemical in 2000, it had approved 454 different products containing diazinon.

The EPA also permitted the chemical to be applied to "non-lactating cattle and sheep," suggesting that it had concerns about the ability of diazinon and its metabolites to accumulate as it moved through food chains, bind to fats, and be excreted in mammals' milk. In a 1956 study, USDA scientists administered radiolabeled diazinon to a lactating cow and found the insecticide in the animal's blood, urine, feces, and milk. But common houseflies in dairy barns were a persistent problem for farmers, and diazinon was effective in killing them. And when cattle were infected with grubs, diazinon again provided relief when administered to the cattle orally, subcutaneously, or directly to the cow's back.[26] A later USDA study found the residues only in the buttermilk fraction of cows' milk.[27] The EPA thus licensed the chemical to be sprayed on pastures, rangeland, corn, soybeans, and alfalfa, all of which may supply animal feeds. In 1986, the agency began a special review to better understand the chemical's threat to birds and two years later, placed special restrictions on outdoor uses other than for home lawns and gardens. Not surprisingly, then, the manufacturers turned to homeowners as the next most likely consumers. By 2000 nearly 83 percent of all diazinon was being used for nonagricultural purposes, predominantly in and around homes and other buildings. Residential application rates were generally allowed to be nearly double those permitted for agricultural purposes. Ironically, the government permitted a more intense use where people lived, worked, played, and went to school than on croplands.[28]

As with DDT, awareness of diazinon's threat to human health grew from understanding its effects on wildlife. EPA scientists found the chemical to be highly toxic to birds, fish, and mammals, and were especially concerned about its effects on endangered species of wildlife, aquatic life, and terrestrial plants. Still, regulators permitted the chemical to be purchased by untrained homeowners in grocery and hardware stores, and manufacturers used aggressive advertising to encourage its intensive use to control a variety of household pests. Each spring homeowners were certain to hear on radio and television a variety of commercials touting the benefits of using diazinon to protect lawns from grubs and other threats, with lawn care companies also promoting its use, in various products, to turn a patchwork of weeds and molehills into the perfect lawn.

PETROLEUM DISTILLATES

Mineral spirits were also an ingredient of the pesticide product released in the Eblings' apartment. Generally considered inert, mineral spirits are also suspected of causing health problems. Mineral spirits are distillates of petroleum and can include more than two hundred different hydrocarbons. Although they are not believed to be carcinogenic, the spirits are solvents, and may cause both acute and chronic health effects. Studies of professional painters routinely exposed to the solvents demonstrate increased incidence of neuropsychological disorders following chronic exposure.[29] Exposure to petroleum distillates like mineral spirits normally occurs by breathing the airborne vapors; painters commonly absorb them directly through the skin. The hydrocarbons are then carried by the bloodstream throughout the body, where some bind to fatty tissues. Neurological symptoms following intense exposures suggest that the chemicals also reach brain tissue. Within fatty tissues the chemicals have a half-life of about forty-eight hours, and are most likely metabolized by the liver and excreted via respiration and urination.[30]

Some types of petroleum distillates are known to affect the central nervous system if exposures are high enough. Following intense short-term exposures, people have experienced symptoms such as headaches, dizziness, nausea, slurred speech, tremors, confusion, unconsciousness, and even coma.[31] Chronic exposures can also take a toll, producing in some occupational settings damage to the peripheral nervous system and ataxia (a lack of voluntary muscle coordination). As with other compounds the EPA has been charged with regulating, the diversity of types of petroleum distillates has created confusion among regulators about their individual or collective effects. The EPA recently reviewed available data for the inert ingredients it permits to be added to pesticides, and has found that there is not enough information to determine the magnitude of the possible health threat. The best one can conclude from the data available is that some pesticide inert solvents may induce adverse neurological effects, some may induce developmental abnormalities, and others may damage the liver; all depend on intensity of exposure.

The Indiana Court of Appeals found in 2000 that the Eblings' apartment was sprayed with chlorpyrifos, diazinon, pyrethrins, piperonyl butoxide, and petroleum distillates.[32] Both organophosphates and petroleum distillates are known to be able to cause nervous system damage at high doses. Their combined effects, however, have never been studied. When the EPA requires pesticides to be tested for their toxicity prior to issuing a registration, only the active ingredient is tested, not the mixture that is normally applied. Nor is it assumed that several different compounds may have been sprayed in the same apartment—a common occurrence since pesticide applicators are not required to know the chemical history of the buildings they spray, nor how long the chemical residues persist.

Although the agency has recognized the need to develop risk estimates for mixtures of organophosphorus pesticides, manufacturers claim that the EPA's methods are not yet precise enough to estimate such risks. This argument was been accepted by federal regulators for nearly two decades and led to regulatory paralysis. The rarely spoken concern of manufacturers is that if mixtures were to be subject to risk estimates, their chemical might be banned because it was found to contribute to risks created by other manufacturers' products.

But synergism among organophosphate compounds is not only possible; it has been demonstrated by the combination of two other organophosphates, malathion and EPN. In another study of various combinations of thirteen organophosphorus insecticides, too, toxicity was dose-additive in twenty-one pairs of compounds and less than additive in eighteen pairs; and a synergistic effect occurred in four cases. Synergism appears to depend on many factors. It is even possible that exposure to one compound can be somewhat protective: when both potent and weak inhibitors compete for the same active catalytic sites on acetylcholinesterase molecules, the less potent pesticide may occupy these sites and lock out the more potent one.[33] The most important point, however, is that we don't have the information we need regarding the dangers, if any, of using these commonly sold mixtures in our homes and workplaces—and the EPA has not required the testing of mixtures to give us this clearer picture.

A further complication is that some organophosphates and carbamates may become transformed in the environment or in the human body to form metabolites that are more toxic than the registered parent compounds. When exposed to heat and air, for example, malathion and parathion degrade into maloxon and paraoxon, both of which are more toxic than the parent compounds. We know this in part because of a maloxon poisoning incident that occurred among malaria workers in Pakistan following their exposure to improperly stored malathion. Another study found that exposure to a combination of parathion and methyl parathion is more likely to induce intermediate syndrome (a complication from organophosphate insecticide poisoning involving muscle weakness and potential respiratory failure) than parathion poisoning alone, even though malathion is a far weaker toxin than parathion.[34]

IS INDOOR SPRAYING DANGEROUS?

When the Ebling children moved to the Prestwick Apartments, no fewer than forty organophosphate insecticides were licensed by the EPA for use in agriculture or indoors.[35] The government had licensed each one individually over nearly forty years, but no agency had asked whether the risks of using these insecticides together might be greater than those for using each one alone. Finally, in 2000, an EPA scientific advisory panel reviewed the question and determined that these pesticides do indeed affect the nervous system in a similar way, and that their dangers should thus be considered at least additive.

While attempting to manage pesticides, the EPA has focused most of its efforts and attention on agricultural uses. Although agricultural uses are important—since the toxins may affect farm workers; contaminate food, water, and soil; and remain as residues in meals—the agency neglected perhaps the single most important source of exposure: indoor environments. Meanwhile, pesticide manufacturers sought and received government licenses to reformulate their crop protection chemicals for an ever-expanding list of indoor uses. As a result of their efforts, more than 100 million pounds of pesticides are now released indoors in the United States each year. By 1999 this figure included nearly 7 million pounds of organophosphate insecticides that were released in homes, schools, hospitals, restaurants, stores, and other building environments to kill termites, cockroaches, ants, rodents, and other pests. In addition, pesticides are deliberate additives to many consumer products such as clothing, carpets, plastics, paints, stains, building materials, play equipment, picnic tables, some detergents, fuels, shampoos, pet products, cosmetics, and pharmaceuticals. Ships, trains, buses, and planes are also commonly sprayed to prevent insects from hitchhiking rides to other areas.

Because Americans spend 80 to 90 percent of their time inside buildings, releasing these chemicals indoors can produce intense exposures, especially if these indoor areas are poorly ventilated. Manufacturers and the EPA had relied on data previously submitted to obtain agricultural registrations to determine whether the indoor and consumer product uses would be safe. Yet the pattern of human exposure that results from indoor applications is completely different from exposures that follow outdoor applications.

When EPA regulators examined U.S. Poison Control Center data carefully in 1999, they concluded, "Diazinon is one of the leading causes of acute insecticide poisoning for humans . . . The majority of incidents occur in the home."[36] In December of 2000, the EPA, recognizing the threat to children's developing nervous systems, cancelled permits for residential indoor uses of chlorpyrifos and diazinon. Yet diazinon sales were permitted to continue through 2002, a date negotiated with manufacturers. Under this agreement, retailers were also allowed to continue to sell the product to homeowners until the end of 2004. Agency regulators have often worried that if the EPA demanded a shorter phase-out period, litigation would continue for years as would human exposures.

Most people have little understanding of the legal and scientific details surrounding pesticide regulation. They are likely to presume that the government would not allow the release of dangerous chemicals indoors; that building managers would do nothing to risk the health of occupants; that those applying the chemicals are well-trained to understand and manage chemical hazards; and that they would not be allowed to remain in buildings—homes, schools, and offices—if the hazards might be serious.

If the chemicals are available in the local hardware store or even the supermarket next to foods, how toxic could they be? During the mid-1990s America's war on indoor bugs had resulted in 19,000 unintentional exposures to organophosphate insecticides that were serious enough to be reported to poison control centers. The EPA estimated that among these, "close to 7,000" involved chlorpyrifos, and approximately a hundred of these had moderate to severe outcomes. The agency also found an average of "116 chlorpyrifos cases per year with moderate to severe outcome reported in residential settings." Chlorpyrifos exposures had the highest rate of severe or fatal outcomes among the thirteen organophosphates reported to the poison control centers. In one case, a twenty-two-month-old toddler drank a cup containing chlorpyrifos, petroleum distillates, and water. Within minutes the child began choking, experienced both gastric and respiratory distress, and had to go to the hospital. The child died ten weeks later.[37]

After reviewing these statistics, the EPA concluded in 2000, "In addition to acute poisoning, chlorpyrifos has been reported to be associated with chronic effects in humans, including neurobehavioral effects, peripheral neuropathy, and multiple chemical sensitivity. Neurobehavioral effects include persistent headaches, blurred vision, unusual fatigue or muscle weakness, and problems with mental function including memory, concentration, depression, and irritability. . . . The main source of serious acute incidents of chlorpyrifos poisoning had been liquids used by homeowners or Pest Control Operators (PCOs) indoors or outdoors, termite treatments, liquid sprays and dips applied to domestic animals."[38]

It is hard to imagine any parent allowing their home to be treated with a pesticide if they understood the chemicals' potential to affect their children's developing nervous systems. Who understood the risks to children posed by organophosphate chemicals, and when did they know them? The National Academy of Sciences certainly sounded the alarm in 1993, when it released a report that highlighted the risks from organophosphate residues in children's foods; and this report was commissioned by the EPA. The NAS study highlighted the rapidly growing literature on the hazards of chlorpyrifos, and recognized the risks posed to infants and children. Manufacturers, too, may well have encountered cautionary results from proprietary studies unavailable to the public. If so, was all of this knowledge shared with EPA regulators, and can labels ever adequately convey the complexity of findings to consumers or applicators?

Most defendants in lawsuits involving pesticides and health loss claims would argue that the illnesses are unrelated to their chemicals. They would also claim that the risks are insignificant when compared with the pesticides' benefits; that the threat is highly uncertain; that there are likely to be other plausible explanations for observed health problems, either environmental or genetic; and that all products were approved for indoor uses by federal regulators after EPA scientists had determined their safety.

Yet given that the NAS had warned the EPA in 1993 about children's excessive risks when exposed to chlorpyrifos residues in foods, wouldn't government and corporate officials assume that indoor exposures to the very same chemicals could be hazardous for children? In 2000, six years after the Eblings moved into the Prestwick Apartments, the EPA required a change in the label for chlorpyrifos so that products including it now must display the warning: "This product is not to be used in and around homes or other residential areas such as parks, school grounds, playing fields."[39] EPA Administrator Carol Browner announced the new rule stating, "This action comes after completing the most extensive scientific review of the potential hazards from a pesticide ever conducted. This action—the result of an agreement with the manufacturers—will significantly minimize potential health risks from exposure to Dursban, also called chlorpyrifos, for all Americans, especially children."[40]

CHAPTER TEN

Without Warning

IN 1999, FIVE YEARS AFTER leaving the Prestwick Apartments, Christina Ebling was hospitalized again for yet another round of intense seizures. Also that year, Cindy Ebling, her mother, described Christina's deteriorating condition to Jim Morris, a writer for *U.S. News and World Report:* "She's gone downhill . . . I found her face-down in her eggs." Morris himself noted, "Christie sat on her bed a few feet away, gaping at a visitor, drooling, and hooting as she struggled to assemble a simple puzzle," and Todd Ebling, sadly recalling his daughter's former intelligence and vitality, reported, "I have a hard time even looking at Christie anymore."[1] The girl's health has continued to decline. In 2008, Cindy was interviewed by Philip and Alice Shabecoff and explained that Christina "doesn't know what to do with the feelings she has. She slaps and bites and pinches. She throws things at her father and pulls out his hair. The last two years have been hell as I watched my little girl progressively go downhill."[2] An Indiana television station also reported that year that Christina had experienced thousands of seizures and hundreds of falls, some resulting in broken bones.[3]

Many Americans would be surprised to know that if they experienced similar serious health problems following exposure to licensed pesticides, the law provides little opportunity to seek personal compensation from the federal government, even if public officials were clearly negligent in approving the pesticide's sale and use. The Congress through legislation has provided compensation for a variety of reasons, but only if authorized by laws such as the Alaska Native Claims Settlement Act, or the Radiation Exposure Compensation Act of 1990.[4] The EPA could be sued if it registered pesticides known to pose uncontrollable danger to human health or the environment, or if the agency failed to enforce the law. Courts, however, would have authority only to demand the setting of standards to protect health, or future compliance, not to provide compensation to those injured. Given this situation, the only recourse for individuals who believe they have been harmed has been to bring a civil action against the private firms that produced and formulated the chemicals, or applied them in their environment. If they were to win the suit, they had to demonstrate, with a "preponderance of evidence," that the injury was caused by the applied chemicals, and that one or more defendants was responsible for the harmful exposures.[5] The history of pesticide litigation to recover damages in the United States dates back to 1884, and has long been a fallback excuse for Congressional inaction. If warnings fail and harm results—outcomes that occur unfortunately too often as poisoning data demonstrate—Congress has long relied on the consumers' right to seek compensation for damages in civil suits brought in state courts.[6]

Manufacturers of all types have worked hard to protect themselves from lawsuits in which plaintiffs claim that products have induced physical harm. One successful corporate strategy has been to encourage Congress to adopt laws that allow manufacturers to sell hazardous products if consumers are warned of dangers via labels. Over the past forty years, then, Congress has relied increasingly on written or symbolic warnings, such as a skull and crossbones, to inform consumers about dangerous products and technologies. Warnings are now required for many products like cigarettes, alcohol, pesticides, pharmaceuticals, medical devices, fuels, paints, cleansers, cars, trucks, boats, aircraft, building materials, tools, and cell phones. The theory behind the law is that consumers may enjoy product benefits safely if they are informed via labels about potential threats to health, property, and environment. In other words, pesticide labels are employed by manufacturers as a defense against claims of liability due to negligence, fraud, and defects in product design, manufacture, instructions, or warnings.[7] Manufacturers of these chemicals have found that they can hide legally behind the logic that if warned, consumers should shoulder the blame for and damages caused by unsafe use. "Express warranties" placed on pesticide labels claim the product is reasonably fit for its intended purpose, but that the buyer assumes all risks.[8]

Indeed, pesticide management in the United States has long relied on this concept of warning consumers via labels. The Insecticide Act of 1910 required truth in content labeling for pesticides to protect farmers against fraud and the dangers of misbranded and misformulated products. By 1947 labeling and registration had become the federal government's primary strategy for controlling agricultural chemicals, one that rested firmly on the principle of caveat emptor, or "buyer beware." This principle has guided the USDA and EPA to allow the distribution of nearly 70,000 separate pesticide products, each with a unique label that provides warnings. Importantly, the label is intended by the EPA to reduce the likelihood that health or the environment will be damaged. Yet these warnings fail to offer sufficient protection for several reasons, including:

Those exposed are often not warned. Warning information is provided only to those who purchase the chemicals, not those who use or apply them, or those who enter chemically treated environments.[9] Apartments, offices, schools, restaurants, airplanes, trains, buses, and many other spaces are commonly treated with pesticides without the knowledge or consent of those eventually exposed. These experiences have led to a number of "failure-to-warn" claims in state courts, brought by those who believe they were harmed after not being sufficiently warned about product dangers. The Indiana Court of Appeals, after reviewing *Dow v. Ebling,* found that the professional applicator "did not provide [the apartment managers] or the Eblings with any of Dursban's EPA-approved warnings and labeling information. Further, although Louisville Chemical provided [the apartment managers] with the EPA-approved labeling for Creal-O, the Eblings were not provided with this labeling before their exposure to that pesticide."[10]

Illiteracy keeps many consumers from understanding the labels. The National Adult Literacy Survey was conducted in 2003 to determine adults' "ability to read, write, and speak in English, and compute and solve problems

at levels of proficiency necessary to function on the job and in society."
Nearly 20 percent of U.S. adults—42 million people—function at the
lowest level of literacy.[11] Of this group, nearly 62 percent did not complete high school, 19 percent reported problems with visual acuity, and
25 percent were immigrants trying to learn English. Almost 44 percent
of those scoring in the lowest literacy category were living beneath the
poverty level, compared with 4 to 8 percent of those scoring in the highest two proficiency levels.[12]

Pesticide labels demand a high level of literacy. They usually include
directions for how properly to mix, apply, and dispose of the product;
emergency medical guidance; and how to calculate the safe and effective
application rate based on the area or volume to be treated (often expressed
as a weight-to-volume or weight-to-area ratio). To fully grasp warnings
and safe-use instructions on pesticide labels, a user would have to have at
least a "level 3" literacy level. Approximately half of U.S. adults between
the ages of 16 to 65, or about 90 million people, are not this literate. The
approximately 73 million children under the age of eighteen could also not
reasonably be expected to have the literacy, quantitative skills, or maturity
to read and comprehend pesticide labels. And the American Foundation
for the Blind estimated in 2004 that nearly 10 million people in the United
States are either legally blind or visually impaired, meaning that they have
"difficulty seeing words even with their eyeglasses on."[13] Thus, nearly
170 million individuals in the United States, more than half the nation's
population, are unlikely to be able to either read or fully understand the
warning labels on pesticide products.

Warnings generally concern only proven adverse effects. Labels for chemical
pesticides are required to warn only about well-demonstrated adverse effects. Although many academics, EPA scientists, and parents who would
like to avoid harmful exposures for children would like to know what
effects would be demonstrated by a preponderance of evidence, rather
than the more rigorous scientific standard of proof, such warnings are
not provided. Cancer risk, for example, need not be included on warning labels for chemical lawn care products as it is for chemicals used to
grow food—even though exposures to lawn care pesticides could easily
be higher than exposures to lingering residues on food.

Claims of safety are often unproven. The federal government does not control how manufacturers proclaim the safety of their products. For example, manufacturers have promoted their chlorpyrifos-containing products with claims such as

- "Proven human safety record"
- "They [pesticides] are essentially environmental medicines"
- "Chlorpyrifos products have been used safely around millions of homes"
- "Chlorpyrifos products are safe for use around adults and children"
- "Every effect that a pesticide could have on our health and environment is tested"

When hazard warnings appear alongside claims of safety such as these, confusion is bound to result. This blurring of messages has not gone unnoticed. In 2003, New York's attorney general announced his plan to sue Dow, asserting that the company repeatedly "concealed information and misrepresented hazards to human health and the environment." One target of the New York complaint was an advertisement formerly on the Dow website that stated: "Are chlorpyrifos products safe for use around children? Used as directed, chlorpyrifos products provide wide margins of safety for both adults and children." The attorney general and the court found these claims to be difficult to reconcile with government-required warnings such as "excessive vapor concentrations are attainable and could be hazardous on single exposure"; "may be fatal if swallowed"; "avoid contact with skin, eyes, or clothing"; and "avoid breathing vapors or spray mist." Dow eventually agreed to pay a $2 million penalty in December 2003, and to discontinue making claims of product safety—but only in New York State. They had pledged similar restraint in 1994.[14]

Information about adverse effects has been withheld. Pesticide manufacturers supply the federal government with toxicity information, and are required to submit any reports of adverse health or environmental effects, known as incident reports, so that the EPA can adjust regulations and label licenses as needed to protect the public's health. DowElanco in 1994 delinquently submitted "incident reports" of individuals who claimed they had been exposed to Dow's chlorpyrifos-containing products. The EPA charged the company with a violation of federal law, and in 1995 the company finally agreed to pay $876,000 as a civil penalty. The EPA noted, "DowElanco reported 249 incidents to EPA well after the thirty-day time period specified in EPA guidance. EPA's review of these incidents revealed that DowElanco had not been reporting adverse effects incidents that it learned of through personal injury claims and lawsuits. After EPA sent DowElanco a 'show cause' letter, the company submitted information on additional incidents."[15] There is convincing evidence that industry disclosure of these reports in a timely manner can result in tighter regulations and reduced exposures. The EPA noted in 2000 that nearly 98 percent of the exposures reported as "incidents" during the 1990s were associated with products that it later chose to ban or further restrict.[16]

Warnings are seldom updated. As older licensed chemicals are more carefully researched, many are found to persist longer and be more toxic than earlier believed, leaving the EPA's earlier regulations and warnings insufficient. Yet changes to rules and labels are often delayed. Although the EPA had been required by the Federal Insecticide, Fungicide, and Rodenticide Act (FIFRA) of 1947 to review scientific evidence on risks to health and environmental quality every five years, these deadlines were rarely met. The law now demands that each chemical be reviewed only every fifteen years. Even so, many products have not been fully reviewed for several decades, while scientific understanding of human and ecological effects is growing faster than ever before. The problem has been compounded by the sheer number of labels to check. Today the nearly 22,000 separate pesticide products on the market each have different labels, with chlorpyrifos products alone accounting for at least 350 of these. This lack of government responsiveness might easily be considered a form of public sector negligence.

When chlorpyrifos was prohibited from residential environments in 2000, label changes to restrict its use, including new label warnings, were phased in over five years. During this period, the EPA did not remove the product from the marketplace or change its chemical formulation. Further, even in those rare instances when products have been banned, manufacturers, distributors, stores, and applicators have been allowed to sell existing stocks with original labels. If an EPA administrator finds imminent danger associated with any product, she or he can ban the chemical and immediately prohibit any future sales, but then the federal government must purchase the product and manage it as hazardous waste, using funds from the public treasury. Consequently, most consumers have remained uninformed of any new restrictions in the marketplace, presuming that what is sold in the stores is safe or it would not be sold. And many stores continued to sell both chlorpyrifos and diazinon products even after new sales prohibitions became effective. Applicators were also allowed to use up existing stocks, all without consumer notification of the new restrictions.

Labels do not adequately explain hazards. The EPA's website for chlorpyrifos contains thousands of pages of scientific documents concerning the health and environmental risks of exposure. A recent search of the National Library of Medicine database for peer-reviewed technical articles on chlorpyrifos returned 1,390 papers. Labels cannot possibly distill this evidence into accurate and meaningful advice to explain how consumers can best avoid dangerous exposures. Labels rarely present quantitative estimates of risk, nor do they compare them to other hazards that might be more easily recognized by untrained consumers. Although the danger to the consumer normally depends on the intensity of exposure to the compound, no information is provided that relates plausible exposure scenarios to different illnesses. And although in some cases the label warnings are simplistic, in others a flood of hard-to-read information is provided. The warnings and directions for proper use for one pesticide I recently examined in my local hardware store contained a "label" that is in fact a folded fourteen-page booklet set in tiny, single-spaced type.[17] The information in the booklet is considered by the agency to be part of the label, and thus a form of risk management. But obviously few consumers read labels and product booklets carefully; consequently most who experience pesticide exposures have little chance of understanding the warnings well enough to prevent harmful doses.

Applicators are often poorly trained. More dangerous pesticides may be purchased and applied only "by or under the direct supervision of specially trained and certified applicators," that is, those who have completed training in the safe handling, application, and disposal of the chemicals. But people who apply pesticides are often untrained in toxicology, environmental chemistry, environmental medicine, exposure, or risk assessment. Moreover, the nearly 1.25 million private and commercial applicators who are certified in the United States may legally delegate authority to apply these chemicals to those who are both untrained and uncertified. Effective control of pesticide hazards demands literacy, experience, education, intelligence, and emotional maturity on the part of the applicator. But the EPA does not monitor applicators to ensure they are following label directions or informing those potentially exposed to the chemicals—and Congress's legal approach toward pesticide risk control relies primarily on self-education, self-regulation, and the self-reporting of incidents.

The government's trust in these weak measures has sometimes led to disastrous results for consumers. In the late 1990s, the EPA discovered that methyl parathion had been applied indoors in hundreds of homes to kill roaches, even though the compound was licensed only for outdoor agricultural uses. In all, nearly 4,500 buildings were reported to have been treated with the insecticide, affecting nearly 18,000 people and 10,000 children. Relocation and decontamination costs exceeded $200 million by 1997. Affected areas were designated as a federal Superfund site (despite the properties being scattered across nine states), a title reserved for the most seriously contaminated sites in the nation.[18]

Residues persist longer than many consumers realize. Apartment managers need not inform tenants that a building has been sprayed with potentially dangerous insecticides or that chemical residues are likely to remain— and most prospective tenants are unlikely to ask questions about prior or future pesticide applications before signing a lease. Occupants of apartments, offices, and schools that are treated by applicator services are sometimes notified so they can leave during the application, but any pesticide residues are generally undetectable. History has taught us that both government and industry have often underestimated the persistence of pesticide residues released indoors, and that the government's understanding of pesticide persistence indoors is remarkably scant and grounded in little experimental evidence.

There would be no need to warn the public if the chemicals instantly dissipated and became nontoxic. Although persistence generally makes pesticides more effective for a longer period, it also can lead to lingering hazards. By 1986, six years before the Eblings moved into their apartment,

Dow had gained approval to market a different formulation of chlorpyrifos named Dursban ME that was "microencapsulated" or coated in a polymer that would gradually break down and release the insecticide. (This "slow release" technology had been developed several decades earlier by the pharmaceutical industry for use in over-the-counter cold capsules.) Dow scientists estimated that a single five-and-a-third-ounce bottle contained nearly 8 billion microscopic spheres. By January 1986, Dow was marketing the product with claims of "low human toxicity" while still selling emulsifiable chlorpyrifos formulations that produced higher environmental concentrations shortly after treatments. Dow advertised that "Dursban ME is designed to deliver peak performance where insect pressure is high, where toxicity, odor or frequency of treatment are concerns . . . [and] the advanced polymer coating provides an additional degree of safety in handling and application." The newer microencapsulated formulation plausibly could lead to lower human exposures indoors when compared with older formulations that release the active ingredient to the environment more quickly. The EPA, however, neglected to compare the likely exposures and effects of the newer formulation to the older one, and not surprisingly granted yet another license to Dow.[19]

Labels often use confusing language. Neither FIFRA nor the EPA specify precise language that must appear on pesticide labels. Perhaps this is one reason that, when the EPA studied consumers' response to labels in the late 1990s, it found considerable confusion. Nearly one-third of those who purchased pesticides for indoor application did not read label information describing the potential health effects of using these products. Moreover, the words used in the labels meant one thing to the government and manufacturers, and quite another to consumers. To describe potential toxic effects, the EPA has long relied on "signal words" such as "danger," "warning," or "caution," each of which refers to a different level of risk. But nearly 50 percent of those surveyed indicated that these terms meant the same thing. And many consumers who did wade through the ingredient section on labels told the EPA that they did not understand the information presented.[20]

Pesticides are often not tested thoroughly. Most pesticides have not been fully examined to know if they threaten developing nervous, endocrine, or immune systems even in test animals. Nor is their persistence within manmade environments well tested or understood. Consequently, government regulators have no way of knowing the intensity of exposure experienced by people living, working, or learning in treated environments. This lack of information also means that required labels cannot provide full and fair warning to consumers.

Labels often use misleading imagery. Labels are part of packaging, and packaging is a primary method for advertising pesticide benefits. But placing warnings about the product next to advertisements of benefits creates an obvious conflict for consumer attention. Colorful images of children running across green lawns, flawless golf courses, blooming flower gardens, and ugly dead bugs often occupy the front of pesticide packages, whereas warning information is placed at the bottom of product panels (and supplemented with information in small black type on the back). One is naturally drawn to the pictures, rather than technical information about hazards and safe use instructions.

Unsubstantiated claims of effectiveness. The EPA stopped evaluating manufacturers' claims of product efficacy in 1976 due to lack of staff time and budgetary constraints. Consequently, for more than thirty years, manufacturers have been free to use labels to advertise unsupported claims of effectiveness. Indeed, all but the most recent history of U.S. pesticide regulation has been premised on the need to balance pesticide benefits against risks. And as we have learned, several National Academy of Sciences committees and many other serious critics have concluded that the EPA has done a poor job of estimating the health and environmental risks of these compounds. But if the risks are not well understood, and the benefits are unsubstantiated, where does this leave pesticide regulation—and consumers?

THE EBLING FAMILY LAWSUIT

The Ebling parents filed a lawsuit in 1997 seeking compensation for the injuries from defendants including Dow Agrosciences LLC, Louisville Chemical Company, Affordable Pest Control Company, and the Prestwick Square Apartments. The Indiana Appellate Court found that all played some role in pest management in the apartment complex between 1994 and 1995. Dow manufactured one of the chemicals, LCC mixed and sold other chemicals, Affordable applied Dow's chemicals, and Prestwick maintenance employees applied Louisville Chemical Company chemicals.[21]

Despite hundreds of civil suits against pesticide companies during the past several decades, few have resulted in the court award of damages to parties claiming pesticide-related injury. Among sixteen lawsuits filed against Dow in 1999, fifteen were dismissed, and one was settled for $1,000.[22] Courts have normally agreed with chemical company arguments that failure-to-warn claims are prevented or preempted by FIFRA's assignment of exclusive labeling authority to the federal government, so the effect of the pesticide on human health is never examined.[23]

Pesticide litigation often follows a pattern. In an effort to understand the facts of a case, defendants' and plaintiffs' attorneys take deposition testimony from family members, employees of the defendant companies, medical doctors, and scientific experts, often generating thousands of pages of evidence and exhibits. Well before questions of case history and fact may be considered by juries, defendants in many pesticide suits have individually attempted to block the lawsuit, arguing that this type of damage claim was implicitly prevented or "preempted" by federal pesticide law. Pesticide litigation is thus an exceptionally expensive and time-consuming process as defendants discourage movement toward a trial, and plaintiffs' financial resources can easily be exhausted before any trial can begin.

The Eblings' suit was originally brought to the Floyd County Superior Court in 1997. In 2000 it was reviewed by the Indiana Appellate Court, and the Indiana Supreme Court heard the case the following year. The family's attorneys presented several reasons to justify their suit, termed "theories of recovery." First, the family complained that because the label information and cautions had not been provided to them, they had not been warned about the dangers of any pesticide applied to their apartment. Second, they claimed negligence—that is, a failure to exercise reasonable care in potentially hazardous circumstances—on the part of the applicator who might have treated the apartment when occupied, might have treated it too intensively, may have contaminated clothing and toys, and may have failed to ventilate the apartment sufficiently after the applications. Relative to this last concern, the Indiana Appellate Court found: "The Dursban label and the Dursban Material Safety Data Sheet warn users that the product can be hazardous to humans if inhaled. Affordable failed to designate any evidence that its employees ventilated the apartment during or after applying the Dursban."[24] Third, they complained that one of the pesticides applied was defectively designed because it had a higher-than-allowable concentration of one regulated chemical.[25] Fourth, the Eblings alleged that the label could not possibly be protective, because Dow had failed to disclose to the EPA several hundred "incident" reports, that is, complaints it received from users, including reports of health problems described earlier. Federal regulations require these reports to be submitted to the EPA in a timely manner, and if the EPA had been promptly informed, the agency might have tightened restrictions on the use of the pesticide.[26]

Even before the case was tried in front of a jury, the defendants challenged the Eblings' authority to bring the lawsuit, arguing that their claims were prevented by FIFRA, the federal pesticide law. The Indiana Supreme Court barely allowed the case to survive: it disallowed the Eblings' failure to warn, nondisclosure, and negligence claims, allowing only the design defect allegation to be heard. To prove this allegation, the Eblings would need to demonstrate that the products had been dangerously overformulated. They would also have the burdens of reconstructing their history of exposure to the pesticides and demonstrating their plausible relation to the neurological illnesses that the children have suffered.

The Eblings' long fight to recover damages remains challenging. Yet this is a civil action and the Eblings must show that a "preponderance of evidence" demonstrates a causal connection between the defendants' products, behaviors, or services and injuries sustained by their children. Preponderance has usually been interpreted to mean that "more likely than not" the defendants were responsible for the plaintiffs' loss. Criminal cases, in contrast, require the causal nexus to be "beyond a reasonable doubt," which is a more rigorous standard and one closer to standards of scientific proof, often defined by a 95 percent or 99 percent confidence that the observed association between two events did not occur by chance.

To estimate the children's exposure is a complex problem. The strength of a causal claim will depend on answers to exacting questions such as

- How were the applied chemicals mixed or formulated?
- What amounts of each chemical were released in the apartment?
- When, where, and how were the chemicals applied?
- Were the applicators trained to use the chemicals applied, and were they licensed to apply them?
- Were the chemicals present in the mixture at the concentrations authorized and advertised on the labels?
- What dose of each chemical did the children experience?
- What doses were received by inhalation, ingestion, or skin absorption (given that toxicity varies by route)?
- Did the periodic spraying lead to an accumulation of residues in the apartment, or in the children's bodies?
- Could these doses have induced the illnesses experienced by the children?
- What warnings were given or read to the tenants?
- Were the tenants present during the spraying? If not, how long after spraying did they enter the apartment?
- Was the apartment ventilated after spraying?
- What was the rate of fresh air exchange in the apartment?

Responses to these questions would provide a credible basis for estimating exposure, and for untangling relative responsibility among the defendants. Importantly, no one measured the chemicals released in the apartment, tracked their dissipation over time, or took blood or urine samples from family members to know the path followed by the residues. Moreover, since the exposures occurred more than a decade ago, exposure histories are costly and difficult to reconstruct. Susceptibility to any dose may vary by age, weight, or if it was experienced along with other illnesses. So it is difficult to know the specific dose that makes chlorpyrifos and other chemicals damaging to an individual's health. Other environmental and genetic factors, such as the inability to detoxify chemical ingredients, may play some role in the onset of disease, further complicating any quest for proof.

The Ebling story also illustrates how powerful commercial interests have shaped the way pesticides have been regulated in the United States. Their narrative may seem to begin when they moved into their Prestwick apartment, but in fact it started decades earlier, when some of the world's largest chemical manufacturers obtained licenses to package and sell poorly tested insecticides to formulators and applicators, and, as the EPA found soon after 2000, these proved to be hazardous to humans. Applicators purchased "restricted use" insecticides too concentrated to be sold to unlicensed consumers. Yet no one followed those applying the insecticides to find out whether they were licensed or to make sure they would follow label directions.

Congress has generally chosen to manage the risks of harmful exposures to pesticides by relying on manufacturers' warnings on product labels—if the consumer buys the product, the reasoning goes, he or she is responsible for faithfully following all directions and heeding all warnings. But, as illustrated previously, labels cannot possibly give sufficient warning if the products have not been fully tested or if those exposed, such as poorly educated adults or children, cannot understand the technical language in the warnings (or cannot read them at all). The warning system is deceptive and defective, and both government and manufacturers are to blame.

Federal preemption is authorized by the Supremacy Clause of the U.S. Constitution, which reads in part: "This Constitution and the Laws of the United States . . . shall be the supreme Law of the Land; and the Judges in every State shall be bound thereby."[27] Laws passed by lower levels of government that conflict with federal law are thus "laws without effect." Preemption may be exercised by Congress when it judges that doing so best serves the public interest. Traditionally, Congress has deferred to states to regulate threats to health and safety; since 1970, however, Congress has exercised its power to preempt state or local regulation in a number of areas, including the control of pesticides. But in this domain, the federal government's approach seems more geared toward helping the pesticide manufacturers than assuring public safety: more than a dozen environmental laws contain provisions that prohibit states from adopting tougher restrictions than the federal government, including pesticide laws. This use of the preemption rule has invited extensive litigation brought by those who believe that the federal rules are too lax and fail to prevent significant damage to human health and the environment.

In 2005, the U.S. Supreme Court decided to clarify whether state damage claims were prevented by the fact that Congress had assigned the authority for pesticide labeling authority exclusively to the federal government. The case, *Bates v. Dow Agrosciences LLC* (2005) was brought by twenty-nine Texas peanut growers who alleged that application of a Dow Agrosciences' weed killer, Strongarm, had damaged their peanut plants during the year 2000 growing season. The Supreme Court concluded that Dow knew or should have known that Strongarm would stunt the growth of peanuts if the soil acidity was above a pH of 7.0. The pesticide label, however, claimed that "use of Strongarm is recommended in all areas where peanuts are grown." When the herbicide was used on the Texas peanut farmland, where soils often exceed a pH of 7.2, it not only damaged the peanut crop, but also failed to control the weeds. By 2001, the EPA had approved a new label for the chemical that included the warning, "Do not apply Strongarm to soils with a pH of 7.2 or greater."

The farmers first sought compensation directly from Dow without success, then notified the company of their intent to bring suit in Texas court. Dow immediately asked the federal district court for a judgment that the farmers' claims were preempted by federal pesticide law. The district court agreed, relying on FIFRA's language, "states shall not impose or continue in effect any requirements for labeling or packaging in addition to or different from those required under this subchapter."[28] The farmers then took their case to the court of appeals, and it ruled that any judgment by a court against Dow would cause it to change its label, and therefore would be a "requirement" expressly preempted by FIFRA. Although this decision grew from other rulings by federal courts of appeals, conflicting opinions had been reached in other courts.

The U.S. Supreme Court had been silent on pesticide preemption until *Bates,* because the majority of lower court opinions prevented many from bringing damage claims based on a failure to effectively warn. The high court ruled 7 to 2 that the appellate court had erred by assuming that a jury award of damages should be considered a state "requirement." Instead the Court concluded that a jury award should be thought of as an "event" or "optional decision" but not a requirement prevented by FIFRA. To assume that the EPA would necessarily force manufacturers to change their label was only speculation. The Texas farmers' property damage claims were not preempted, nor should other similar claims of health loss: "Congress surely would have expressed its intention more clearly if it had meant to deprive injured parties of a long available form of compensation. Moreover, this history emphasizes the importance of

providing an incentive to manufacturers to use the utmost care in dis-
tributing inherently dangerous items. . . . It seems unlikely that Congress
considered a relatively obscure provision like Section 136v(b) to give pesticide manufacturers virtual immunity from certain forms of tort liability.
. . . We have been pointed to no evidence that such tort suits led to a
'crazy-quilt' of FIFRA standards or otherwise created any real hardship
for manufacturers or for EPA."[29]

The *Bates* decision should reverse a trend in federal appellate courts'
decisions and open the way for damage suits to be heard in state courts.
Fraud, negligent use, failure to warn, inadequate hazard testing, design
defects, and manufacturing defects now may provide theories for seeking
recovery through state tort litigation. Rather than shutting down these
civil suits due to federal preemption, the Supreme Court openly encouraged these civil actions as a backstop for failed federal pesticide regulation. But the scientific complexity, cost, and decades-long delays common
to litigation make it a poor safety net at best.

The Ebling history also brings to light poorly conceived government
priorities. The vast majority of EPA regulators' time and resources is
spent managing chemicals released to outdoor environments, especially
those used on foods. In particular, almost all of the EPA's pesticide regulatory efforts have focused on foodborne residues that linger following
crop treatments. These are a real concern, especially for those who apply
them, and for those who may be at risk for drinking contaminated water.
But indoor exposures to pesticides may be far higher than doses that one
might experience at the dinner table. In a few years, we may well look back
to our current era and wonder why we did not address the risks posed by
indoor chemicals; they will surely be considered one of the nation's most
significant sources of health risk, and one of the most neglected targets
of twentieth-century environmental law.

Alexis de Toqueville argued that local government is most responsive
to and reflective of local will and values. A decentralized system of governance in his view could best tailor its rules to consider how problems
vary from place to place. And because the very definition of a problem is
grounded in human values, local governance could assign priorities to
problems or issues it cared most about. In the area of pesticide regulation,
however, both the U.S. Congress and many state legislatures increasingly

have adopted laws that limit lower levels of government from tailoring regulation to their local conditions and values. These decisions have been promoted by those who hope to maintain a consistent and less costly regulatory system. But neither criterion is likely to protect health or the environment.

The stringency of California air quality regulations when compared with federal regulations demonstrates that the freedom to adopt new forms of control on a less than national scale can benefit the country, if not the world. The Clean Air Act of 1970 and the Air Quality Act of 1967 before it prohibit states from adopting vehicle emission standards different from federal requirements, with the exception of a waiver granted by Congress only to California. Congress approved this strategy "to foster California's role as a laboratory for motor vehicle emission control, in order to continue the national benefits that might flow from allowing California to continue to act as a pioneer in this field."[30] The act also gives other states the right to adopt the California standards, if those states do not meet the federal air quality standards for carbon monoxide, nitrogen dioxide, ozone, lead, sulfur dioxide, and particulate matter. Today no fewer than forty states fall into the "nonattainment" category, making them eligible to adopt the more stringent California standards.

When Congress chooses to prevent lower levels of government from enacting regulations more stringent than federal requirements, state and local governments lose their ability to manage problems that are distinctive to their regions. When authority is centralized, too, knowledge is less likely to reach individual consumers. California's history of successful innovation in the area of air-quality control makes sense. Those bearing the highest risk have the most incentive to adopt innovative solutions, and often they have the best understanding of how their effects are likely to vary according to local ecological conditions (consider the example of the Texas peanut farmers who lost their crop after spraying pesticide on a particular soil type). Local applicators, not EPA regulators, know best what is sprayed when, where, and at what concentrations. And local and state governments are far more likely to recognize and address successfully variations in ecological conditions, as well as any conflicts that might arise between those who wish to use pesticides and those who hope to avoid them.

The Eblings' predicament illuminates a troubling dilemma. Technological hazards such as nuclear power, weapons, chemical wastes, pesticides, and highly contaminated sites demand highly specialized expertise to manage them safely. But society has no efficient legal or economic system to diffuse that expertise among states or local governments. Preemption centralizes authority and expertise in a manner that at first seems to be efficient yet it leaves many, including the Eblings, to face dangers to their health and environmental quality that vary by household and individual. Allowing local and state governments to exercise control will be one important first step toward making our use of pesticides safer, both indoors and out. And such an approach will be much more democratic than the centralized system in place now. Today those bearing the risks simply do not have the knowledge to participate in choices about how their environments are influenced by hazardous technologies such as pesticides. Nor do they have the financial resources to buy that expertise. Under these conditions democracy is eroded, as is our right to know and control chemical dangers in everyday life.

CHAPTER ELEVEN

The DDT Dilemma

IN 1972, WILLIAM RUCKELSHAUS, the first administrator of the EPA, decided to ban the pesticide DDT (dichloro-diphenyl-trichloroethane) twenty-seven years after it had been first sold. Ruckelshaus based his decision on convincing evidence that the pesticide harmed wildlife and caused cancer in laboratory animals. Rachel Carson was one of the first to express alarm over the effect of the chemical on wildlife: its use had pushed many species of large raptors, such as eagles and ospreys, near extinction. Anxiety grew primarily from DDT's persistence and its tendency to increase in concentration as it moved up food chains from predator to predator, until humans, and finally breastfeeding infants, were exposed. At the peak of its use, everyone in the nation carried residues of the insecticide. In 2005, more than thirty years following the U.S. ban, the FDA reported that DDT was still the most commonly detected pesticide in the U.S. food supply.[1]

The initial enthusiasm for DDT had stemmed primarily from its exceptional effectiveness in combating malaria and other insect-borne illnesses. The World Health Organization has long estimated that DDT has saved tens of millions of lives over the past fifty years by reducing the incidence of malaria. In fact, it announced in 2007 that DDT would again become one of its primary chemical weapons to fight the illness, especially in Africa.[2] How should a nation balance the public health benefit of DDT against the risks it poses to human health and the clear damage it causes to wildlife? And what does it mean that most people in the United States still carry residues of the chemical in their tissues?

Malaria has killed well over 100 million people during the twentieth century—a number roughly equivalent to all casualties from warfare during the same period. The World Health Organization estimates that each year a million people worldwide die from malaria, most of them young children. Among those half million who develop the most severe cases of cerebral infections, 50,000 to 100,000 die, and tens of thousands are left with permanent neurological damage. There is little dispute that the severity of the illness is increasing, as drug-resistant strains of the parasite become more prevalent. And nearly 125 different species of mosquito have demonstrated resistance to at least one pesticide. Not surprisingly, where insecticide resistance occurs, rates of illness and death are highest.[3]

Poverty is perhaps the best predictor of malaria; 90 percent of all deaths from the disease occur in Africa, and a map of malaria endemicity could be used equally well to represent the poorest nations in the world. Effective control of malaria is expensive. Officials must discover which insects carry the disease, understand their habitats, apply insecticides, and provide rapid access to quality medical care when people fall ill. Financial resources needed to accomplish these tasks are rarely available to most living in the poorest nations. The ability to fend off the illness, too, depends in part on having a healthy immune system: pregnant women, small children, the undernourished, or those suffering from other illnesses are often the first to succumb and are more likely to live in the most impoverished parts of the world.

To break the cycle of malaria transmission, mosquito bites need to be prevented. The use of pyrethroid-impregnated bed nets during nighttime biting periods has proven to be effective in reducing prevalence rates in some regions. Screening of doors and windows may also help prevent mosquitoes from entering homes, workplaces, and schools, yet many in the poorest parts of the world live without windows and doors. People in these areas also often live, work, and play near slow-moving water such as that found in marshlands, irrigation swales, or roadside ditches, which are prime mosquito breeding areas. The mosquito habitats can be treated with insecticides, but it is difficult and expensive to do so; moreover, routine outdoor application often creates locally resistant strains of mosquitoes.

Malaria is a complex disease that varies regionally within the world's tropical and semitropical climates. It is caused by the single-celled *Plasmodium* parasite, transmitted by female *Anopheles* mosquitoes. There are nearly 430 different *Anopheles* species in the world, and thirty to forty of these can transmit malaria with varying effectiveness. Some species feed on humans, while others feed on animals. Some feed indoors, and others outdoors. Some feed at dusk and dawn, while others feed at night. All of these factors influence the success of transmission and probability of epidemics.

Much of the United States and Europe was prone to malaria outbreaks before and during the twentieth century. Eradication in the United States did not occur until the 1950s following the intensive use of DDT, which was sprayed over tens of millions of acres of rivers, lakes, ponds, streams, and wetlands, both coastal and freshwater. Every acre treated by aircraft was sprayed with a combination of one pound of DDT and one gallon of kerosene. The same landscapes were sprayed repeatedly, at a cost that could not be borne by the poorest African nations, then or now. Some hailed DDT as being as important a breakthrough as penicillin, and the man who discovered the compound, Swiss physician Paul Müller, was awarded the Nobel Prize in Medicine in 1948 for his contribution to global public health. Yet although the toxic mixtures undeniably killed many mosquitoes, they also wiped out many insects, birds, fish, and mammals, leaving landscapes with detectable DDT residues nearly a half century later. Without question, DDT and other insecticides have changed the biological diversity of the United States.

The successful reduction of illness and deaths gave the chemical a powerful image of sanitation, health promotion, and safety. In addition, DDT has long been inexpensive relative to other chemicals. By the end of 1944, E.I. du Pont de Nemours had dropped the price of DDT from $1.60 to $1.00 per pound and by 1949 it was selling for $0.32. Sales of the compound skyrocketed.[4] Also around this time, Westinghouse Electric mixed DDT, pyrethrum, petroleum distillates, and Freon in pint-sized aerosol spray canisters, making the compound easy to carry and apply indoors.[5]

Following World War II, hundreds of new commercial DDT formulations were designed and added to paints, wallpaper, wood finishes, rugs, and fabrics used for clothing and drapes. Within several years DDT was being sprayed on hundreds of crops, on forest lands to control gypsy

moths, and along rivers, lakes, and coastal marshlands to control mosquitoes. In some regions it was applied by aircraft to entire towns, as it
was in Cambridge, Maryland, in 1945. Many homeowners also applied
the chemical outdoors on their lawns and gardens and indoors on surfaces such as walls, windows, doors, and screens to kill nuisance insects.
DDT's low cost, persistence, and low level of acute toxicity made it very
attractive to producers, consumers, and public health officials alike. In
fact, USDA officials repeatedly expressed their concern that manufacturers were not putting enough DDT in products for them to be effective—a
worry that in 1947 led government regulators to pass new rules regarding
labels and claims of product effectiveness.[6] In the USDA's annual *Yearbook on Agriculture* for that year, USDA scientists recommended using
the chemical not just for crop protection, but also within homes, schools,
offices—literally anywhere that insects might be found: "There are several
DDT sprays that may be used in the home for the application of a DDT
residual deposit. The most suitable one is a 5-percent solution of DDT
in deodorized kerosene." The author suggested painting DDT on walls,
windows, doors, and screens, because "it is perfectly safe to use if a few
simple precautions are used."[7]

The author of the yearbook entry had no medical training, and even
as of 1950 only one toxicologist had been employed by the branch of
USDA responsible for pesticide registration. The USDA had become less
a regulator than a cheerleader for those designer chemicals that appeared
to promote health and profits.

SOUNDING THE ALARM

In August 1945, three pharmacologists working at the FDA reported in
the journal *Science* that dogs fed DDT stored the insecticide in their body
fat. The researchers also found that when the dose was continued, the
DDT concentration in fat accumulated. Other mammals, they cautioned,
may also preferentially store the chemical in fat tissues. But their most
alarming finding was that milk derived from lactating dogs contained
"appreciable levels" of DDT.[8] Government scientists immediately grew
concerned over the safety of the nation's milk supply.

Another study published in *Science* in 1945 demonstrated that rats
fed DDT could pass lethal doses of the chemical to their young via lactation. The authors noted that DDT concentrated in milk fat, and that both

residue levels and toxic symptoms increased as the dose continued, even if at a constant level. They warned, "The data strongly suggest the need for more intensive research on the toxicity of milk from dairy cows ingesting DDT residues either from sprayed or dusted forage plants or from licking themselves after being sprayed or dusted with this insecticide."[9] The *New York Times* in 1946 also covered the story: "DDT spraying of pastures or woodlands where dairy cattle graze may poison users of their milk."[10]

These extraordinary findings suggested that if a person regularly ate foods contaminated with low-level residues, DDT concentrations would continue to increase in body fat. This was the first articulation of the concept of "bioaccumulation," whereby persistent chemicals gradually build up in tissues as they are passed to predators along food chains. The chain of DDT's movement began with the pasture hays sprayed with the insecticide and continued to cattle consuming the hay; to beef and dairy products, including milk; to the human dinner table; and eventually to human fat tissues.

The rate of residue buildup depended in part on the consistency of exposure, and children consume few foods more than milk. Chronic exposure to children was assured if animal feeds, dairy barns, and cattle were routinely sprayed. Even if some farms did not spray, milk is often distributed through collectives that gather it from regional dairy farms lying within "milksheds," a term coined by the AEC, which was worried that radionuclides would rain from the sky as fallout and contaminate pasturelands, dairy products, and human tissues.

DDT is stored in fat tissue with a half-life of four years, while the metabolite DDE has a half-life of six years.[11] Once inside the human body, the longevity of exposure depends on a chemical's rate of degradation into other compounds (and this can vary from individual to individual), the propensity of the body to store it in certain kinds of tissues, such as bone or body fat, as well as its rate of elimination from the body via excretion or respiration. Persistent chemicals include chlorinated pesticides—DDT, aldrin, dieldrin, chlordane, heptachlor, toxaphene—and some industrial chemicals such as PCBs, combustion by-products including dioxins and furans, and flame retardants such as polybrominated diphenyl ethers (PBDEs). These all tend to bind to fatty tissues, and will concentrate if exposure is chronic, or routine, even if environmental levels are extremely low. It pays to carefully avoid exposure to these persistent, fat-loving chemicals, especially if they contaminate foods and water that are regularly consumed, or environments that are frequented.

USDA scientists also recognized that no effective method existed to remove the residues from contaminated plants, produce, or animal products. For example, it was impossible to spray a dairy barn with DDT and then somehow remove the residue from foods. The chief of the USDA's entomology division cautioned a Congressional committee in 1947 that DDT especially concentrated in butterfat, so the chemical should not be used on crops used to feed beef or dairy cattle.[12] Recall Willard Libby at the AEC wondering how to filter strontium-90 from milk, rather than proposing underground nuclear weapons testing to prevent further food contamination.

The National Audubon Society had also begun calling attention to the potential toxic effects of DDT. As early as 1949, society members were routinely reporting dead birds, fish, crabs, and bees following the spraying of DDT in these creatures' outdoor environments.[13] Following extensive DDT spraying to combat Dutch Elm disease, gypsy moth infestations, Japanese beetles, and fire ants, honeybees, along with other beneficial insects that pollinate and keep parasites at bay, were recognized to be especially susceptible. Later it became clear that DDT poses a higher threat to some species than others. Goldfish, for example, die at lower doses than frogs, which in turn are more sensitive than rats, cats, and rabbits.[14]

WEIGHING BENEFITS AND RISKS

Debates over the dangers of DDT pitted those concerned about this mounting evidence of toxic effects on wildlife (and potentially humans) against those eager to win the war against insect-borne diseases like malaria and promote higher agricultural yields. The agricultural industry had a lot at stake in this argument. As early as 1945, the USDA reported that beef cattle treated with DDT gained on average fifty pounds over those left untreated (and thus prone to horn fly infestations). Other studies in Illinois claimed to prove that DDT was responsible for a 10 to 20 percent increase in milk production. The American Chemical Society supported this assertion, arguing that insecticide use controlled parasites and thereby increased annual milk production by several million quarts, and beef production by a billion pounds per year, creating a combined increase in farmer income of $800 million.[15]

What logic could justify removing these immediate benefits in order to protect hypothetical risks to human health? This comparison of tangible and certain benefits with unproven health effects illustrates the uphill battle that environmentalists and others fought during that era as they sought to more strictly regulate pesticide use in the United States. Looking back from our twenty-first-century perspective, however, many of the voices of this early movement seem prescient. As early as 1949, for example, the American Medical Association warned that if the chemical industry did not create a credible and sound toxicity testing program, the government was likely to adopt "hastily conceived laws and regulations."[16]

EARLY (AND DELAYED) ATTEMPTS AT REGULATION
One of the more persistent environmental debates of the twentieth century concerned the safety of pesticide residues in foods. By 1950 nearly three hundred different crops were treated by DDT alone, and DDT was only one of hundreds of new pesticides that had been combined to formulate tens of thousands of new pesticide products used in the agriculture industry. By 1955 DDT mixtures, with USDA approval, had been incorporated into hundreds of different pesticide products and was being used to protect hundreds of different species of plants, including crops, shrubs, flowers, and trees. The scale of DDT's use in agriculture was startling by any measure. As of 1950, 363 million acres in the United States were planted with major crops such as corn, soybeans, wheat, cotton, rice, alfalfa, and tobacco, and the USDA recommended using DDT on most of them. When the acreage of specialty crops is added to that of these major commodities, the total acreage of U.S. farmland that the USDA recommended for DDT treatment was nearly 400 million.

Public concern and interest in pesticide residues in foods had begun in the early twentieth century when lead, arsenic, mercury, and fluoride compounds were used intensively as biocides. These elements were well known for not only their acute and chronic toxicity, but also their persistence. DDT has a much lower level of acute toxicity, partially explaining the enthusiasm of agricultural scientists. It was found to be effective almost whenever it was applied; if not, the dose was adjusted, it was mixed with different solvents or additional pesticides, or it was applied with different technologies until the desired effect was obtained.

The primary legal strategy to manage pesticide risks in the twenty-first century was established by the Insecticide Act of 1910. This law was intended to discourage fraudulent formulation of pesticides, and to require the accurate labeling of the contents of pesticide products.[17] The statute set ingredient standards for pesticides, and any product that was mislabeled or impure was prohibited from manufacture, transport, or sale. Farmers were the primary beneficiaries of the law, because it gave them some protection from fraudulent or deceptive marketing claims by manufacturers. The intent was truth in advertising, with no hint of a concern to protect human health or the environment.[18]

Only four years earlier, in 1906, Congress had passed the Pure Food Law, a response to untested preservatives and colors, as well as biological contaminants. Upton Sinclair's *The Jungle,* an exposé of the Chicago meat packing industry that same year, had awakened the public to the larger issue. It seems the "good old days" were a time of remarkable ignorance about the effects of chemicals in and on food. Formaldehyde, for example, was commonly used to preserve milk.[19] But the Pure Food Law was at least a step in the right direction. Unfortunately, the Insecticide Act was not similarly grounded in concern for public safety. It did not provide the government with the authority to control levels of commonly used pesticides such as lead, arsenic, and nicotine in foods. Moreover, it would be nearly four decades before Congress once again sought to regulate pesticides.

The year was 1947, and nearly 30,000 separate pesticide products were available in U.S. markets. Their manufacturers had for years enjoyed encouragement by the federal government rather than regulation. But Congress, at long last, felt compelled to review governmental oversight of the enormous pesticide industry. It passed the Federal Insecticide, Fungicide and Rodenticide Act (FIFRA) in that year, requiring that all pesticides sold in interstate or international markets first be registered with the USDA. It also required labels to include the manufacturer's name and address, the product's ingredients, and warning statements to protect applicators, as well as a list of potentially affected species of plants and animals. The USDA continued to be in charge of pesticide licensing for the next twenty-three years, although it had no power to set residue limits in food, drinking water, or other sprayed environments, and never banned a single pesticide.

The list of pesticide residue limits or tolerances for foods grew slowly, beginning in 1955. The FDA directed its attention to a list of persistent organic pollutants including aldrin, dieldrin, BHC, chlordane, DDT, methoxychlor, and toxaphene. Lead arsenate, calcium arsenate, and fluorine compounds, however, were allowed to remain as residues on dozens of foods. All DDT residues were limited to 7 parts per million, with no consideration given to the fact that some foods such as grapes and apple juice are consumed more frequently, and in greater amounts, than others such as peppers and blueberries—especially by children. Government regulators ignored the possibility of a cumulative dose of pesticide that might result from FDA-approved DDT residues on dozens of foods. They did not consider that some people might be more susceptible to toxic effects than others. And they did not include in their assessment many highly consumed foods, such as milk, meat, apple juice, corn, and dairy products.[20]

The mixture of pesticides allowed on single foods such as apples was then a toxic brew that included BHC, chlordane, DDT, 2,4-D, dieldrin, fluorine compounds, lead arsenate, methoxychlor, naphthalene, nicotine, parathion, and toxaphene, among others. The 1954 Miller Amendment to the Federal Food, Drug, and Cosmetic Act (FFDCA) provided the FDA with the authority only to set limits on raw agricultural products, leaving processed foods unprotected. Thus the FDA during its first four years of tolerance setting (1954–58) restricted its attention to crops rather than processed foods such as oils, juices, dried fruits, and dairy products. By avoiding these foods, which many people consumed frequently, the FDA gave the chemical companies, farmers, and food supply an air of wholesomeness and legitimacy that they hardly deserved. Although the Miller Amendment required that the FDA consider health and safety when setting tolerances, it did not demand that tolerances be set to assure health protection.

DDT IN MILK

The possibility that animal feeds could concentrate chemical residues in meats and fats was well recognized in the late 1940s. Yet the FDA did not have the legal authority to control residue levels in feeds until the Food Additives Amendment to the FFDCA, known as the Delaney Clause, was adopted in 1958. This clause considered pesticides to be food additives in need of tolerances, but only under the arcane circumstance that the residues appear in processed foods at concentrations higher than in parent crops. If DDT existed in milk and butter at the same concentration, no separate tolerance would be required for butter. And since butter is not a "raw agricultural commodity," no other tolerance would be needed. This proved to be a serious oversight in the Delaney Clause: it allowed the FDA to avoid residue testing, and then to conclude that no limits needed to be set for dairy products produced from milk. In another example, the FDA considered fruits, vegetables, and nuts similarly—thereby avoiding the setting of tolerances on processed foods made from them such as juices, oils, and peanut butter. Yet these were precisely the foods delivering the largest doses of DDT to the population. And due to children's unusually high intake of milk, dairy products, and juices, they received especially high doses of chlorinated insecticides and their metabolites. The inadequacy of the Delaney Clause seems obvious by the FDA's choice for its first attempt to regulate pesticides as "food additives": seemingly out of touch with the urgency of the problem for children, the agency initially set tolerances for DDT and DDD residues in pet foods.[21] It would be five more years before similar restrictions would be passed for cow's milk, a staple of children's diets.

Limits for DDT residues in animal meat fat were not established until 1961, sixteen years after the USDA had found DDT residues in the fat of cattle.[22] A specific tolerance or residue limit for DDT in milk was not established until 1967, twenty-two years after the USDA first warned that the insecticide could be easily passed from cattle feed to finished milk, and five years before its ban. The milk tolerance level was proposed not by the FDA, but by the State of California, early evidence that the state's environmental and health instincts were stronger than the federal government's. The manufacturers of DDT requested that California's petition be reviewed by a science advisory committee, and the FDA agreed. Nearly two years later, the committee agreed with California's proposal, and the FDA adopted tolerances at levels proposed by California in 1967, a full twenty-two years after the chemical had been released for agricultural use.[23] The delay in regulation became a pattern repeated for nearly 10,000 other pesticide tolerances set by the federal government.

It is not unusual even today for cattle to be sprayed or dipped with pesticides. In the 1950s, cattle were routinely lowered into a vat of DDT to kill external parasites, and the insecticide quickly passed through the cowhide into the bloodstream, binding chemically to fat molecules in tissues throughout the body. USDA scientists knew soon after World War II that cattle would quickly absorb DDT from dips as well as dairy barn sprays, which were commonly done to prevent the nuisance insects and disease-carrying flies in manure from biting the cattle.

Although dipping and spraying were the most obvious ways in which cattle would absorb DDT, residues in the cattle's food supply were also sources of the pesticide. The USDA, with some foresight, recommended that farmers avoid the use of DDT on grain, forage, or other crops used as animal feed even in 1945, the year the chemical was introduced into U.S. agriculture. But even though their recommendations were grounded in a concern over DDT residues in milk, the department did not directly warn the public about DDT in the milk supply or establish regulations that would keep DDT from concentrating in dairy and beef products.[24]

In fact, three years would pass before the USDA focused serious attention on the safety of the nation's milk supply. In April 1949, it asked manufacturers to cease recommending that DDT be used directly on the hides of dairy animals for parasite control, in feeds they might consume, and within dairy barns for insect control. The suggestion was published in the *U.S. Federal Register* four years later: "Insecticides containing DDT *should not* be used on dairy animals, or on forage to be fed dairy animals or animals being finished for slaughter. They *should not* be used in dairy barns pending the carrying out of adequate tests which show that under the proposed conditions of use, they will not cause contamination of milk."[25]

At the time, the USDA had only the authority to control pesticide labels and the uses recommended on them. Farmers could still purchase the product and use it any way they pleased, since no regulations limited DDT levels in any foods. Not until 1954, nearly a decade following the release of DDT for agricultural uses in the United States, did the FDA obtain the statutory authority to regulate pesticide residue levels in individual foods. It then took several decades more before it set tolerances (residue limits), all while failing to test for DDT's presence in the food supply, tests needed to ensure compliance. Worse, the residue limits eventually set had no relation to health protection, because toxicity studies had not been required from manufacturers; that is, due to a lack of information, the FDA was unable to identify what threshold dose would be dangerous.

In 1962, the FDA acknowledged that corn feeds could transfer residues to cattle tissues and milk. It then adopted a regulation discouraging the feeding of DDT-contaminated feeds to cattle: "Animals that consume corn forage, corn fodder, corn silage, corn stover, or sweet corn cannery waste containing DDT may accumulate considerably more of the chemical in their fat than is present in the feed itself, and a long time may be required on a diet free of DDT to reduce excessive levels of DDT to the tolerance level. Unless the person who raises meat animals is in a position to determine the magnitude of DDT residues in these corn feed products and to insure that the conditions of feeding are such that the residues in meat from such animals will be within the established tolerance, these products from DDT-treated corn should not be used in the feeding of meat animals."[26] The language was so indirect that it had little effect. The agency might have written "No cattle may be fed products containing any DDT residue." And it might have adopted the regulation fifteen years earlier, when DDT's concentration in the human food chain was first discovered.

Corn was hardly the only contaminated animal feed. Once apples are peeled, cored, and mashed to make sauce, the remnants, with their lingering DDT residues, are mixed with other materials to produce animal feeds. The FDA was well aware that wherever DDT was used, it left a trail of residues including its metabolites, DDE and TDD, for many years. It acknowledged as much in its 1954 published list of pesticides used on fruits and vegetables, which included residue concentrations commonly found on each food.[27]

Although many dairy farmers fed their cattle from their own pastures, treated with insecticides to reduce insect and fungal damage, specific tolerances for hays were not set until 1962. The cows' diets were often supplemented by pomace from citrus pulp, apples, tomatoes, and grapes, all crops that were commonly treated with DDT. Other inexpensive (and thus popular) supplements included animal feeds made from plant remnants like the hulls of rice, cotton seeds, and sunflower seeds, which commonly had higher pesticide residues than the interior seeds, fruits, and grains.[28] As farms became larger and more specialized, many relied less on pasture grasses for feed, and more on mixtures of grasses, grains, processing wastes, and roughage from off-farm sources. Significantly, residue levels in feed mixtures from diverse sources are very difficult to predict and costly to test.

WAKE-UP CALLS

In 1952, L. G. Cox, a Beech-Nut Packing Company executive, testified before Congress that the company had spent nearly $650,000 since 1946 attempting to keep pesticides from its baby food and peanut butter product lines. Beech-Nut was forced to reject apples from New York, squash and celery from Florida, and peaches from Pennsylvania, leading it to contract directly with growers and specify pesticides they could use. Cox argued that the very young are commonly more susceptible to adverse effects from DDT and other related chemicals, and that newborn babies may already have detectable levels of residues in their tissues, even as they receive more from their mother's breast milk. All of these claims would be proven in the decades to come.[29]

Cox judged the FDA's efforts to be inadequate and its regulatory authority insufficient. He called for six improvements: complete toxicity testing prior to marketing pesticides to protect the farmer, the food processors, and consumers; the establishment of residue limits or tolerances that would be health protective; the granting of tolerance-enforcement authority to the FDA; the collection of residue data from food samples; public warnings and seizures of products when tolerances were exceeded; and publication of residue limits in the *Federal Register*.[30] More than forty years would pass before Cox's suggestions became federal law.[31]

Another warning was sounded by Edward Laug, an FDA scientist who in 1951 was the first to publish findings that DDT residues are commonly found in human breast milk. All samples tested contained DDT and DDE, a metabolite or degradation product, suggesting widespread exposure from residues in the human food supply. It is curious that Laug's research went unreported by the major media of the time, especially given sustained public and FDA interest in cow's-milk contamination. The absence of public attention to the research is especially difficult to explain given that the results may well have been anticipated: similar DDT residues had been found in 1945 studies of cattle, dogs, rats, and goats.[32]

Since 1950 nearly three hundred synthetic and natural compounds have been detected in human milk, including pesticides, metals, solvents,

PCBs, and combustion by-products, including dioxins.[33] Over time, dozens of studies have demonstrated a gradual decline in some persistent compounds including DDT, explained by multinational pesticide bans or use reduction. Other molecules, including flame retardants and plasticizers, however, are increasing in concentration. Since few chemicals are banned worldwide, and most national bans are phased in over long periods, residues remain. In fact, manufacturers of chemicals banned in the United States have often redirected their sales to poorer nations abroad, making these poorer nations a global dumping ground for pesticides phased out by the rich.

In some tropical nations, DDT is still used to spray the interior of homes for malaria control. (Most outdoor spraying for malaria has ceased due to problems of cost and ineffectiveness; insects quickly become resistant with this kind of spraying.) KwaZulu mothers from northern South Africa have been exposed to relatively high levels of DDT from such interior spraying of their homes, leading to some of the highest recorded concentrations of total DDT in breast milk fat. Curiously, residues of DDT in milk declined among KwaZulu mothers as they had more children, while the children who had been breastfed continued to have high levels. Yet despite these findings, pediatricians overwhelmingly agree that the benefits of breastfeeding—which include a stronger immune system and better nutrition for the baby, and psychological benefits for both baby and mother—far outweigh concerns over exposures to industrial, agricultural, and pharmaceutical residues.[34]

EARLY SUBSTITUTES FOR DDT

The U.S. ban of DDT announced by EPA administrator Ruckelshaus in 1972 soon led to the use of an alternative insecticide known as dicofol, which replaced some of DDT's use on citrus and cotton crops. Dicofol was first licensed by the USDA in 1957. But it is made from DDT feedstock and, as the EPA discovered in 1984, contained 9 to 15 percent DDT, meaning that thousands of pounds of DDT were in fact still being released legally into the environment. By the turn of the century, approximately 860,000 pounds of dicofol were being used on nearly 720,000 acres of cropland.

Moreover, dicofol is structurally similar to DDT, and concentrates in apple pomace, citrus oil, mint oil, dried tea, citrus pulp, cottonseed oil, raisins, and plum prunes. In 1998 the EPA concluded that the chemical was likely to concentrate in adipose or fatty tissue, and it is excreted in the milk of tested animals. This meant that the EPA would need to set tolerances to limit dicofol in animal feeds.[35]

That same year, the EPA published a list of dicofol residues anticipated in 124 different foods. Residues were believed highest in foods heavily consumed by children, including milk, meat products, oranges, raisins, and some stone fruits. Six years later, in 2004, the agency finally proposed to limit dicofol residues in animal feeds, milk, meat, and poultry products and in 2007 it finally revoked these tolerances, completing the ban on DDT, dicofol's parent, fifty-two years after it was first used in the United States.[36]

Another substitute for DDT was also suggested in 1972: methyl parathion, a chemical relative of nerve gas agents that in high doses can be highly toxic to the central nervous system. Twenty-seven years later, in 1999, the EPA announced that it was banning future uses of methyl parathion on foods that made up a large percentage of children's diets, while maintaining its license for use on cotton, wheat, and corn.[37] The chemical has poisoned many people, and is especially dangerous when used indoors; as mentioned earlier, the EPA had to relocate nearly 2,300 people living in nine Southern and Midwestern states after the insecticide was applied illegally in homes, schools, and businesses.[38] The dicofol and methyl parathion experiences demonstrate the dangers of bureaucratic inertia and suggest that banning one technology may lead to the use of a poorly tested substitute that poses very serious risks. Anticipating the cost, effectiveness, environmental fate, and toxicity of likely substitutes is essential for avoiding a never-ending cycle of poor choices for pest control.

The World Health Organization announced in September 2006 that DDT will again play a major role in the international efforts to combat malaria. Although less persistent but more costly pesticides are available, the international donor community has not provided the necessary funds to purchase and apply them.[39] Indoor spraying will certainly increase concentrations in human tissues, although care will be taken to minimize harm to wildlife.

The choice by WHO and African public health officials to accept these risks in order to save lives is a very different one than allowing it to be used for agricultural purposes. The risk of contracting malaria is considerably higher than the danger of experiencing a toxic reaction to DDT sprayed in mosquito-abatement programs, and these programs are vital to the prosperity and well-being of entire cultures. Still, I am not an advocate for the use of DDT anywhere in the world because it builds in food chains and in human tissue. The United States would certainly not spray indoor environments with DDT, nor should we recommend this policy abroad. Wealthier nations have the capacity and, I believe, the responsibility to provide aid to purchase less persistent and less toxic chemicals, products that would sustain both health and environmental quality in impoverished and malaria endemic areas.

CHAPTER TWELVE

What Is Acceptable Risk?

FOR NEARLY TWENTY-SIX YEARS, the EPA was charged with balancing the economic benefits of pesticide use against their risks to health and environmental quality. During this period, it neglected the review of chemical mixtures, largely avoided product bans, slowed the pace of regulation, and sped the movement of new chemicals to the marketplace with incomplete testing. Some of these outcomes grew from the agency's restricted budgets, insufficient statutory authority to limit risk, and incomplete knowledge. But the EPA routinely has been slow to react to scientific consensus that pesticide risks are greater than earlier believed.

The EPA has had successes: in particular, it has banned some chemicals believed to be especially dangerous, including DDT, aldrin, dieldrin, chlordane, heptachlor, EDB, and 2,4,5-T. Among all pesticides registered during the twentieth century, however, fewer than 3 percent have been banned following government recognition of their potential to endanger humans or other species.[1]

Significantly, each ban followed debates that often lasted more than a decade, meaning that human exposures continued long after scientists recognized their dangers. Moreover, persistent chemicals may create an exposure problem that lingers long after bans are implemented. DDT, for example, is still one of the most commonly detected pesticide residues in some foods, although concentrations are very low. DBCP, EDB, and aldicarb have all been banned, yet their residues still contaminate aquifers that lie beneath fields treated decades ago.[2]

Many different types of bans are possible. Production, formulation, sales, export, and use may each be prohibited independently. Yet when the EPA attempted to stop a compound's use in the United States, it allowed phase-out periods that permitted the sale of existing chemical stocks, often for years. The EPA also has allowed domestically banned products produced in the United States to be exported to unregulated foreign markets.

The remaining 97 percent of pesticides not banned have been managed by labeling requirements, concentration limits, specific use licenses (outdoors versus indoors), or guidelines for methods of application. Some individual pesticides are present in more than a thousand separately licensed products, and the EPA may choose to restrict some of these products but not others. Regulations that restrict use are more difficult and expensive to monitor and enforce than outright prohibitions that are politically more difficult to accomplish. Consequently, the agency has routinely requested producers to voluntarily abandon a small percentage of licensed applications (often those with little market share) to avoid product bans. This strategy has tended to reduce litigation, and allow manufacturers to help judge how risks might be reduced in the most cost-effective way for them. It also sidestepped the odd requirement, long demanded by federal law, that manufacturers be compensated by the government for loss of property value resulting from such bans.[3]

During the twentieth century, several waves of pesticide products were developed and used in the United States. The initial wave included many toxic metals (including lead, arsenic, and mercury); the next brought about persistent organochlorines such as DDT; another wave followed that included acutely toxic organophosphates and carbamates; and finally came the new synthetic pyrethroids. As each new wave provided substitutes for older chemical designs, government regulators believed they were encouraging the use of newer, less risky pesticides. The public was surprised to find that each class of chemical carried different types of risk that were often misrecognized (or possibly misrepresented) when they were initially licensed.

CHANGING PUBLIC PERCEPTIONS

Several committees of experts convened by the U.S. National Academy of Sciences have criticized the EPA over the years for neglecting the risks of pesticides to human health. One committee's 1987 report, *Regulating Pesticides in Foods: The Delaney Paradox*, concluded that the EPA's approach toward pesticide regulation had permitted significant public exposure to carcinogens in the diet.[4] The negative news has had an effect on public perceptions of food safety. A second NAS committee reported in 1993 that variations in diet could lead to differences in pesticide exposure, and raised special concern over residues in the diets of children. They found that children are more likely than adults to be harmed from pesticide exposure, due to their immature organs and the body's rapid growth during this stage, and that they are often more exposed than adults. Furthermore, the committee found that these concerns had not been considered by either the USDA or the EPA when nearly 10,000 food tolerances were established, or when 70,000 pesticide products were issued licenses.[5]

These reports stimulated broad public concern over the quality of the international food supply. Public dissatisfaction with pesticide residues detected in foods helped organic products become the fastest growing sector in the food industry. And numerous reports of pesticides in drinking water have helped fuel phenomenal growth in the bottled water industry.

The reports also stimulated Congress to pass the Food Quality Protection Act (FQPA) in 1996.[6] FQPA became the most health-protective environmental law adopted during the twentieth century. Key innovations within the statute include:

- A precautionary general safety standard requiring the EPA to find "reasonable certainty of no harm" before licensing any pesticide.[7]
- A shift in the burden of proof so that instead of the government's having to demonstrate danger of registered pesticides, manufacturers are required to demonstrate their safety.
- A requirement that the EPA explicitly find that each pesticide given a permit does not pose significant health risks to children.
- A mandate that the EPA consider how individuals may be exposed

to "aggregate risk" from pesticides; that is, a potentially unsafe total exposure to a single chemical from contact with foods, water, air, lawns, indoor environments, and pets.

- The end of the EPA's usual practice of "one-at-a-time" chemical review; instead the cumulative risk—that is, the risk of being exposed to multiple compounds from either a single mixture or multiple sources—must be assessed.[8]
- An acceleration in the pace of review, so that all tolerances are reviewed to determine compliance with the new standards by 2006, with those chemicals and uses that pose the greatest risk being checked first.
- The requirement of "an additional tenfold margin of safety for the pesticide chemical residue and other sources of exposure" for infants and children, unless another margin is deemed to protect the health of this vulnerable age group.[9]

The pathbreaking passage of FQPA has clearly posed an extraordinary challenge to the EPA. It has been difficult and expensive to transform thousands of regulations—nearly all created by favoring economic benefits over health risks—to ensure the protection of children's health. As we will see, the EPA's choices to apply or reduce the tenfold safety factor when judging the reliability of data have clouded independent scientists' ability to determine the risk that remains. But FQPA is certainly a step in the right direction, and provides a model for the management of many other types of industrial chemicals that are perhaps equally dangerous, but untested and unregulated.

ON BEING "REASONABLY CERTAIN" OF SAFETY

Over the past decade, scientists have become increasingly concerned that pesticides may interfere with normal growth and development in fetuses and children. Previously, the EPA simply had not asked the necessary questions or demanded answers before licensing tens of thousands of products. Nor had the USDA. Importantly, the government had also neglected to explore the effects of pesticides on developing nervous, immune, endocrine, and respiratory systems. Moreover, toxicity to the developing nervous system has rarely been tested in animal studies. Even so, in 1998 the EPA found that among nine—only nine—pesticides tested for developmental neurotoxicity, six produced effects in young animals at lower doses than in the adults.[10]

Then, in 1999, Theodore Slotkin at Duke University found that the insecticide chlorpyrifos inhibited neural cell replication in rat brains: "When cells were allowed to differentiate in the presence of chlorpyrifos, cell replication was inhibited even more profoundly and cell acquisition was arrested." The EPA also concluded that although no threshold of toxicity was discernable in a developmental neurotoxicity test that demonstrated structural alterations in brain development, "a clear qualitative difference in response (susceptibility) existed between adult rats and their offspring."[11]

Susceptibility to neurotoxicants appears to be related to the stage of nervous system development. Since human brain growth occurs most rapidly until age six, early childhood exposure to neurotoxicants deserves careful consideration. Studies on lead, irradiation, fetal alcohol syndrome, and oxygen deprivation generally indicate that the developing brain is more vulnerable than the mature brain. In 1993, the NAS Committee on Pesticides in the Diets of Infants and Children may have anticipated Slotkin's results when it concluded: "The data strongly suggest that exposure to neurotoxic compounds at levels believed to be safe for adults could result in permanent loss of brain function if it occurred during the prenatal and early childhood period of brain development. This information is of particular relevance to dietary exposure to pesticides, since policies that established safe levels of exposures to neurotoxic pesticides for adults could not be assumed to adequately protect a child less than 4 years of age."[12]

These findings also motivated the EPA in 2000 to request developmental neurotoxicity data for 140 additional pesticides, data that were to begin to be submitted by chemical companies and be peer reviewed over the following several years. Yet by 2008 the EPA still had not interpreted these essential test results, nor had it requested developmental neurotoxicity data for hundreds of other pesticides.

Some of these other untested pesticides are known to have caused developmental immunotoxicity in rodents. In one study of rodent exposure to chlordane, for example, immunity was depressed for 101 days following birth. When adult mice were exposed to the same level as pregnant mice, immune effects were either lower or not observed.[13]

Many synthetic chemicals, including some pesticides, mimic human hormones. Both the Food Quality Protection Act of 1996 and the Safe Drinking Water Act Amendments of 1996 require that pesticides be tested to identify their potential to block or mimic human hormones. In 1998, in a further positive step, the EPA's Endocrine Disruptor Screening and Testing Advisory Committee proposed protocols for the screening of approximately 15,000 chemicals to understand their potential effects on the human endocrine system. Nearly a decade later, however, these protocols have not yet been applied to judge the potential for most pesticides to alter normal growth and development by affecting the human endocrine system.[14]

A serious additional gap in our knowledge about chemical pesticides stems from the limited toxicological testing of chemical mixtures. Humans are routinely exposed to mixtures of pesticides through residues in food, water, and different indoor and outdoor environments. How should exposures accumulated from food, water, and indoor sources be considered together? How should risks from different chemicals be added if they contribute to the same illness? And how should the relative potency among chemicals be accounted for when judging the collective risk? These questions have stalled the EPA's analyses of mixtures for nearly fifteen years.

The National Academy of Sciences warned the EPA in 1993 that it should consider the probability that children are exposed to mixtures of organophosphate insecticides that pose a collective threat. As mentioned earlier, the FQPA requires the EPA to regulate groups of chemicals that may contribute to the same adverse health effect via a "common mechanism," and mandates that maximum allowable exposure to pesticides by children meet the new safety standard of "a reasonable certainty of no harm." The EPA Scientific Advisory Panel for Pesticides, a panel of scientists convened by the International Life Sciences Institute, as well as other academic, government, and private-sector scientists, concurred with the opinion that organophosphates should be regulated together to protect against their cumulative effects. Yet it was still thirteen long years before the EPA completed its first review of these organophosphates, a review that considered only exposures and risks from individual chemicals.[15]

These delays and inconsistencies mean that today pesticide toxicity evidence remains unreliable in numerous respects: animal tests are used to infer human risks; short-term tests are used to predict long-term health loss; "safe doses" are predicted from concentrations that produce the lowest-known adverse effect; susceptibility of the nervous, immune, and endocrine systems among the young has not been tested for most pesticides; and the effects of mixtures are neither tested nor known. Given the FQPA's requirement that all tolerances be reviewed by 2006 to determine compliance with the new standards, the government's pace of scientific inquiry seems inexplicably—and irresponsibly—slow.

COMPOUND EXPOSURES: FOOD AND WATER

By 1999, in response to requirements in the 1996 Food Quality Protection Act, the EPA had set 9,700 separate tolerances to limit pesticide residues in food. This exceptionally large number grows from combinations of nearly four hundred different pesticides and almost a thousand different foods. For example, nearly 150 distinctive tolerances limited residues in milk, while 1,300 limited residues in animal feeds as a way of keeping residues from finding their way into milk.[16]

The EPA estimates national exposure to pesticide residues from nearly a thousand different foods by combining data on detected concentrations of the chemicals on the foods themselves with information drawn from food intake surveys. But information about what people ate in the mid-1990s may not be accurate for today's food consumers. Dietary patterns change rapidly in response to shifts in tastes, lifestyles, marketing techniques, and food processing technologies. Juice boxes introduced in the early 1990s, for example, increased children's consumption of single fruit juices (and created the added challenge of figuring out pesticide exposures from juice mixtures that often came from several foreign nations). The processing of normal-sized carrots into "baby carrots" and the packaging of these in individual serving bags convenient for lunch boxes similarly increased carrot consumption among children.

Food intake, and thus pesticide exposure, also varies by age, seasonal availability of different foods, income, regional differences in availability, and ethnic preferences, but these differences not well captured by current survey designs and limited sample sizes. In particular, the diets of children change rapidly between birth and age six, demonstrating the need to sample children year by year. Averaging their intake over five-year periods, as the EPA does, is likely to lead to underestimates of exposure to at least some pesticide residues.

Another gap in the EPA's knowledge comes from a lack of attention to processed foods, which have become an ever larger part of the American diet. In particular, processed foods like fruit juices, milk, raisins, and vegetable oils—all favorites of children—are tested for residues far less frequently than are raw foods such as fruits and vegetables. In addition, market-basket surveys, a common way of collecting field trial data on food consumption, usually do not discriminate between treated and untreated pieces of food. Because some fat-soluble pesticides concentrate in oils and fats, while other water soluble compounds concentrate in dried foods such as raisins, this seems an unfortunate oversight that is also likely to lead to underestimates of pesticide exposures.

Some foods are consumed unblended—apples, peaches, steaks—whereas others, such as peanut butter, applesauce, hamburger, and oils, are eaten as mixtures. The government does conduct some tests on blended foods: it creates slurries from different pieces of food, then samples the slurry to obtain an average residue level. Using this methodology, however, means that the highest levels of residues in the slurry, as well as in the marketplace, are overlooked.

Pesticide residue levels in meat and poultry are influenced by numerous factors including how and whether pesticides are applied to animal feeds and directly to animals for veterinary purposes, how long after the pesticide is applied the animal is slaughtered, and if the particular pesticide or toxic metabolite used can bioconcentrate. The USDA is still in charge of monitoring residues in meats, poultry, and eggs, whereas the FDA is responsible for all other foods.[17] But as late as 1999, the USDA's Food Safety and Inspection Service was using a testing method incapable of detecting thirty-three organophosphate insecticides.[18] So although the findings of this inspection service may give the public the impression that the foods we eat are safe, in fact their information is woefully incomplete.

Water is the "food" most consumed by children. Pound for pound, children and especially infants consume much more water than do adults. In addition, if tap water is contaminated but used to reconstitute common dried or concentrated foods like fruit juices or infant formula, both of which are frequently given to children, exposures can be increased.

Pesticide residues in drinking water are poorly managed in the United States. The Safe Drinking Water Act requires testing only for chemicals that are "listed," not for all chemicals that pose a health threat. Among

1,200 active ingredients now registered by the EPA, only twenty-six have established "maximum contaminant limits" set pursuant to the act. These limits are not set exclusively to protect public health; they consider, for example, the economic feasibility of detection and removal, a factor that tends to vary by the size of the community served by the aquifer. Among forty organophosphates registered for use, none have established maximum contaminant limits, even though nearly 100 million pounds of these insecticides are released to the environment annually.[19] The absence of any legal requirement in the United States—at any level of government—for testing water from wells serving fewer than fifteen service connections (households) or twenty-five individuals further increases uncertainty regarding the extent of residues in water and human exposures.

Although water intake is far easier to predict than food consumption, pesticide residues in water are poorly understood due to a variability in chemical concentrations over time and in long- and short-term weather conditions, as well as the expense of monitoring and analysis needed to capture this detail. If water supplies are contaminated with pesticides, exposure will normally occur through drinking the water; eating foods prepared with the water; inhaling volatile pesticides in steamy showers, baths, and kitchens; and absorbing them through the skin during washing. Those most at risk are likely to be those who derive water from shallow wells on or near lands that are treated routinely and intensively with pesticides such as golf courses, croplands, and dense residential communities. If soils are sandy and porous, the groundwater table is high, or if the well is old, the risk of contamination increases.[20]

THE CHALLENGE OF INDOOR PESTICIDE USE

Early in 2007, nearly 107 separate pesticide active ingredients were registered for use in indoor environments. Many interior environments are treated routinely with pesticides, including homes, offices, schools, stores, industrial sites, recreational settings, restaurants, cafeterias, hotels, theaters, hospitals, and public vehicles. The diversity of settings obviously makes it difficult for the EPA to estimate how one might accumulate exposures moving through daily life. And there is no legal requirement to inform occupants of the chemicals that have been applied, their potential health effects, or their rate of dissipation, all of which are necessary to know when it is safe to reenter the structure following treatment.[21]

The agency has attempted to identify the range of human behavioral differences that may affect pesticide exposure in its *Exposure Factors Handbook,* and *Child-Specific Exposure Factors Handbook.* But considerable uncertainty still surrounds the agency's exposure and risk assessments due to the lack of research into both how people use pesticides and how they live in and around treated environments.[22]

Children's exposures to pesticides have long been a concern for those interested in limiting contact with these compounds. Young children spend more time indoors within residential settings than adults, and this time is usually spent on or near floors, where pesticide residues, dust, molds, and other contaminants settle. Young children touch surfaces that may be treated with pesticides more frequently than do adults, and they tend to put their hands and objects in their mouths, crawl on floors, and wear fewer clothes than adults, especially in warmer climates.

Untrained homeowners are permitted to apply many pesticides, and the mixing, storage, and disposal of pesticides can all potentially expose the applicator and others in the household to high concentrations. Clothing worn by professional applicators will normally carry residues indoors, leading to additional exposures of residents. The extent to which eating and cooking surfaces, toys, furniture, and clothing are protected during spraying; the material composition of clothing, carpets, and furnishings; and the extent of the area sprayed all affect potential exposure levels.

The allowed interval between spraying and permitted reentry may also have an important effect on exposure levels. Farmworker exposure is normally governed by field "reentry intervals," but products registered for homeowner application are not; in fact, many products may legally be sprayed when rooms are inhabited. In the absence of contrary evidence, the EPA has long assumed that indoor residues dissipate and pose no significant health threat. Recent experiments, however, have shown that residues may persist for months and years following application. Residue levels are influenced by structural characteristics such as the design and location of the heating and ventilation system (especially fresh-air exchange rates) and its quality of filtration, the location of windows and doors, and the tendency of homeowners to reduce ventilation to save energy. For some chemicals, residue levels in the air continue to rise for days following application—and are highest near the floor.

Intense chemical exposures may also occur within vehicles such as buses, trains, planes, and ships, which are routinely treated with pesticides. The Airline Flight Attendants' Association has brought several lawsuits against airline companies for spraying planes prior to transoceanic flights, yet little is known about the magnitude or seriousness of residue persistence and movement, or the intensity of passenger exposures. Factors that are likely to affect exposure intensity include the amount applied, the application method, the number and kind of inert ingredients used, the materials treated, how long the passenger spends in the vehicle and how much contact he or she has with treated surfaces, and the vehicle's ventilation and air-filtration systems. The EPA admits knowing little about the issue, although it has licensed dozens of chemicals to be used inside vehicles.[23]

Several thousand pesticide products are available for lawn and garden use by homeowners. These are sold in hardware stores, building supply outlets, supermarkets, and drugstores. As with indoor applications, those who apply them outdoors face potential exposure while mixing, applying, cleaning up, and storing the pesticides, as well as when reentering treated areas; and each step carries additional risks if label warnings and directions are not followed perfectly or if other mistakes occur. In addition, children's exposure to pesticides applied outdoors may be much greater than adults': they love to roll, play, and sit in the grass, and they enjoy touching and smelling vibrant ornamental shrubs and flowers, which are often sprayed more than other plants.

Hundreds of pesticide products are also available for animal treatments, and may end up contaminating children. These include pesticide-impregnated collars, dusts, sprays, liquids, ointments, and shampoos. Once applied, these chemicals dissipate at rates and ways that are not well understood. Children may be more exposed than adults to pesticides applied to animals because they are more likely than adults to touch and hug pets, and a child's breathing zone is closer to that of pets. Those who allow treated pets to sleep on their furniture and beds, adults and children alike, will also experience higher exposures.

Many insecticides are registered for mosquito control and may be applied in a manner that results in direct contact by adults and children. The spraying of Shea stadium to control the spread of West Nile encephalitis during a Mets baseball game in 2000 is an example of avoidable human exposure.[24] News reporters have captured trucks driving through

Manhattan in New York City, spraying pedestrians walking along side-walks.[25] In Connecticut, children waiting for a school bus were sprayed from aircraft in an attempt to contain Eastern equine encephalitis, also transmitted by mosquitoes.[26] And the proximity of many schools to agricultural lands that are treated have led to numerous complaints about sprays drifting into school facilities.[27]

Restaurants, cafeterias, and food warehouses are routinely treated with pesticides. In June 2000, the EPA estimated that 24 percent of all food-handling establishments were treated regularly with the insecticide chlorpyrifos, which has since been banned from use in homes or schools. Human contact within these facilities could occur through food contamination, inhalation of airborne particles, and dermal absorption following contact with treated surfaces.

Nearly one hundred active pesticide ingredients—including bactericides, algaecides, and antifungal agents—are currently used legally to kill microbes in swimming pools and spas. Releasing biocides in pools kills bacteria, viruses, fungi, and parasites that otherwise could produce serious illness. Those who treat pools to improve water quality, however, are rarely trained to understand the potential for exposure and associated health risks. And even if they were trained, there is little empirical evidence of whether and where residues persist, and what the predominant routes of exposure are. For children, frolicking in the pool may mean absorbing pesticides through the skin, by swallowing water, or by breathing the air in an indoor pool room. For those children and adults who train competitively and spend hours daily in the pool, significant exposures could result.

RECENT STEPS TOWARD TIGHTER REGULATIONS

Since 1996, the EPA has cancelled all future uses of thirteen organophosphates, including uses of methyl parathion on all fruit trees, all household uses of chlorpyrifos and diazinon, crop uses of both phosmet and azinphos-methyl, and other uses of propargite.[28] All of these compounds are organophosphate insecticides, and the new prohibitions will reduce the probability that children will experience dangerous exposures. But even in this bold action, there is evidence of the EPA's bargaining with powerful manufacturers, who want to continue to sell their products in as many forms, and for as many uses, as possible. In 2000, for example, chlorpyrifos was prohibited from future use in homes and schools, but allowed for use on some crops and other types of buildings. It was also disallowed for future use on tomatoes, but only restricted to prebloom applications on apples and grapes. Overall, among 9,700 food tolerances, the EPA revoked nearly 1,900. Manufacturers voluntarily withdrew many of these from the market due to lack of profitability, or their unwillingness to pay the expense of further environmental and health testing, but the effort is still notable, and laudable: it constitutes the first serious and comprehensive overhaul of federal pesticide regulations since the EPA was created in 1970. The pragmatic logic employed since 1996 has reduced children's health risks while preserving some uses profitable to the manufacturers and growers, and should be considered a partial success.

Health advocates still argue that protections are superficial, and point to the EPA's chemical-at-a-time approach to risk reduction, rather than more thorough consideration of the mixtures of chemicals we all face routinely. Such concerns are understandable. To be fair, the agency, in its final cumulative risk assessment for organophosphates released in July 2006, does develop cumulative exposure estimates from numerous sources including food, drinking water, and residential applications. It also begins to explain the dangers of different chemicals that may threaten human health via a common mechanism of toxicity. (Indeed, the assessment controversially uses a single measure to describe the collective or "cumulative" risk posed by forty different pesticides that can harm the human nervous system via the same biochemical mechanism. That is, the EPA relied exclusively on cholinesterase inhibition to judge organophosphate neurotoxicity, ignoring research that reported decreased brain development.[29]) But the government did not release the assessment until nearly a decade after Congress had mandated the review, and did not begin to consider this type of cumulative risk until 1996, nearly thirty-six years after Congress had given the EPA regulatory authority over pesticides.[30]

The most important decision regarding pesticide regulation during the presidency of George W. Bush concerned the collective or "cumulative" risk posed by 33 different organophosphate pesticides capable of harming the human nervous system. The EPA in 2006 made the choice to dismiss the extra tenfold margin of safety for a number of profitable insecticides, despite its requirement in the FQPA intended to protect against unforeseen risks to infants and children. The statute demands that the government must find it has "reliable" evidence of risk before it may reduce the safety margin. The practical effect is an increase in allowable levels of insecticide exposure. The agency officials argued that they had reliable evidence of toxicity, yet neglected the common unreliability of available exposure information that is a fundamental determinant of overall chemical risk.

The agency found in 2006 that its risk estimates were reliable for twenty-two of the organophosphates. As mentioned earlier, all uses were cancelled for thirteen of the compounds. But for eleven others, including chlorpyrifos, the EPA reduced the margin of safety—thereby reversing the more cautious approach of the agency under the Clinton administration and, significantly, protecting many of the manufacturers' registered uses. Agency regulators claimed that they reduced the margin of safety not because they thought that chlorpyrifos was less potent than earlier believed, but that they better understood how dangerous childhood exposures could be prevented while protecting uses where exposures are either unlikely or more easily avoidable. They also noted that indoor uses have been prohibited since 2000 and restrictions on food applications (for apples, grapes, and tomatoes) have reduced children's exposures. As the authors of the report put it: "Taking all of these factors into account, EPA finds that there is a reasonable certainty of no harm to all major, identifiable population subgroups from cumulative exposure to the OP's [organophosphate pesticides]." One might ask, however, whether the goal of protecting 99.9 percent of the population is sufficient, when it means that as many as 23,000 children will still be acutely exposed.[31]

In 2006 the EPA also completed a cumulative risk assessment for three triazine herbicides, including atrazine, simazine, and propazine, along with their chlorinated metabolites. Atrazine, which has been registered in the United States since 1958, is the most heavily used herbicide used today. Currently nearly 75 million pounds are applied to U.S. lands each year, and the chemicals have contaminated groundwater supplies in many parts of the South, Midwest, and California, as well as aquifers beneath agricultural lands in Europe. The EPA believes that triazines have the potential "to cause neuroendocrine developmental and reproductive effects that may be relevant to humans. Specifically, they may disrupt part of the central nervous system (the hypothalamic-pituitary-gonadal or HPG axis), potentially causing changes to hormone levels and developmental delays." Yet the agency lowered the FQPA safety factor from ten to three, thereby increasing acceptable exposure levels. In doing so, they neglected their own external scientific advisory board's warning: "Because of the rapid developmental brain changes alluded to above, the influence of atrazine on neurotransmitters in the hypothalamus and on GnRH may well have a differential, permanent effect on children."[32] In addition, even after these alarming reports, the agency recommended "monitoring of groundwater supplies" rather than suspending registered uses that were contributing to groundwater contamination.

One additional action the agency undertook in 2006 to reduce children's exposure and risk was the required phase-out of pressure-treated lumber containing copper-chromated-arsenate (CCA), a chemical that often leaches to the surface of treated wood used to fashion playground equipment, picnic tables, and decks. Those playing or eating on these surfaces commonly experience dermal and oral exposures, and arsenic has been detected in the soils beneath these structures. Arsenic is a well-known poison that was used by assassins as early as the fifteenth century. It affects most human organ systems and functions, and can cause peripheral neuropathy, renal and kidney failure, and at high enough doses, death.

The new regulation did not affect CCA-treated wood already in use, and it does not effectively address the potential for serious exposures to sawdust or fumes generated by burning older materials. CCA-treated wood is now considered hazardous waste but is still often mixed with discarded household items, creating a problem if placed in landfills or incinerated. Thus although the agency has taken a valuable step toward protecting children's health from inorganic arsenic, we now need to contend with a massive hazardous waste management problem. The history is reminiscent of the waste problem created when building materials were coated with leaded paints.[33]

There is little doubt that children are better protected from pesticides today than they were prior to passage of the Food Quality and Protection Act of 1996. The most significant decisions by the EPA since then have prohibited use of the more potent products indoors and on foods children generally eat most. Yet surprisingly, all of the data gathering, rulemaking, litigation, and expenditure of hundreds of millions of dollars to regulate the industry have resulted in relatively minor changes to how most of these chemicals are allowed to be used. In addition, a small percentage of the registrations and tolerances that the EPA has adjusted are likely to significantly lower the danger to human health. This fact brings up a tantalizing possibility: perhaps despite the numbing array of chemicals released to the environment as biocides, only a few of these compounds are responsible for much of the overall risk. If these can be identified with compelling data from reliable scientific studies, benefits from the safer pesticides may legitimately be protected, even as the environments in which we and our children live, work, and study are made safer.

PART FOUR BREATHING TOXIC AIR

CHAPTER THIRTEEN

Airborne Menace

NEARLY 86 PERCENT of the world's energy is derived from fossil fuels, or hydrocarbons found in the earth. The combustion of oil, gas, diesel, natural gas, fuel oil, and coal is responsible for most of the nation's outdoor air pollution problem. Two-thirds of all petroleum in the United States is used to move vehicles, while producing a highly diverse mixture of hazardous contaminants including particulates, ozone, nitrogen oxides, and volatile "air toxics" such as benzene. Adding to the hazard are emissions from incinerators, construction equipment, coal-fired power plants that generate electricity, two-stroke engines such as motorcycles and lawn mowers, and increasingly, wood-burning stoves.[1]

The Clean Air Act of 1963 required the EPA to establish and enforce national standards to protect public health from outdoor air pollutants. The original goal of the act was to establish and achieve these standards in every state by 1975. Many areas of the country, however, were unable to meet the standard by then, so the act was amended in 1977 to set new dates for meeting the "attainment goals." By 1990 further amendments had been made to address air quality problems including ground-level ozone, stratospheric ozone depletion, and air toxics. Despite these regulations, more than half of all U.S. residents live in counties that currently do not meet federal standards for ozone and fine particulates. Nearly 15 million experience excessive carbon monoxide exposures, and more than

28 million are exposed to levels of larger particles (ten micrometers or less) at levels exceeding federal standards.

The situation for the nation's children is especially troubling, and may be getting worse. By 1994 several scientists had found that severe asthma occurs more commonly than mild asthma among children living in areas that exceed federal air quality standards. In 2000, the EPA estimated that 9 million children were living in areas where ozone standards were not met; 3.5 million children were living in areas where the particulate standards were exceeded; 2.8 million children were living in counties where the carbon monoxide standard was surpassed; and 1.4 million children lived in counties where the air limit for lead was not met. Seven years later, in 2007, no fewer than 20 million children were living in areas of the United States that failed to meet at least one of the federal standards for air quality.[2]

Those most exposed live in the most densely populated areas and where combustion of oil and gas is concentrated. Southern California is the most severely polluted area in the nation, while nearly one-third of the U.S. population lives in the densely populated Northeast Corridor from Washington to Maine, where air quality in the urban centers routinely exceeds allowable limits.

Moreover, the standards and limits may not be strict enough to protect public health, even where they are followed. Researchers have found evidence of harmful effects at levels near or below limits for ozone, for particles smaller than 10 and 2.5 micrometers, and for nitrogen dioxide.[3] Monitoring is incomplete as well. Current air quality laws require that the states monitor only some pollutants at a few sites. Living in a zone that meets "national standards," then, hardly guarantees that one's air is free of dangerous levels of those hazardous chemicals that have not been monitored—or are not yet regulated.

OZONE

Almost half of the U.S. population lives in areas where ozone levels have exceeded federal limits. Ozone is created when ultraviolet light interacts with nitrogen oxides and with volatile organic compounds created by hydrocarbon combustion—especially the burning of petroleum-based fuels such as gasoline. Ground-level ozone is the principal component of urban smog. The association between high ozone levels and increased asthma-related emergency room visits for children is well established. Other experimental studies have shown that ozone can interact with allergens, exacerbating breathing-related symptoms. Ozone can also aggravate chronic lung diseases, reduce the immune system's ability to fight off infections in the respiratory system, and worsen heart disease.[4]

One of the largest ozone pollution studies ever conducted found a connection between short-term changes in ground-level ozone and mortality in ninety-five large urban areas, which together are home to 40 percent of the U.S. population. The study linked day-to-day variations in ambient ozone levels and daily mortality in urban areas, finding an association in most cities between higher daily ozone levels and deaths due to cardiovascular and respiratory distress, especially among the elderly. The scientists who led this pathbreaking study estimate that a reduction of 10 parts per billion, or about 35 percent of the average ground-level ozone on any day, could save an average of four thousand lives a year in those urban areas.[5]

PARTICULATES

Fine particles floating in the air can be very harmful to breathe, depending on their size. The EPA estimates that nearly 90 million people live in counties that violate federal particulate standards for some portion of the year. Particulate matter (PM) includes particles found in the air (such as dust, dirt, soot, smoke, or liquid droplets). Particulates are emitted from vehicles as well as from power plants, factories, construction sites, tilled fields, unpaved roads, stone crushing operations, and wood-burning fireplaces and stoves. Significantly, particle pollution concentrations depend in part on the weather. Stagnant and humid air masses allow the chemicals to accumulate, while swift-moving weather systems and storm fronts tend to blow the particles away or wash them from the atmosphere.

Two sizes of particles are regulated by federal law, PM-10 and PM-2.5, and each are associated with different health effects. PM-10 refers to coarse particles with a diameter of 10 micrometers or less, which have been known to aggravate asthma. Sources of PM-10 include crushing or grinding operations, and dust from paved or unpaved roads.[6] PM-2.5 is defined as those fine particles less than 2.5 micrometers in diameter, and these result from fossil fuel combustion (motor vehicles, power plants, and wood burning), some industrial processes, and incineration. Whereas larger particles are commonly filtered in the upper nasal passages, these smaller particles can more deeply penetrate lung tissue and so pose the more serious threat to human health.

Nearly 90 percent of particles emitted as diesel exhaust are considered to be "ultrafine," that is, less than 1 micron in diameter.[7] These PM-01 can pass through the lungs directly into the bloodstream, where they thicken the blood and increase the chances of inflammation, thus escalating the risk for heart attack, stroke, and respiratory problems.[8] Analysis of data from an American Cancer Society study of long-term exposure to fine particulates found that for each increase in fine particulate air pollution of 10 micrograms per cubic meter of air, the risk of dying from lung cancer increased by 8 percent.[9]

Particulate matter is especially harmful to people with lung diseases, children, the elderly, and people with heart disease. Many studies have shown that exposure to fine particulate matter in vehicle exhaust, even briefly, can irritate the lungs, particularly in asthmatics, and increase the potential for later reaction to other allergens. In one such study, volunteers with mild allergic asthma were exposed to vehicle exhaust while sitting for twenty minutes inside a car that was traveling through a tunnel. Those exposed to PM-2.5 at levels common during rush hour traffic had an increased reaction to an allergen administered four hours after the tunnel exposure. Vehicle exhaust therefore may make groups with preexisting respiratory diseases more likely to suffer additional health loss when exposed to other irritating chemicals.[10]

HAZARDOUS AIR POLLUTANTS

The 1990 amendments to the Clean Air Act addressed a large number of air pollutants believed to harm human health or the environment. Emissions standards for these "hazardous air pollutants" were established to limit the release of some of them from specific industrial sectors, even as other industrial emissions continued to be ignored. Other failings of the amended law include basic terminology. The effort to control these pollutants relies on the use of the "best available control technology" that a particular industrial sector can afford, whereas ideally the companies involved would be required to protect human health with "an ample margin of safety." Even the name for this group of chemicals is misleading, since other chemicals outside the category are also hazardous.

Moreover, solving the problem of industrial air pollution will not be enough to give us clean air. According to the EPA, about 40 percent of toxic emissions in the nation's air comes from mobile sources including cars, trucks, buses, and farm equipment.[11] In 2005, American families spent the second largest proportion of their income on transportation (housing was first). The U.S. fleet now exceeds 240 million vehicles— one for everyone in the nation eligible to drive. As of 2008, no fewer than 22 percent of the vehicles in the world were registered in the United States and in one year those vehicles travel, on average, three trillion miles as they burn 175 billion gallons of fuel. In that year, the U.S. fleet was predicted to consume twice as much petroleum than was produced domestically—a situation that causes its own complications as U.S. consumers draw from the reserves of foreign countries.[12]

U.S. residents today are less self-propelled and more reliant on engines and fossil fuels than any other nation in the world, and we live in close proximity to sources of combustion, especially vehicles and heating systems that consume fuel on-site (consider, for example, that many people bring their vehicles literally into their homes by parking in attached garages). But the situation is not just local; it's global. Your vehicle's contribution to the emissions problem may seem incidental, but the energy-intensive lifestyle of high-income nations is quickly being adopted by developing nations, where the global fleet is now growing most rapidly. Our society's love affair with vehicles is apparent even in our architecture: late-twentieth-century residential designs commonly devoted more indoor square footage for vehicles than bedrooms. When the emissions from your vehicle are multiplied by those from nearly 1 billion others in the world, chronic background or "ambient" pollution becomes the norm.

The most intense exposures to air pollution result from emissions in residential, educational, and occupational settings and in transit. North Americans, for example, spend an average of nearly ninety minutes a day in vehicles, sitting only feet from their own vehicle's tailpipe. And because vehicles are concentrated in travel corridors, we also experience the exhaust produced by others traveling on the same roadway.

The Texas Transportation Institute has estimated that 3.7 billion hours are lost to traffic delays each year (assuming only one person per vehicle), resulting in nearly $67 billion in lost economic activity and more than

5 billion gallons of wasted fuel per year. The longest average commuting times are experienced in urban areas such as New York (116 minutes a day), Chicago (106 minutes), and Los Angeles (58 minutes).[13] These cities also have the largest populations in the United States and experience the nation's highest levels of air pollution. Confinement within slowed or stopped vehicles in these areas not only increases collective misery on the roadways, but also exposes drivers to more vehicle exhaust.[14]

Fuel efficiency standards and emissions-control technologies have been in force for nearly three decades, and have reduced emissions modestly. The EPA estimates that new cars today are capable of emitting 90 percent fewer air toxics per mile than older vehicles. Yet the national vehicle fleet collectively produces more of some pollutants such as nitrogen oxides than ever before because people drive vehicles longer before purchasing new, less polluting models; the number of vehicles has steadily grown as the population and average income have increased; and individual vehicles are being driven farther each year.[15]

In fact, vehicle miles traveled in the United States have been on the increase since the 1990s. A shift in driving patterns is contributing to the trend, with the most prominent of these being an increasing average distance between home and work. Suburbs have grown rapidly, and public transportation has been in decline. Off-road vehicle use has also increased.

The increasing popularity of sport utility vehicles and other light trucks (a classification including minivans, pickup trucks, and sport utility vehicles) has had a profound effect on national fuel consumption. Although passenger cars still account for most of the vehicle miles traveled, between 1990 and 2000 the percentage of automobile miles traveled per year declined (from 66 percent to 58 percent), while miles traveled by light trucks increased.[16] These larger, heavier vehicles achieve poorer fuel economy while producing more pollution per mile than lighter passenger cars.

The fuel efficiency of U.S. vehicles, measured in miles per gallon, has improved since 1978, when standards were first implemented. Between 1978 and 1988, the average fuel economy of a new passenger car increased from 19.9 to 28.8 miles per gallon, while light trucks improved slightly, from 18.2 miles per gallon in 1979 to 21.3 in 1988. Since 1988, however, the average fuel economy of the entire fleet of vehicles sold in any one year has remained flat. In particular, the average fuel economy for long-haul trucks has remained the same since the 1970s, at about 5.5 miles per gallon.[17] And they don't have to be moving to pollute. Every year, long-haul trucks consume an average of more than 838 million gallons of fuel while idling overnight.[18]

Toyota and Honda have incorporated devices in automobiles that automatically stop the engines when drivers brake with the gearshift in drive. When the driver removes her or his foot from the brake pedal, the engine restarts. Idling cutoff devices have also been installed in trucks and buses produced by Hino Motors, Isuzu Motors, and Nissan Diesel.[19] The Energy Conservation Center of Japan tested two cars of the same make and model, one with an automatic idling cutoff and the other without. They found that the vehicle with the idling-prevention device used about 3 percent less gasoline on open roads between cities where there were few stops, but in cities the savings was much greater—a little over 13 percent.[20] Clearly, using less fuel has positive implications for air emissions as well.

MISSTEPS IN MEASURING AIR POLLUTION DATA

The definition of acceptable or legally compliant air quality depends on what is measured, where and when it is measured, and how the data are interpreted. Of the thousands of pollutants commonly released into the atmosphere, the EPA routinely measures only six types. If and when regulated pollutants are measured, readings are taken from fixed monitoring stations and considered chemical by chemical. In addition, the EPA permits states to average levels of some pollutants over extended periods of time and across both urban and rural areas when determining their compliance with federal standards. The effect is that high pollution episodes are often hidden in the data, and the states are deemed to be compliant.[21] Indeed, what we know from the government about air quality in the United States has little relationship to the mixture of pollutants we breathe as we move among indoor and outdoor environments in our daily lives. Moreover, contrary to the EPA's claims, there are many reasons to assume that air quality is declining where children and other susceptible members of the population spend their time.

Measuring air quality is very difficult because it is continually changing, not only breath to breath, but also with variations in weather. We do know, however, that it is poorest where traffic is heaviest and most constant: that is, where highway traffic slows due to congestion; at or near construction sites; and where trucks, buses, and cars tend to idle, such as at schools, hospitals, shopping centers, truck stops, warehouses, ports and shipping facilities, rail stations, and bus terminals. Air pollutants also tend to concentrate where diesel-powered vehicles are used within enclosed spaces such as highway tunnels, warehouses, hockey rinks, ships, or mines. In addition, gas-powered lawn and garden equipment can generate significant exhaust.

The implications for health are enormous. Medical and public health scientists have found that short-term exposures to air toxics, even those lasting only hours or minutes, are potentially dangerous to susceptible populations, that is, those with cardiovascular, respiratory, and other illnesses. If outdoor pollution is averaged over extended periods to include the nighttime and weekends when industrial and vehicle emissions are greatly reduced, pollution levels will be reported to be lower than if averaged over the normal workweek. If higher urban pollution levels continue to be averaged with concentrations reported by cleaner rural sites, government reports of safety will remain seriously exaggerated. Perhaps the most important lesson is that little relation exists between the quality of air that governments monitor and report and the quality of the air you breathe as you move through daily life.

POLLUTION IN INDOOR AIR

U.S. residents spend roughly 85 to 90 percent of their time in transit or indoors, within homes, schools, offices, and recreation and shopping centers. This is the untold story of air pollution, for although most of the EPA's attention to air quality has focused on the burning of fossil fuels, indoor air is often more polluted than that outside.[22] Many factors contribute to poor indoor air quality, including tobacco smoke; cleaning products; pesticides; formaldehyde from adhesives in many types of non-solid wood furniture; paints and other wood finishes; cleaning agents; waxes and polishing compounds; fragrances; plasticizers in wallpaper, rugs, and fragrances; components of building structures (such as sealants, plastics, adhesives, and insulation materials); animal and insect allergens; and molds and fumes from household gas appliances. Carbon monoxide, fine carbon particles, and polycyclic aromatic hydrocarbons emitted from poorly vented fireplaces, wood stoves, furnaces, water heaters, kerosene heaters, and idling vehicles in garages also may pose a serious threat to health indoors.

The 1974 Energy Policy and Conservation Act encouraged building standards to promote energy efficiency, and to reduce the exchange of indoor and outside air. But tighter, more energy-efficient structures often have one-tenth the air exchange rates of older structures that have windows, doors, and walls that are less well insulated and sealed. Windows that do not open, and heating and cooling systems that recycle air rather than exchange indoor and outdoor air, often lead to the accumulation of chemical and biological agents that can trigger or exacerbate asthma. Poor ventilation within the walls can lead to a damp, dark environment ideal for the growth of molds and fungi. Biological material such as animal and bird dander, residues from dust mites and cockroaches, fungi, infectious agents including bacteria or viruses, and pollen also cause respiratory distress in many and can concentrate in homes without adequate ventilation.[23]

The EPA has estimated that 30 to 50 percent of all structures in the United States have damp conditions favoring the growth and buildup of microbes that can harm health. Humidity levels above 50 percent, commonly found in areas such as bathrooms, damp or flooded basements, and within and around wet appliances such as humidifiers, washing machines, and dishwashers, are common breeding grounds for fungi and molds, with some mold species capable of producing hazardous mycotoxins. Carpeting in humid environments, as well as leaking roofs and poor drainage, can also contribute to their growth, while humidifiers, poorly vented heaters, and air conditioners increase the chances that moisture will form on interior surfaces. Site conditions can also contribute to interior moisture. Structures built in former wetlands or in floodplains, coastal communities, and other areas with high groundwater tables are more likely to nurture molds. Dense foliage and shrubbery immediately adjacent to structures also reduce air movement and contribute to moisture buildup, and buildings sited on south facing slopes tend to be drier than those on north-facing hillsides.

Nitrogen dioxide is another common indoor air pollutant. Produced during combustion processes, its indoor sources include gas stoves, gas heaters, and tobacco smoke. In the home, improperly vented and used home appliances fueled by natural gas are the most common source of nitrogen dioxide poisoning. High levels of nitrogen dioxide can cause lung injury and a decrease in pulmonary defense mechanisms due to the oxidizing effects of the gas. Emissions from gas stoves have been associated with respiratory symptoms and reduced lung function in both healthy and asthmatic children.[24]

Environmental tobacco smoke contains several hundred recognized toxic substances, including numerous carcinogens. Tobacco smoke increases the severity of asthma; moreover, childhood exposure to tobacco smoke may cause asthma.[25] Nearly 48 million Americans smoke tobacco products, including 4.5 million adolescents between the ages of eleven and seventeen, whose lungs have not fully matured. Every day, nearly two thousand children under the age of eighteen become regular cigarette smokers.[26] More than one-third of U.S. children in 2000 lived in households where residents or visitors smoked regularly indoors.[27] As of 2007, the Surgeon General was reporting that nearly 60 percent of U.S. children ages three to eleven were still being exposed to secondhand smoke in the home.[28]

Children living in households with smokers have a higher risk of asthma, bronchitis, ear infections, and pneumonia.[29] The most dramatic declines in children's health occur when the mother smokes. Studies have shown that smoking is associated with clinically significant asthma in children, and that the more the mother smokes, the more likely it is that the child will have asthma: in one study in particular, heavy maternal smoking more than doubled the chance of having an asthmatic child.[30] A pregnant mother's smoking habit is also known to interfere with her fetus's normal lung development.[31] Childhood exposure to secondhand tobacco smoke, regardless of the source, is associated with an increase in asthma attacks per child, earlier symptom onset, an increase in medication use, and prolonged recovery from attacks.[32]

Tobacco smoke is considered a volatile organic compound, and it is not the only one polluting indoor air in the United States. By 1989, the EPA had identified more than nine hundred volatile organic compounds in indoor air, where they normally persist at higher levels than outdoors.[33] These chemicals are found in building materials, paints, furnishings, adhesives, cleaning agents, solvents, perfumes, cosmetics, clothes that are dry cleaned, tobacco smoke, fuels, and other combustion by-products including auto and diesel exhaust. The indoor burning of fuel oil, natural gas, propane, kerosene, and firewood all release particles and gases into the air, especially when furnaces, heaters, and fireplaces are not regularly cleaned and tuned up.

Formaldehyde is a volatile organic compound and recognized human carcinogen that is released to indoor air from a variety of sources, including insulation and construction materials, chipboard, plywood, water-based paints, fabrics, cleaning agents, and disinfectants. Cigarette smoke, vehicle exhaust, and poorly ventilated heating systems are additional sources of formaldehyde.[34] By 2000, the U.S. National Academies of Science had concluded that "indoor VOC's [volatile organic compounds] and formaldehyde may cause asthma-like symptoms." Higher prevalence rates of asthma and chronic bronchitis were found in children from houses with higher formaldehyde levels than in those less exposed, with the most dramatic difference noted for those children exposed to both secondhand tobacco smoke and formaldehyde. In a study of mobile homes and homes with particleboard flooring, formaldehyde was found to be associated with increased severity of dry and sore throats, bloody noses, sinus irritation, sinus infection, cough, headache, chest pain, and other symptoms. Similar conditions have been reported by those housed in trailers provided by the Federal Emergency Management Agency following hurricane Katrina in New Orleans.[35] Exposure to chemical emissions from indoor paint, which generally contains formaldehyde, has been also associated with asthma, with some volatile organic compounds from indoor paint causing inflammatory reactions in the airways.[36]

Fragrances are not usually thought of as contaminants, but they do generally contain volatile chemicals that may increase the prevalence and severity of asthma among sensitive individuals. Fragrances are commonly added to paints, cosmetics, pesticides, cleaning solutions, some fabrics and furnishings, and increasingly, magazines.[37] In one study, 90 percent of a group of sixty asthmatics surveyed reported that their symptoms had flared after smelling a fragrance, and nearly 40 percent had visited emergency rooms after such an incident. In the United States, the chemical content of fragrances does not have to be listed on the product label; those experiencing an adverse reaction have no way of identifying and avoiding the offending compound.[38]

Indoor ozone "air purifiers" have been sold since the mid-1990s and often are purchased by those with asthma or allergies who hope to improve the air quality in their homes, workplaces, and cars. They continue to be very popular, with nearly a hundred different models available for sale in the United States in 2008. Ironically, however, ozone is well known to worsen respiratory and cardiovascular illnesses, and some of the "filters" raised ozone levels indoors to above the federal maximum outdoor limit. In 2007, the California Air Resources Board, recognizing the danger to the nearly 800,000 people in the state who had purchased these products, banned ozone air filters outright. Yet as of this writing, no similar ban or restriction has been ordered by any federal agency, including the EPA, Consumer Product Safety Commission, or FDA—all of which have the authority to regulate devices such as these that claim to improve health.[39]

Indoor air quality is rarely regulated in the United States. Exceptions include restrictions on the use of tobacco products in some occupational, commercial, and public settings, and bans of some pesticides for indoor use. The quality of the air inside vehicles has also been generally neglected by U.S. environmental law; only limits on the use of tobacco products in public vehicles such as buses, trains, and aircraft, and certain restrictions on applications of pesticides, have been successfully pursued. Consequently, no fewer than 107 different active pesticide ingredients are allowed by the EPA to be used indoors—even in schools while children are in classrooms.[40]

Since 1976, the EPA has been empowered by the Toxic Substance Control Act to test and regulate consumer products that could introduce hazardous air pollutants to indoor air. The agency, however, has rarely tested or prohibited the use of commercial products other than pesticides. The inability or unwillingness of the agency to take action demands an explanation. Many environmentalists and some government scientists blame federal regulators for overlooking new evidence that airborne chemicals induce human illness. Federal officials have countered by pointing to a lack of resources, scientific uncertainty, and fears of litigation. And EPA administrators may well consider the overwhelming task of managing industrial and vehicle emissions a reasonable excuse. Whatever the rationale, the facts are clear: a vast majority of chemicals in indoor environments (including within vehicles) remains unregulated under federal, state, and local law. Moreover, if history is any guide, the situation will not get better soon. Generally it takes about ten years after the EPA becomes aware of a danger for it to revise a standard for a single air pollutant such as ozone or fine particles—and even when tougher standards are set, manufacturers are granted a four-year period before the new rules apply. Without a comprehensive new approach, globalization of trade, technological innovation, and government neglect together will assure that the already serious threats to health from air pollution, both indoors and outside, will continue to increase.

CHAPTER FOURTEEN

Who Is Most at Risk?

AIR POLLUTION NOW kills more Americans than breast and prostate cancers combined, and about as many people die premature deaths associated with particulate matter pollution as are killed in traffic accidents. Fine airborne particles are responsible for tens of thousands of premature deaths in the United States each year.[1] At special risk are tens of millions of Americans who suffer from serious illnesses such as asthma, chronic obstructive pulmonary disease, cardiovascular disease, diabetes, and lung cancer—as well as children, the elderly, those with compromised immune systems, and those with certain genetic traits. Also during the past decade, scientists have confirmed a relationship between two forms of air pollution, ozone and particulate matter, and increased rates of mortality, especially among those with known cardiovascular disease.

Children's vulnerability stems in part from their rapidly growing lungs. In the developing fetus, more than forty different types of pulmonary cells rapidly divide and differentiate as they change from a more primitive, in utero state into a fully functional lung. Throughout early childhood, the lungs continue to develop more—and more complex—branches. Even in adolescence, lungs continue to grow and differentiate; in fact, the lungs are not fully mature until the late teenage years for females, and the early twenties for males.[2] Lungs work by drawing air deep into their branches, where several hundred million tiny sacs called alveoli replenish blood with oxygen while removing carbon dioxide. Pound for pound, children breathe nearly 50 percent more air than is inhaled by adults. During strenuous exercise and energetic play, children's breathing rates and air-volume intake increase rapidly, often to five times above normal resting rates, thus also increasing the inhaled dose of any air pollutant present.[3]

ASTHMA

Asthma is a chronic inflammatory disorder of the airways in which muscles surrounding airways constrict and become inflamed, thereby reducing airflow to the alveoli, and thus the amount of oxygen that can be derived from each breath.[4] Children and the elderly are commonly more susceptible to asthma than adults; and some children are more susceptible to chronic airway sensitization and restriction than other children.

Asthma has reached epidemic proportions in the United States and is now the most prevalent chronic disease among children, as well as a common affliction of adults. The National Center for Health Statistics estimates that nationally 31.3 million people will be diagnosed with asthma in their lifetime. The most recent estimates published by the U.S. Centers for Disease Control and Prevention indicate that 13 percent of children under the age of eighteen in the United States have been diagnosed with asthma.[5] And between 1980 and 1995, the asthma prevalence rate for children ages five to fourteen increased 74 percent.[6] Today nearly 8.2 percent of non-Hispanic white children experience the illness, while 12.7 percent of non-Hispanic black children, and 19.2 percent of Puerto Rican children, also suffer from it.[7] Asthma prevalence is highest among urban children, and the disease is the primary cause for childhood hospitalization in urban areas.[8]

A variety of factors, alone or in combination, can trigger an asthma attack in predisposed individuals. These include household dust mites, indoor or outdoor air contaminants, allergens, food, exercise, respiratory infections, and cold weather. Air pollution in particular is known to increase the frequency and severity of asthma attacks; the EPA has found that children with asthma are 40 percent more likely to have an attack on days when the outdoor air has high levels of contaminants. Wood smoke, tobacco smoke, volatile substances, motor vehicle exhaust, ozone, sulfur dioxide, nitrogen dioxide, fumes from heating and cooking appliances, pesticides, paint fumes, and synthetic fragrances all have been associated with asthma flare-ups in peer-reviewed studies.[9] It has also become clear that biological matter, including dust mites, pet dander, molds, pollen, cockroach feces, viruses, and bacteria may increase the severity and prevalence of the disease, even though they do not appear to cause it.[10] Compounding the problem for those most vulnerable, more particulates are deposited and retained in the lung tissues of severely asthmatic children than in the lungs of mildly asthmatic children; this means that those who suffer most from asthma may be receiving the highest delivered dose of particulates and other hazardous pollutants.[11]

As concentrations of particulate matter rise, the prevalence and severity of asthma increase. Short periods of high exposure can be especially dangerous. In one study, children experienced an increase in severe asthma symptoms, with one-hour exposures at the maximum allowable PM-10 levels more strongly associated with increased symptoms than twenty-four hours at only mean levels. Breathing fine particulate matter has been strongly associated with an increased use of medications, more hospital admissions and emergency room visits for asthma attacks, and premature mortality.[12]

Children living near roads with high traffic volume are more likely than those residing near quieter roads to have more medical care visits per year for asthma and a higher prevalence of most respiratory symptoms.[13] Traffic-related air pollution in Austria, France, and Switzerland is estimated to be responsible for 290,000 episodes of bronchitis in children, and 0.5 million asthma attacks.[14] The lungs of children living in areas with higher ambient or background levels of PM-10 are known to grow more slowly; they also have higher concentrations of carbon.[15]

Heavy truck traffic near residences and schools and close proximity to the freeway may also pose a special danger for children. Most diesel particulates are considered "ultrafine" in size, and the smaller the particulates, the more likely they are to cause asthma, as well as more severe symptoms of the illness.[16] In addition, ozone is the pollutant most consistently associated with newly diagnosed asthma cases.[17] Even low levels of particulate matter and ozone in ambient air may increase symptoms of asthma in children.[18] Together, these facts explain why even short-term exposure to vehicle exhaust may harm asthmatics, and in particular, children who struggle with the illness.[19]

In 2000, the U.S. Centers for Disease Control estimated that the cost of asthma in the United States exceeded $14 billion a year. Costs include premature mortality, medications, medical care, hospitalization, and lost time at work and school; in 2005, nearly 4 million children missed 12.8 million school days due to asthma, often resulting in loss of parents' work time.[20] Of these costs, treatment is the greatest. In 2001, the estimated cost of treating children under the age of eighteen was $3.2 billion. Asthmatic children required 2.7 times the prescription drugs, and experienced twice as many in-patient care days and 65 percent more non-urgent medical care visits, as asthma-free children. The poorest tend to receive the worst medical care, in part because they are less likely to have regular physician checkups or to have help adjusting their medications. They often experience the most severe episodes, which may lead to emergency room visits, hospitalization, and even death.[21]

CHRONIC OBSTRUCTIVE PULMONARY DISEASE

Chronic obstructive pulmonary disease (COPD) refers collectively to emphysema, chronic bronchitis, and in some cases asthma. An estimated 10 million adults were diagnosed with COPD in 2000, but data from a national health survey indicate that as many as 24 million Americans are affected. The fourth leading cause of death in the United States today, COPD is projected to be the third leading cause of death for both males and females by the year 2020. COPD is also a primary cause of illness and disability in the United States. The Centers for Disease Control estimated that in 2000, COPD was responsible for about 120,000 deaths, 725,000 hospitalizations, 1.5 million visits to hospital emergency rooms, and 8 million additional cases of hospital outpatient or personal physician treatment. Asthma and COPD are different. The inflammation in asthma can be triggered by contact with irritating substances and improves when treated. In contrast, the inflammation in COPD is not triggered by allergies and does not respond well to anti-inflammatory medication. The lungs of patients with COPD generally show evidence of permanently damaged and blocked airways.[22]

Vehicle emissions are particularly harmful to COPD patients. Lung or pulmonary function is a measure of how efficiently the lungs inhale and exhale air, and transfer oxygen to the blood stream. As airborne fine particle levels increase, the lung function in patients with COPD declines. In addition, it has become well known that ozone can exacerbate COPD symptoms. And both air pollutants lead to higher rates of hospital admissions.[23]

CARDIOVASCULAR DISEASE

Cardiovascular disease is the leading cause of mortality in the United States, and is responsible for approximately 660,000 deaths per year. Rates increase with age and males experience greater death rates than females. Risk factors traditionally associated with cardiovascular disease include a family history of the disease, increasing age, gender (male), race (black), cigarette smoking, high blood cholesterol levels, high blood pressure, diabetes, physical inactivity, and being overweight or obese.[24]

Fine particles were recently discovered to inflame the heart and change its rhythm. Particles can pass through the lungs directly into the bloodstream, thickening the blood and increasing the likelihood of vessel inflammation. Human exposure to fine carbon particles reduces heart rate variability, a sign of health that is known to decline with acute stress and age.[25]

The risk of dying early from cardiorespiratory illness and lung cancer is higher in more polluted areas, and the relationship between increased exposure to particulate matter and adverse cardiovascular effects has been well documented in epidemiological studies.[26] Fine particles have been linked to cardiovascular symptoms, cardiac arrhythmias, heart attacks, and premature death from heart disease. Long-term exposure to fine particulate air pollution at levels that occur in North America have been associated with an increased risk for cardiovascular mortality: the danger of dying from a heart-related illness increases by 12 percent for every 10-micrometer increase in airborne particulate matter, with heart attacks accounting for most of this increased mortality rate. Other causes, such as heart failure and fatal arrhythmias, also increase.[27]

Heart attacks are more common among people exposed to heavy traffic and those living near polluted roadways. In one German study of 691 patients with cardiovascular disease, examination of diaries revealed an association between exposure to traffic and the onset of myocardial infarction within an hour afterward: the risk of a heart attack tripled. Time spent in cars as well as on public transportation were associated with increased risk of cardiovascular events; they were 2.6 times more common for people in cars, and 3.1 times higher for people taking public transportation, when compared to those not in transit. Another study reported that those living near polluted roadways were twice as likely to die from a heart attack when compared to those living in cleaner areas.[28]

CANCER

Cancer is the second leading cause of death in the United States, causing nearly 559,000 people to lose their lives each year. In 2004, the EPA estimated that vehicle emissions account for as many as half of all cancers attributed to outdoor sources. Several other studies have found that vehicles are responsible for most of the excess cancer risk associated with exposure to toxic air pollutants in urban areas. Epidemiological studies over the past fifty years have found associations between ambient air pollution from incomplete combustions of fossil fuels and increased rates of lung cancer. Long-term exposure to fine particulate air pollution, too, including vehicle emissions, has been associated with lung cancer mortality in many studies. Several studies have also linked some childhood cancers to motor vehicle exhaust and the proximity of vehicle traffic to the child's home.[29]

Fine particulates often bond with carcinogens such as benzene; 1,3-butadiene; and formaldehyde, creating a toxic mixture that can then be inhaled.[30] The contaminant 1,3-butadiene is a component of diesel fuel and its exhaust, and was recently found to be both cytotoxic and genotoxic to human bronchial epithelial cells. And polycyclic aromatic hydrocarbons, found in diesel emissions, are among the most potent carcinogens and mutagens known.[31]

In response to the developing body of scientific evidence about the dangers of these particulates, the International Agency for Research on Cancer (IARC) classified diesel exhaust as a probable human carcinogen in 1989.[32] The State of California Scientific Review Panel on Diesel Exhaust concluded: "A level of diesel exhaust exposure below which no carcinogenic effects are anticipated has not been identified" and classified diesel exhaust as "known to the State of California to cause cancer" in 1990. In 2000, the U.S. Department of Health and Human Services National Toxicology Program followed suit, classifying diesel exhaust as "reasonably anticipated to be a human carcinogen," due to the increased risk of lung cancer found in most human studies of long-term occupational exposures to diesel exhaust. Even more recently, the California Scientific Review Panel labeled diesel exhaust the sixth most potent carcinogenic substance after dioxins, chromium IV, inorganic arsenic, and benzo(a)pyrene.[33]

Also in California, the South Coast Air Quality Management District estimated that 71 percent of all cancer risk from air pollution is derived from diesel exhaust, with high concentrations in outdoor air causing an additional cancer risk of 1.4 cases per 1,000 people. Another study found that motor vehicles and other mobile sources accounted for about 90 percent of the cancer risk in Southern California (with about 70 percent of all risk attributed to diesel particulate emissions and about 20 percent to other air toxics such as benzene, 1,3-butadiene, and formaldehyde). Still other reports have linked various childhood cancers to motor vehicle exhaust and proximity to traffic.[34]

DIABETES

Nearly 7 percent of the U.S. population, or around 21 million people, suffer from diabetes, with the annual cost of the disease estimated to be $167 billion.[35] People with diabetes are more susceptible to illness from air pollution.[36] The American Lung Association recently added diabetics to its list of groups most at risk from particle pollution, based on their vulnerability to fine particulates.[37] Epidemiological studies have found that the ability of vessels to control blood flow is impaired in diabetic adults who are exposed to traffic pollution, resulting in an increased risk of a heart attack or other cardiovascular emergency.[38] This fact is especially significant given that adults with diabetes are about two to four times more likely to die from heart disease than adults without diabetes; a full 65 percent of deaths among people with diabetes are due to heart disease and stroke.[39]

In general, diabetes affects the old more than the young: just 2 percent of the U.S. population between the ages of twenty to thirty-nine have the illness, with the prevalence rising to 21 percent for those over age sixty.[40] But the incidence of type 1 diabetes is disproportionately increasing among infants and very young children, and some scientists believe that the development of this syndrome in children may be associated with air pollution. In particular, independent researchers have found that increased ozone exposure may be a "contributory factor" to the increased incidence of type 1 diabetes and that exposure to PM-10 may be a "specific contributory factor" to the development of type 1 diabetes in children under the age of five.[41]

DANGERS OF SHORT-TERM EXPOSURES

Most people believe that short-term exposures to high levels of air pollution are not dangerous. Federal standards allow the averaging of daily (twenty-four-hour) particulate levels over three years, and these averages are then compared with federal limits. This method assures that high pollution episodes of short duration will not be recognizable—that is, brief periods of high pollution levels are thus obscured from the data, which seemingly demonstrate that states and urban areas have pollution well under control. But there is growing evidence that adverse health effects among asthmatics can occur following shorter duration exposures to some pollutants, and these episodes are concealed by the government's averaging rules. For example:

- In a 1998 study, one-hour and eight-hour maximum PM-10 levels were found to have larger effects in inducing asthma symptoms among children ages nine to seventeen than did twenty-four-hour average levels, and the effects were strongest among children who already frequently had symptoms.[42]
- A 2000 study of 133 Seattle children found a significant association between asthmatic symptoms and short-term particulate and carbon-monoxide levels.[43]
- Volunteers with mild allergic asthma were exposed to vehicle exhaust in a car within a Stockholm road tunnel for twenty minutes. Subjects exposed to PM-2.5 at levels above 100 micrograms per cubic meter of air had an increased reaction to an allergen administered a full four hours after the tunnel exposure. The authors of the study, published in 2000, concluded: "Exposure to air pollution in road tunnels may significantly enhance asthmatic reactions to subsequently inhaled allergens."[44]
- A 1999 German study of asthmatic children concluded: "Exposure to traffic flow and in particular, truck traffic and diesel exhaust leads to significant increases in respiratory symptoms and decreases in lung function."[45]

- Lung function among children in the Netherlands was associated in 1997 with the intensity of truck traffic near residences and schools: "Cough, wheeze, runny nose, and doctor-diagnosed asthma were significantly more often reported for children living within 100 meters from the freeway. Truck traffic intensity and the concentration of black smoke measured in schools were found to be significantly associated with chronic respiratory symptoms."[46]
- In a 1995 study, when asthmatic volunteers who were exercising strenuously were exposed for just ten minutes to sulfur dioxide, a component of diesel exhaust, they experienced more severe asthma symptoms than usual.[47]
- A 1997 study of eighty-nine asthmatic children in the Czech Republic found that the effects of air pollution on asthmatic children with respiratory infections may be greater than on those without infections.[48]
- The EPA concluded in 2000 that short-term exposures to diesel particulate matter can produce allergenic effects, and that these effects are caused both by inhaling carbon particles and by absorbing gaseous compounds through the lungs.[49]
- According to a study published in 1996, human volunteers exposed to diesel exhaust for one hour experienced increased airway resistance and irritation of the eyes and nose.[50]

Since 1990 many scientists have warned of grave dangers posed to human health by air pollution, especially for children, the elderly, and those with underlying serious health conditions—a group that now includes nearly 100 million Americans.[51] These studies are difficult to reconcile with our common acceptance of daily exposure to thousands of outdoor and indoor air pollutants. Cardiovascular disease, cancer, asthma, diabetes, immunological responses, and mortality are clearly associated with the kinds and levels of airborne pollution experienced routinely in the United States (Figure 14.1).

The EPA spends more than 99 percent of its air-quality improvement efforts trying to understand the behavior and effects of only a few pollutants in outdoor environments. The agency neglects thousands of contaminants now suspected of harming human health, and disregards the

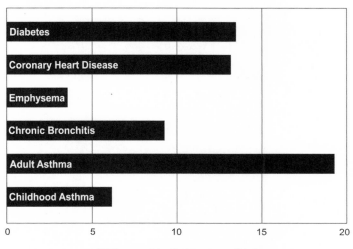

Millions of Individuals at Risk

FIGURE 14.1. Susceptibility to health loss from air pollution in the United States. Tens of millions of U.S. residents suffer from illnesses that many scientists believe worsen following exposure to air pollution. *Sources:* Environment and Human Health, Inc., U.S. Census, U.S. Department of Health and Human Services, U.S. Centers for Disease Control and Prevention.

fact that most of us spend nearly 90 percent of our lives in buildings and vehicles, where the air quality has rarely been monitored or regulated by any government. The nation's health is suffering significant, preventable losses, and children, the elderly, and those already burdened by serious illnesses are paying the dearest price.

CHAPTER FIFTEEN

The Trouble with Diesel

THE CENTERS FOR DISEASE CONTROL and Prevention estimated in 2007 that more than 6 million children under the age of eighteen, or 9 percent of all children in the nation, have asthma, now the most prevalent chronic disease among the young. Although children make up only 25 percent of the U.S. population, 40 percent of all asthma cases occur in children. Nearly 160,000 children are hospitalized for asthma annually; in fact, it is the primary reason that children living in urban areas are admitted to the hospital.[1] A 2002 study estimated the costs of pediatric asthma attributable to environmental contaminants to be approximately $2 billion per year.[2]

Since the mid-1980s, asthma prevalence in the United States has increased rapidly among all races, both sexes, all age groups, and in all geographic regions—although urban children (ages five to fourteen, especially African Americans) continue to have the highest rates of the syndrome. Between 1980 and 1995, the rate for children ages five to fourteen increased an astonishing 74 percent—a rate that the Centers for Disease Control and others have concluded is too fast to have resulted from any genetic change in susceptibility; something else is to blame.[3]

While working in New Haven, Connecticut, over the last twenty years, I have repeatedly heard New Haven teachers and school administrators express concern that asthma rates were high and had increased rapidly during the 1990s. Connecticut then had no registry for childhood respiratory diseases, and in 2001 Environment and Human Health, Inc. (EHHI), a Connecticut nonprofit research group, surveyed school nurses in the state to determine how many students had registered asthma medications with their offices. Prevalence among school districts ranged between 3 and 14 percent, while some individual schools in poor urban communities reported rates as high as 22 percent. One effect of this study was the creation of a school-based asthma registry in 2003.[4]

What changes in the children's environments, or perhaps their behavior, might explain the growing illness rates? We suspected that something has changed in the chemistry of the air or children's behavior, or perhaps both. These became our primary hypotheses, although other types of chemical exposures might have previously stimulated children's immune systems to become especially sensitive to pollutants. The EPA has resisted this line of argument, however, by claiming that air pollution has declined in the United States in response to federal regulations limiting industrial emissions and to improvements in engine design that have reduced vehicle emissions. The government's data are collected from fixed sampling sites scattered around each state, and do suggest a decline in carbon monoxide, lead, sulfur dioxide, nitrogen oxides, some types of particulate matter, and ozone. Why, then, would respiratory illness still be rising in prevalence and severity among children in the United States, and what relation might exist between the chemicals tested at government sampling stations and the air inhaled by individual children as they move through their daily routines?

Several differences in childhood behavior during the past several decades suggest a possible answer. Children are spending less time outdoors; instead of playing outside, they are increasingly engaged with electronic media and games, so that now nearly 85 percent of their time is spent indoors. Children are less physically active, as demonstrated by a 2008 study by the National Institutes of Health that found children were engaged in measurable physical activity only forty-nine minutes on weekdays, and thirty minutes on weekends. And children are spending more time within vehicles: EPA found in 2002 that a child's average time in any vehicle was approximately 75 minutes. Among those children surveyed, 10 percent (7 million) spent 2.5 hours per day in vehicles, and 5 percent were in transit nearly three hours daily. (Time spent within vehicles likely has increased since this study, in part due to suburbanization and increasing traffic density.)[5]

HOW ARE CHILDREN EXPOSED?

To attempt to explain the rising prevalence of respiratory disease in children, David Brown, a toxicologist formerly with the Centers for Disease Control and Prevention, Nancy Alderman of EHHI, and I designed a study in which measurements of the air quality near each child participant would be taken every ten seconds from the time they left their homes for school until they returned late in the afternoon. Participating children pulled an intravenous cart that supported a microwave-sized carbon detection meter and carried a spherical stainless steel vacuum canister the size of a basketball to collect air samples to test for diverse volatile chemicals. We then strapped a meter to their shoulders to measure fine particles, and pinned a formaldehyde detector to their sides.

We expected to find higher levels of respiratory illness in schools with poor ventilation, molds, and older heating and ventilation equipment. We speculated that volatile organic compounds would be concentrated in chemistry labs and photo labs, that the use of cleaning supplies and pesticide applications would release these pollutants into the air, and that newer schools would be cleaner than older ones due to updates in design standards for heating and ventilation. We also wondered if variability in school maintenance practices such as floor waxing, painting, use of cleansers, fragrances, and pesticides would be measurable in our study, and whether the use of unventilated basement rooms for classes in over-crowded schools might cause intense chemical exposures.

We were surprised, then, to find that the data did not support any of these hypotheses, and that the air in many schools had low concentrations of many of the chemicals we measured. Instead, the air surrounding our student subjects proved to be most polluted during trips to and from school. All of the students we studied experienced periodic high concentrations of very fine particles, carbon, and volatile organic compounds. When we matched the pollution log to the children's activity log, peak levels of chemical exposure coincided with rides on diesel school buses. Children seemed to be inhaling the most pollution while riding in the buses.

Diesel exhaust is a complex mixture of particles and hazardous volatile chemicals. Emissions include gases formed from combustion (nitrogen, oxygen, carbon dioxide, and water vapor), and from incomplete combustion (benzene, formaldehyde, 1,3-butadiene, and polycyclic aromatic hydrocarbons). Significantly, the composition of diesel exhaust varies by engine type (heavy engines pollute more than light engines), engine age (older vehicles are more polluting), fuel used (high-sulfur fuels produce more particles than low-sulfur fuels), operating conditions (acceleration, deceleration, idling, and uphill and downhill runs all increase emissions when compared to constant speeds on flat terrain), and vehicle load (as load increases, so does pollution).

Diesel emissions contain a higher concentration of particulates, especially those of small diameter, than gasoline emissions. Importantly, nearly all diesel exhaust particles are less than 10 microns in diameter, while almost 94 percent are less than 2.5 microns, and 92 percent are less than 1 micron. These are very fine particles that penetrate most deeply into the lungs of children, who have small airways. In addition, these fine and ultrafine particles act as a nucleus and attract other hazardous particles and gases including carbon monoxide, formaldehyde, sulfur and nitrogen oxides, and PAHs (polycyclic aromatic hydrocarbons), which can then stick to the particles and be inhaled.[6] These smaller particles have a larger surface area, and may be capable of delivering a higher dose of toxic gases to the lung than coarser particles. This may threaten the health of children with smaller airways and those with reactive airway disease.[7] Among the toxins that can stick to these very small particles, formaldehyde is notable. The CDC in 2008 reported that formaldehyde exposure can cause asthma among those sensitized: "Previously sensitized individuals can develop severe narrowing of the bronchi at very low concentrations (e.g., 0.3 ppm). Bronchial narrowing may begin immediately or can be delayed for 3 to 4 hours; effects may worsen for up to 20 hours after exposure and can persist for several days." Formaldehyde also is known to cause cancer in laboratory animals and may cause cancer in humans.[8]

If the EPA has neglected research on indoor air quality, it has almost completely overlooked potential air-quality issues within vehicles, including diesel-powered school buses. We joined students on buses along both urban and rural routes, during high and low traffic conditions, during different types of weather, and on different buses. Untangling the source of the pollutants we measured became a serious challenge. Could pollution

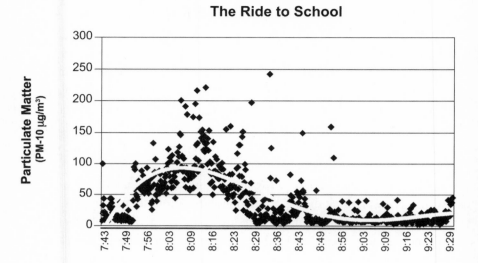

FIGURE 15.1. Diesel exhaust in school buses: the ride to school
Exposure to high levels of diesel particles (PM-10) from exhaust inside school buses
has been demonstrated on both rides to (15.1) and from (15.2) school. These data were
repeated for dozens of children, and show that the most intense exposures to carbon

from high traffic volume have caused our meters to jump, instead of the
bus itself? Could the particles we measured have been generated from
coal-fired power plants, the burning of home heating oil, other diesel-
fuel-burning vehicles, or industrial emissions?

The importance of these other sources is certainly plausible, so we designed a test bus route in a remote area where we could not detect carbon
or particulate matter from any source. Within seconds after school bus
engines were started, our meters registered carbon particles commonly
associated with diesel engine emissions. We measured chemical levels
near the bus exterior, and at various locations inside the bus. The highest
concentrations of particles were found within or near the school buses,
or when walking along routes with high traffic volumes. Particle levels
rose when students boarded the buses, and declined after they exited.
Somehow particles and gases were being drawn inside.[9]

Readings were within the range of particle concentrations reported
by state background monitors but only before and after the bus ride,

The Ride Home

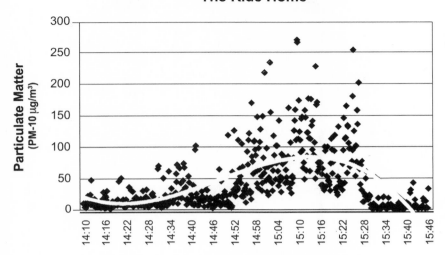

FIGURE 15.2: Diesel exhaust in school buses: the ride home

particles during the entire school day occurred while riding school buses.
Sources: Environment and Human Health, Inc., John Wargo, David Brown.

when engines were turned off. Once the engines were started, the levels jumped several times higher than state reports. The magnitude of the difference depended on the averaging time; the longer the period, the less the difference. Children were generally being exposed to bursts of intense concentrations of particles and gases at bus stops—concentrations that tended to decline as the buses gained speed between stops and ventilation improved. If the bus was stalled in traffic, however, interior levels tended to accumulate quickly.

Switching to more sensitive equipment capable of detecting finer particles also produced surprising results. The highest concentrations detected within buses were ten times higher than background levels. When buses stopped and idled, diesel exhaust often entered bus interiors through opened doors and windows, depending on weather, wind direction, the configuration of the windows, and the location of the exhaust pipe. When diesel exhaust concentrations were highest, we didn't need monitors to tell us that pollutants were present inside the bus: we could detect them by sight (smoke was sometimes visible), by smell, by a lingering acrid taste, and by a burning sensation at the back of the throat.

As buses moved through the test route with the windows closed, the carbon levels often increased at stops. Exhaust is emitted from the rear tailpipe, and some portion appears to be pulled along behind the bus, depending on wind and traffic conditions. When buses stop and the door is opened, some of the exhaust may enter the bus along with the children. When windows were closed, carbon levels increased with each stop. When windows were open, however, the levels of carbon and particulates fell between stops. Buses traveling in traffic experienced higher concentrations of interior particulates than buses moving along routes with no diesel traffic. Significantly, particle levels increased rapidly when following other diesel vehicles, such as when buses left schools in caravans. None of the buses we studied had air filtration equipment capable of diminishing fine carbon particles.

Diesel exhaust may enter the bus interior from unsealed engine compartments, leaking exhaust systems, windows, or doors. Managing interior pollution levels by opening and closing windows is possible, but hardly practical. Not surprisingly, a cloud of exhaust followed each bus, catching up to it when it stopped to pick up children; then the open door diverted the air flow, directing it inside. Exhaust also entered through open windows as the bus came to a stop.[10]

Idling practices affected concentrations of both particles and carbon within the passenger cabins. When buses arrive in a line to pick up students or drop them off, interior particle levels rise quickly, often within a minute of arrival. In some cases they were found to be ten to fifteen times higher than background levels recorded by state monitoring programs. When buses idle while queued, tailpipes are often only six feet away from the open doors of the bus behind. Particle levels were especially high when idling, queued buses were loading and unloading students at schools. The pollution can even follow the students into the school. Several school administrators and teachers complained that bus queuing and idling normally produces diesel odors within schools; fumes were drawn indoors through the school's air-intake vents located nearby.

Even within buses, diesel exhaust is often not the only pollutant to which children are exposed. Background pollutants emitted by motor vehicles and those imported from distant power plants, incinerators, and urban traffic often contribute to the concentrations found within buses. This is especially the case in urban areas such as the U.S. Northeast Corridor, near heavily used roads and highways, and in close proximity to "non-road" sources of diesel emissions, especially construction sites, rail yards, and ports such as Long Beach in California. Average background levels of fine particles (PM-2.5) detected in New Haven, for example, already exceeded the federal standard during this study, adding to concentrations experienced by children on bus rides. Readings demonstrated that a several-hour bus ride would be nearly equivalent to the average particle exposure one could expect over an entire day in New Haven.

Clever statisticians and regulators can disguise intense but short episodes of high pollution by averaging them with readings detected over longer periods such as late at night or on weekends, when traffic and industrial activity is lower than during the work week. The highest levels generally occur from 7 A.M. to 7 P.M., and the remainder of the day normally produces lower readings. As the averaging period increases, the concentrations reported as averages diminish, causing periodic jumps in particulate and carbon concentrations to become invisible to state regulators—and giving a misleading impression of safety.[11] For although most governments assume that intense short exposures do not harm respiratory health, it is clear from the scientific literature reviewed earlier that a decline in health among susceptible individuals should be anticipated.

The high exposures to air pollution that we found on school buses represent a significant risk for children, if only because transporting children this way is such a regular and lengthy part of many children's school experience. Each school day nearly 600,000 buses transport 24 million students to school in the United States, traveling a combined 4 billion miles each year. On average, the bus rides last thirty minutes in each direction or one hour per day per student for 180 days per year. Over the typical thirteen-year period of schooling, this adds up to an average 2,340 hours of bus time for each child. And this is only the average; within some large rural communities some children ride buses for nearly three hours a day. Students with disabilities, too, generally travel longer distances, since they are often sent to specialized regional schools.[12]

All of us breathe diesel exhaust every day of our lives—even indoors. The California Air Resources Board estimated that average indoor concentrations of particulates from diesel emissions were 1.6 micrograms per cubic meter of air throughout the nation. Whenever diesel vehicles are used in enclosed structures such as warehouses, airports, construction sites, or hockey rinks, pollution levels can become dangerously high. Buildings near highways or transit centers, such as bus stations and storage lots, may also draw diesel fumes indoors.

The air we breathe while commuting by car or driving around town is also often hazardous. In 1998, the California Air Resources Board conducted a study of air contaminants within cars. It found levels of benzene, formaldehyde, carbon monoxide, toluene, and other pollutants that were often two to ten times higher than levels measured at a nearby fixed monitoring site. The group also found that pollution levels within vehicles increased as traffic became more congested, with concentrations doubling when the vehicle followed diesel-powered trucks and buses. The International Center for Technology Assessment, upon concluding a review of twenty scientific reports on in-vehicle air contamination, declared that pollution levels within vehicles were significantly higher than levels found along roadsides.[13]

TACKLING THE PROBLEM

How can we reduce children's exposure to diesel pollutants? The federal government could require and fund the retrofitting of existing school buses with particle traps and catalytic converters designed to reduce emissions. In areas of the nation beyond compliance with current federal pollution standards, including the New York metropolitan region, Atlanta, Houston, and Los Angeles, the government could provide funds to replace older, dirtier buses with low-emissions vehicles. It could also establish health-protective air quality standards for public transit vehicles, and provide support for compliance by supplying air filtration equipment, especially in urban areas where traffic intensity is highest. Federal outdoor pollution limits could be further restricted to account for probable exposures within vehicles. The government could also monitor the air quality of individuals as they move through daily routines, as we did in the school bus study, instead of relying only on fixed outdoor monitors. Finally, state governments could establish laws that limit idling to save fuel and reduce pollution.

Although progress has been slow, the government has made some real improvements over the past decade. In 2000, the EPA adopted new emissions requirements for heavy-duty diesel engines. The sulfur content of fuel has been reduced from 500 to 15 parts per million, and refiners and retailers are required to provide the less polluting ultra-low-sulfur fuel by mid-2006. Trucks and buses manufactured since 2006 have engines that produce 90 percent fewer particulate emissions than previous models. And by 2010, nitrogen oxide emissions are required to be 95 percent lower than current levels. The EPA estimates these changes will result in 17,600 fewer cases of acute bronchitis in children every year.[14]

The data and interpretations of the school bus research were challenged by many but not by diesel engine manufacturers anxious to sell more engines that emit fewer pollutants. Moreover, both Health Canada and the State of California replicated our findings of intense within-bus exposure to diesel emissions. In response to the results of such studies, California in 2006 approved a bond issue to spend $200 million to retrofit and replace older buses. In 2003 the federal government allocated $16 million to the EPA to provide state grants for a similar purpose.[15]

The EPA has also joined with the Justice Department to fine corporations that have violated federal air quality standards, with the proceeds allocated to retrofit buses with particle traps and catalytic converters and to purchase ultra-low-sulfur fuels. Toyota Motor Company, for example, was fined $20 million in 2003 for the failures of on-board emissions diagnostic devices installed in 2.2 million vehicles between 1996 and 1998. These funds are being spent to retrofit nearly three thousand school buses. The Virginia Electric Power Company was fined $2 million and the Mystic River power station in Everett, Massachusetts, owned by Exelon Mystic LLC, was fined $3.5 million. All of these fines are being used to retrofit buses in the Boston area.

The quickest way to save fuel and reduce emissions is to stop unnecessary idling. The Connecticut legislature passed a law limiting school bus idling to three minutes; three years later, an evaluation demonstrated nearly 70 percent compliance. Similar statutes have since passed in twenty additional states, including California, Massachusetts, and New York.[16]

These policies do represent some progress, but more could and should be done. Perhaps the most striking conclusion of our research was that children's exposure to vehicle exhaust remains virtually unmonitored and unregulated where it counts—that is, where they breathe at home,

school, and while traveling. The EPA's reliance on fixed outdoor monitors, its averaged data, its neglect of intense short-term exposures, and
its failure to manage chemical mixtures leave many who are susceptible
due to previous illness at special risk. The trends described here may
not explain the rise in asthma prevalence beyond a reasonable doubt,
but they do provide a sufficient reason to restructure air pollution law:
vehicle manufacturers should be required to prove that their emissions
are safe for all. The Supreme Court agreed in 2001 that the Clean Air Act
demands the EPA set standards "allowing an adequate margin of safety
. . . requisite to protect the public health." The agency's failure to meet
the statute's mandate continues to take a serious and preventable toll on
children's health.[17]

PART FIVE LESSONS LEARNED AND
EMERGING THREATS

CHAPTER SIXTEEN

Forgotten Lessons

THE PRIMARY ANTAGONISTS in these histories of nuclear weapons testing, military training, pesticides, and diesel emissions have been large public and private institutions: the Departments of Defense and the Interior; the USDA, EPA, and FDA; and major chemical companies. All have played powerful roles in shaping our nation's chemical environment. Further, despite the very different types of risk posed in each case, there are striking similarities or patterns in how threats to the environment and to human health were discovered and either addressed or ignored. What are these lessons, and how might they guide government policymakers, corporate officials, and consumers to create a healthier and more sustainable world?

NUCLEAR WEAPONS

The discovery of hazards created and dispersed by nuclear weapons testing had a profound effect on the future of environmental science and law. Nuclear weapons were obviously intended to produce catastrophic environmental damage, but the Atomic Energy Commission's discovery of global fallout and ubiquitous human exposure was a surprise. Consequently, although each of the nearly five hundred atmospheric tests was first and foremost an experiment in nuclear destruction, gradually government scientists were compelled to expand their observations to include the persistence, movement through the atmosphere, unsuspected ecological pathways, variable human exposures, and unanticipated health effects of the radionuclides that had been produced. As the destructive force of weapons grew, AEC officials gradually realized their failure to predict global circulation of fallout and universal exposure. The enormous scale of state science necessary to understand these effects became a model for later attempts to comprehend climate change and many other hazardous chemicals that circulated globally.

Public understanding of the problem was impossible at first due to the absence of environmental data, and later the government's determination to maintain exclusive control over both nuclear technology and the science used to understand its effects. This monopoly on intelligence created a potent ability to shape public opinion by restricting the release of information. It also empowered the government to weave its own narrative about the extent of fallout, health implications, and the relative threat of nuclear fallout when compared to natural sources of radiation. Secrecy effectively shut out the public from understanding and debating these risks; it also left public officials unaccountable, while preventing any impartial evaluation of scientific findings.

Most details regarding early health and environmental testing were classified and thus not publicly available until the 1990s, when the U.S. Department of Energy and the National Cancer Institute documented the extent of contamination, human exposure, and cancer risk attributable to the weapons testing program. The institute graphically displayed the variability of the threat by color-coding counties on a national map based on the concentration of fallout detected in the soils. Before release of the risk estimates, the government had reasoned that public understanding of the threat would jeopardize national security. Public panic, officials believed, could easily have led to economic collapse of markets for products or property where radionuclides concentrated. Certainly national leaders also wished to avoid being held accountable for their failure to prevent universal exposure.

The government's race to know what happened to radioactive fallout taught much about chemical pathways linking the atmosphere, oceans, soils, plants, animals, and our bodies. This ecological form of thinking in turn helped society to understand risks associated with other technologies circulating globally, including pesticides, flame-retardants, mercury, air pollutants, and persistent compounds.

Persistence is an important indicator of potential hazard. Strontium-90's half-life of twenty-seven years means that exposure threatens genetic damage for entire lifetimes. Chronic exposure to persistent compounds may lead to their accumulation, as was the case with strontium-90 in human bone and DDT in body fat. Chemicals that are immediately damaging and quickly dissipate are perhaps the most dangerous because there is no warning and often little trace of their source. Persistence, by contrast, allows scientists the opportunity to track chemical movement and fate, and often motivates testing for health effects. Tracking of radionuclides after weapons testing, for example, led to the unexpected discovery that the food supply had become the primary pathway for human exposure to radioactive particles. AEC scientists carefully documented the movement of these particles from blast sites to dinner tables, and by the early 1950s scientists could measure contaminants in human tissues such as bone, urine, blood, and fat, correlating the levels to fallout patterns. Recognizing the vulnerability of the food chain became one of the most important paradigms for future efforts to control pesticides, mercury, dioxins, PCBs, and plasticizers such as bisphenol-A. All may migrate from the environment to human tissues, hitchhiking as food residues. It is sobering to realize, however, that the cost of understanding the effects of a globally dispersed technology, and of developing such paradigms, is beyond the capacity of most nations and all but the largest corporations. Dow Agrosciences, for example, spent nearly $100 million to study the effects of only one insecticide. This expense inhibits research and explains much of society's ignorance about the fate and effects of toxic chemicals and other contaminants.

The dangers of nuclear weapons testing were underestimated and misunderstood, but concerns were voiced. In 1956 the National Academy of Sciences concluded: "How much of this radioactivity will actually reach the population depends on how successfully it can be kept out of the great network—ocean and air currents; food and water supplies—which connect man to his surroundings."[1] This was an obvious and authoritative warning that the only way to successfully prevent exposure was to cease atmospheric testing.

But although blast damage was obvious near test sites, fallout was invisible and its global spread was imperceptible. Most hazardous substance problems are similarly undetectable by nonspecialists because they cannot be sensed by sight, smell, sound, touch, or taste. In the case of weapons testing, once inexpensive detection technology such as radiation badges, X-ray films, and Geiger counters made it possible for those without technical expertise to recognize the scale of the threat surrounding them, they challenged government claims of safety and purity. When a hazardous technology and the ability to detect it are both held exclusively by governments or corporations, however, the public cannot challenge such claims.

With no other evidence to consider, the AEC initially presumed that fallout was uniformly distributed beyond the test zone, which encompassed an area two hundred miles around the blast sites. This led them initially to believe that human exposures were minimal in different parts of the nation and world. By the mid-1950s, however, hotspots of fallout were being detected thousands of miles from the explosions; these hotspots had created unexpectedly higher exposures among populations as they consumed dairy products and meats from the contaminated areas. Several public agencies, including the Public Health Service and Weather Bureau, developed credible challenges to the commission's safety estimates by identifying this sort of concentrated fallout. Their challenges underlined the importance of separating the roles of technology development and testing from those of environmental monitoring and regulation. The conflict of interest created when the roles are joined under a single authority is obvious, and can easily work against the public's interest. Although the Atomic Energy Commission established a separate office to regulate the civilian use of atomic energy in 1957, it remained responsible for the study and regulation of its weapons programs—and the public continued to be left in the dark about the dangers of fallout and the extent of contamination at testing sites. The consequences of concentrating the powers of technological innovation and regulation in a single agency can be devastating, and given the legacy of the AEC, are well known. We need to remember this lesson as we consider how the EPA has come to rely on the private sector to monitor the levels and effects of its own polluting technologies.

Early in the testing program, the AEC had also assumed that all people were equally susceptible to health damage from radionuclides. It wasn't until several years later that scientists learned that children's rapid rates of bone formation had led to accumulations of strontium-90 to levels that in many cases exceeded those of adults. In 1955, the Department of Health and Human Services established a committee to explore the susceptibility of children to radiation from fallout, and to examine why the youngest are more susceptible to high doses of radiation (in this case, from X-rays) than adults. In 1956, the British Medical Council recommended lowering allowable U.S. levels of radiation tenfold, because of children's rapid absorption of the isotope and their heightened sensitivity to its ill effects. And during the 1950s and 1960s, the growing anxiety of Presidents Eisenhower and Kennedy over rising levels of radionuclides in the world's children and fears for their health and that of future generations played a prominent role in reversing international nuclear policy. This concern for the most vulnerable became a model for later inquiry into the effects of childhood exposure to pesticides, lead, air pollution, mercury, and more recently, ingredients in plastics.[2]

The nuclear weapons testing program also provided the first environmental model of cumulative irreparable damage.[3] By 1956 the U.S. National Academy of Sciences had concluded, "Anything that adds radiation to this radiation from naturally occurring background rate causes further mutations, and is genetically harmful. There is no minimum amount of radiation which must be exceeded before mutations occur. Any amount, however small, that reaches the reproductive cells can cause a correspondingly small number of mutations. The more radiation, the more mutations. The harm is cumulative."[4] Although this conclusion came too late for millions of people worldwide who were affected by increases in radioactive toxins in their environments, the message looking forward is clear: the public needs protection from unwarranted, unnecessary, and even accidental increases in environmental radiation.

During the atmospheric nuclear weapons testing era, messages to the public were crafted to be reassuring, not informative. Political and military leaders often compared natural hazards such as X-rays, cosmic radiation, and radon with fallout that they deceptively claimed was less threatening. They argued that any risks from testing were worth the benefits of having superior weapons that would supposedly deter nuclear war. They also compared costs to benefits when weighing the importance of developing new and more powerful weapons—a logic that resurfaced in later twentieth-century debates over nuclear power, pesticides, hazardous waste site management, air and water pollution control, and pharmaceutical regulation. The cost-benefit approach, however, rarely considered the non-uniform distributions of the gains and losses—for example, unpredictable hotspots of fallout or the extraordinary concentration of radionuclides in children. Nor did it acknowledge that those reaping the benefits were rarely those who bore the greatest risks.

Since the U.S. bombing of Japan at the end of World War II, nuclear weapons have not been used in an offensive campaign. But the costs of deploying them for deterrence, in terms of both dollars and loss of health, have also been enormous and are often discounted simply as the price of security. Several hundred thousand innocent people died in Japan while many more suffered lifelong illness and economic devastation. In 1995, the National Cancer Institute estimated that between 11,000 and 212,000 additional cancers are expected in the U.S. population due to the weapons testing program; and many production sites such as Savannah River in Georgia, Hanford in Washington, and Rocky Mountain Arsenal in Colorado remain among the most seriously contaminated sites in the world. The costs to restore these sites to a safe and habitable condition were estimated in 2006 to be nearly $500 billion.

Consider what the world might have done with the $5.5 trillion expended by the government to create, store, and deploy nearly 65,000 nuclear weapons held by both the United States and the Soviets during the 1980s. Imagine the improvements in health, education, environmental protection, parks, transit, technology, sustainable development, and foreign aid that might have changed the course of civilization if these resources had been redirected for the greater good. As Richard Rhodes wrote in 2007, "Far from victory in the Cold War, the superpower nuclear-arms race and the corresponding militarization of the American economy gave us ramshackle cities, broken bridges, failing schools, entrenched poverty, impeded life expectancy, and a menacing and secretive national-security state."[5]

Unlike the other historical examples reviewed in this book, radionuclide contamination was caused by a single manmade source: the atmospheric testing of nuclear weapons. When tests ceased in 1963, the fallout, as well as its associated health hazards, gradually declined. Like the national ban on lead in gasoline, and the consequent declines in lead levels in children's blood, once the cause and effects were identified, the problem could be addressed. Unfortunately the problems of climate change, pesticide contamination, and hormonally active chemicals in plastics cannot be countered as easily.

A final cost of the weapons program has been a continuing and growing threat of proliferation. The technology is so desirable to governments of all kinds that exclusive control has been impossible to contain. Seven nations now acknowledge possession: the United States, Russia, Great Britain, France, China, India, and Pakistan, with most of the 20,000 weapons in existence held by the United States and Russia. Israel also probably has nuclear weapons, although it will neither confirm nor disavow an arsenal. North Korea tested its first nuclear device in 2006, while many experts fear the same capacity in Iran.

This history also demonstrates two distinctive forms of risk: the potential for massive and immediate destruction of human life, and the potential for chronic environmental contamination that would threaten health for generations. The two threats became inseparable and once the public understood this, continued testing became intolerable. The eventual diffusion of knowledge from the AEC scientists to the lay public also left an indelible impression on the structure of environmental science and law.

THE PESTICIDE PARADIGM

Although many know Rachel Carson's *Silent Spring* as a work decrying the hazards of pesticide use in the United States, few realize that the 1962 book grew from Carson's perception of the growing ecological threat created by the U.S. and Soviet nuclear weapons tests. Carson kept files documenting rising strontium-90 concentrations in crops and human tissues through the late 1950s and early 1960s, then worried about genetic damage and cancer in humans.

Pesticides have long been released to the environment to protect plants, animals, and even humans from vector-borne illnesses. After their initial use, residues continue to travel through air, water, and food chains. Between 1945 and 2008, billions of pounds of pesticides have been released annually to indoor and outdoor environments, usually to improve agricultural productivity and protect human health. Everyone in the United States lives, works, studies, plays, or rides in environments where pesticides are applied, and residues are detectable in the tissues of most people on the planet. For this reason, and because of international trade in tens of thousands of pesticide products and the deliberate addition of pesticides to hundreds of thousands of consumer and industrial products, this technology has been exceedingly difficult and expensive to detect and regulate. The ever-changing combinations of chemicals that humans experience are beyond the capacity of any government to comprehend or contain.

The way intelligence has been gathered on the environmental and health effects of pesticides has complicated the problem: this information has been produced almost exclusively by manufacturers seeking government licenses. Government reliance on private science has created obvious conflicts of interest. The process discourages industry disclosure of damaging effects and demonstrates well the need for a government-sponsored independent testing regime to assure full disclosure and transparency. Like the AEC's tight control over intelligence concerning the effects of nuclear weapons testing, pesticide manufacturers' directed and secretive "science" has obstructed the public's right to know.

The failure by government to ask obvious questions has also stood in the way of public intelligence regarding pesticide risks. The EPA and the USDA before it have neglected to demand or collect the data necessary to understand a pesticide's hazards before allowing it to be sold in global markets. Although the federal government has been regulating pesticides for more than sixty years, no government agency has yet done the testing needed to understand where pesticide residues linger, how people are exposed, and what dangers they pose to the health of all people, but especially children, infants, and fetuses.

The histories of the AEC and the EPA have their differences. Within ten years of the initial 1945 Alamogordo test, the AEC understood well the risks created by exploding nuclear devices in the atmosphere. The EPA, in contrast, spent much of its first thirty-five years misunderstanding the effects of pesticide mixtures. One likely explanation for this distinction is that the AEC effort evolved as highly centralized state science that quickly became global in scale, whereas the EPA effort depended on highly decentralized and incrementally produced corporate science.

Just as the AEC gradually recognized that radiation posed a greater threat to children than adults, EPA scientists eventually became convinced that children are more susceptible and exposed to pesticides than others. This realization, however, was long in coming. For nearly half a century between 1947 and 1998, neither the EPA nor the USDA asked if young people are especially vulnerable to pesticide hazards before offering licenses to manufacturers. Eventually political leaders chose not to tolerate the presence of radionuclides in children's bodies, while allowing thousands of pesticide combinations to accumulate in their tissues. Why would the government not infer that if animals were building pesticide residues in their milk, humans might do so as well? Why would society tolerate children's exposure to pesticides, but not to radiation?

Pesticide law now relies on safety factors to manage environmental dangers. The practice emerged from the need to address the question of what level of exposure the public should be allowed to experience. EPA experts have tried to identify the threshold level beyond which harm is likely to occur. The answer is crucial if the government is to license drugs, set pesticide residue limits in foods, limit contaminants in drinking water, and identify acceptable levels of ozone or particles in the air. If adverse effects are found in animal studies above a certain dose, what relevance does the threshold have for humans? Many uncertainties surround the answer, and the government's approach has been to employ a safety factor to buffer against any underestimate of danger.

In fact, safety factors have become keystones of corporate arguments to limit rather than ban marketplace hazards. They can use them to justify the sale of very dangerous products, avoiding their prohibition. Bans still may become necessary, however, if normal exposures exceed government limits. For example, in 1996 Congress raised the pesticide safety factor from 100 to 1,000 to offer children additional protection. As a result, hundreds of pesticide tolerances in foods consumed frequently or in large amounts by children had to be revoked or reduced.

Since 1954 the government has defined the task of regulating pesticides mostly as one of managing residues in foods, in part due to amendments to the Food, Drug, and Cosmetic Act made in the mid-1950s and in part because of bureaucratic inertia. This focus has distracted federal officials from the more dangerous conditions created when pesticides contaminate drinking water, workers labor in recently sprayed fields, sprays drift from farms to nearby lands, and pesticides are released indoors. Each of these scenarios occurs tens of millions of times each year, and has produced intense exposures that far exceed those created when residues persist in foods. In particular, the EPA knows least about exposures to pesticides within indoor environments, even though people spend most of their time indoors and their exposure is likely to be far higher there than outdoors. Although a hundred different pesticides were still permitted by the EPA to be used indoors in 2008, the fate of most of these chemicals remains poorly tested.

In addition, among nearly a thousand active pesticide ingredients now registered by the EPA, only twenty-six are limited in drinking water. The absence of a comprehensive surface- and groundwater testing program has guaranteed government and public ignorance of pesticide levels in these important sources. Federal regulators also know almost nothing about the presence of residues in wells that are exempt from testing requirements—but that provide water each day to 60 million people in the United States.

The EPA's traditional chemical-at-a-time testing protocol guarantees that the potential for additive, synergistic, and even antagonistic effects among mixtures of compounds will remain unknown. As recently as 2006, the agency assessed risks posed by a group of organophosphate insecticides and concluded that since individual chemical risks were insignificant, then exposure to mixtures of the same chemicals would be acceptable. This logic neglects many instances of unpredicted chemical interactions and synergy, some among insecticides.

To manage the risks of pesticide exposures, the government has long relied on consumer product warning labels. These certify to the public that if the product is used properly, the risks may be safely controlled. This approach, however, has failed to prevent significant damage to the environment and human health. Tens of millions of users have limited vision that precludes them from reading the small sized print on pesticide labels. Others do not take the time to read labels, or don't have the educational background needed to understand them. Some are careless and do not mix chemicals precisely, apply the chemical as instructed, or properly dispose of unused portions. In fact, most pesticide poisonings occur among children, who could not be expected to understand how best to avoid health risks. Despite these serious limitations, labeling persists as our nation's dominant legal strategy to protect against dangerous commercial chemical exposures.

The ideal regulatory regime would encourage the rapid substitution of less hazardous new products for older, riskier ones. Yet the EPA routinely licenses new chemicals in the absence of full hazard testing, which can create false impressions of safety. DDT, for example, was indeed an effective substitute for pesticides made from metals such as lead, arsenic, and mercury, all recognized early in the century to pose a serious threat to human health. But when concerns arose about DDT, it was replaced with methyl parathion, which is now banned. Also missing from current policy is a way quickly to remove products from the marketplace when they are discovered to be dangerous. Instead, credible evidence of hazard has often been followed by decades-long delays and debates over the quality of the data, acceptable exposures, and the potential of proposed schemes to manage risk. The result is a regulatory paralysis that normally allows exposures to continue if the technology in question is highly profitable. Litigation similarly slows government decision-making, extending market life for profitable but risky technologies. Even regular reviews of pesticide tolerances have become bogged down; the EPA often fails to meet its deadlines, but no penalty is applied. In 2006, the agency institutionalized the delay by adopting a regulation requiring that chemicals be reviewed only every fifteen years. At this rate, an entire generation could easily grow up being exposed to a chemical known to be dangerous.

The government and the pesticide industry have long presumed that risks are knowable and manageable, and that human exposure can be minimized using a variety of educational strategies. Yet the history of pesticide regulation teaches that effective protection against dangerous exposures can occur only if information is available and understood by all who might be exposed. Without a government willing to mount the large-scale independent scientific inquiry needed to identify and control the dangers of pesticide residues, and willing to disentangle influential business interests from regulatory decisionmaking, people will continue to poison themselves and their environments simply because they don't know better.

LAND RESTORATION

U.S. national security has long relied on the capacity to launch a strong and rapid offensive, one that is intentionally destructive to the environment and human life often on a massive scale. Military operations normally target highways, power plants, pipelines, bridges, water supplies, and sewage facilities. If they are to succeed in a real conflict, soldiers and sailors must first practice using the weapons they will employ on the battlefield. Maintaining military readiness with live-ammunition training, however, has created some of the most contaminated landscapes and seascapes in the world. The intent of this training is to destroy and disable, and historically there has been little concern for the places that are bombed, shot at, and invaded.

Prior to 1970, the U.S. Department of Defense was free to ignore the environmental consequences of its weapons development and training programs. The passage of the Clean Air Act, Clean Water Act, National Environmental Policy Act, Endangered Species Act, and hazardous site restoration laws, however, changed the political landscape by requiring that environmental conditions be monitored and reported, by increasing accountability to the public, and by creating opportunities for litigation.

When Congress adopted these laws, neither the EPA nor the Defense Department understood the scale of environmental damage incurred by former military activities. The Defense Department simply had not paid attention to the problem of chemical management at any of its facilities. Live-ammunition training and weapons testing facilities were especially contaminated. Base commanders licensed, subsidized, or directed the release into the environment of hundreds of millions of pounds of munitions at some sites, as well as fuels, solvents, pesticides, radioactive materials, chemical and biological weapons, and many toxic metals. Fleets of ships were intentionally scuttled at sea by the Navy, leaving them to rust slowly while leaking fuels, solvents, and munitions. Between 1951 and 1966, weapons and radioactive wastes from nuclear-powered vessels were also routinely dumped at sea. More than 150,000 containers were tossed overboard with little regard for how radionuclides move through marine food webs to human dinner tables. Highly radioactive wastes produced at the Hanford weapons production facility in Washington, for example, were dumped into pits and tanks that still leak into the Columbia River watershed. Radioactive materials were also intentionally released to the atmosphere at Hanford without community knowledge or consent.[6]

For its part, the EPA for the first two decades of its existence was overwhelmed by private-sector air and water pollution problems along with its inherited responsibility to regulate pesticides. Although in 1986 Congress enacted the Superfund law, which directed the agency to prioritize and restore the nation's most hazardous sites, the EPA still had neither the expertise nor the resources to assess damage at Department of Defense facilities. Nor had Congress allocated the funds needed to restore them.

Since 1990, Congress has directed the Defense Department to close many bases and facilities as recommended by the U.S. Base Realignment and Closure Commission. This effort was motivated by the expectation that defense costs will decline in the transition to a more technologically sophisticated and less labor-intensive military force. It also grew from the belief that most current facilities are designed to prepare for conflict among Cold War superpowers, whereas warfare in the future would more likely involve civil and sectarian conflicts in which nonuniformed opponents would need to be confronted, often in urban settings.

Congress expected that the closed bases would be transferred to other public or private institutions, but understood little about their environmental conditions or lingering health hazards. The discovery of concentrated hazardous chemicals in soils, water, and structures created the unexpected need for exceptionally costly restorations. Full restoration of these sites has never been considered due to cost; instead military leaders and the EPA negotiate "how clean is clean enough." The choice of cleanup level often depends on the area's anticipated future uses: residences or schools, for example, would require a higher standard than industrial facilities that could be capped with concrete. But because markets will drive what happens to the site in the future, there is no guarantee that the anticipated use will end up being the only use—and there will be little incentive to remember or warn others about imperceptible chemical residues.

As discussed earlier, the people of Vieques have endured government confiscation of property, severe environmental damage, and economic stagnation due to the military's use of the area as a training site. The U.S. Navy in particular consistently neglected to monitor the movement of chemicals it released to the Vieques environment during bombing exercises. The Navy freely admits it released tens of millions of pounds of hazardous materials within the island's environment, yet no one knows precisely where, or what happened to them (unless they are unexploded ordnance and happen to be visible). Nor did it monitor the islanders' health. Nearly sixty years after the United States acquired the site, the Navy finally began to develop scientific estimates of exposure, using the expertise of large national defense contractors.

Government agencies and corporations normally avoid intensive environmental and health monitoring because it is very expensive. In addition, this type of intelligence may lead to causal claims of responsibility. Without detailed information, the Viequense and others could not understand past exposures and how these might relate to illnesses among the islanders. The absence of environmental monitoring and poor recordkeeping (regarding the types, location, and timing of chemical releases) also left the Centers for Disease Control and Prevention with the impression that there was no significant exposure or threat to health.

The nature of the problem made it easy to ignore, at least at first. Except for obvious problems, such as unexploded bombs lying on the Vieques landscape or beneath its coastal waters, the dangers are imperceptible. Failure to conduct routine tests for the presence of chemical residues in the environment, in island animals, and in the Viequense themselves also contributed to a collective ignorance that has worked to the Navy's narrative advantage.

Secrecy has also been a factor. Finding out what chemicals have been released, which sites are contaminated, and who has been exposed has been impossible due to classified information and restrictions on site access. The Defense Department has argued, with merit, that disclosure of the amounts and types of weapons released during training exercises could damage U.S. security. Community activists have countered that their health and environmental security have been diminished, along with their property values, all with no compensation.

Perhaps the core problem has been that the Navy never internalized a sense of environmental stewardship normally associated with property ownership. Routine changes in command, troops, and technology make it difficult to assign individual accountability or responsibility for the toxic mixture that remains on the island and in nearby coastal waters. For example, when a former Marine who had trained on Vieques was asked why he ordered unused boxes of live munitions to be buried on a beach following a mock assault, rather than ferrying them back to their waiting offshore troop ship, he explained that his team was late moving to a different training site. In the haste to make a deadline during a mock war experience, munitions were buried that could leach contaminants into the environment for years to come. No one bothered even to record where the boxes were buried.

The extraordinary expense of removing dangerous chemicals from soils, groundwater, and the ocean floor have shaped and limited the government's restoration goals. Despite the deceptive symbolism of Vieques's federal "wilderness" designation, neither the EPA nor the Navy intend to return the island to a natural, wild state. Instead, technical experts will define "cleanup targets," "acceptable levels of exposure and risk," and "future use and site access restrictions," all designed to facilitate transfer of the lands to other responsible parties while limiting costs. The Department of Defense spent nearly $42 billion on its environmental programs between 1997 and 2007, and nearly $4 billion in 2008, yet 90 percent of those funds have been spent to restore fewer than 5 percent of the seriously contaminated military and energy sites.[7]

Vieques residents often express the belief that the Navy is responsible for a variety of illnesses prevalent among longtime residents. Proof that the Navy's wastes have caused disease remains elusive, however, mostly because exposures are difficult to reconstruct with precision, and plausible alternative causes are difficult to discount. It is far easier to demonstrate the absence of causal evidence than its existence. No funds have been set aside for health surveillance, epidemiology, or medical care, meaning that these questions are likely to remain unanswered. Islanders have been intelligent, articulate, and aggressive in their requests to the Navy for island restoration and restitution. Yet the cost of scientific and legal expertise limits the capacity of the local community and their ability to influence cleanup decisions. Significantly, too, Puerto Rico has no voting representation in Congress.

Information and support are supposed to be more readily available. Congress created the Agency for Toxic Substances and Disease Registries (ATSDR) in 1980 to examine whether hazardous chemicals found at Superfund sites have created exposures and illness. Any citizen can petition the ATSDR to request an evaluation, and Congress intended that the agency would provide scientific support for communities surrounding Superfund sites worried about prevalence of illness. When the ATSDR investigated Vieques, it found that exposures to chemicals released by the Navy were detectable in fish commonly consumed by islanders, but not in sufficient amounts to induce illness. They also concluded that it was safe to eat fish without restriction, despite the EPA's warnings that levels were sufficient to threaten the health of pregnant women and women of childbearing age. But the ATSDR sampled so few fish—including some obtained from the local fish market that may have been caught far from Vieques—that no sweeping conclusion of safety is justified. Moreover, although many studies are flawed by small sample sizes, the agency concludes that their absence of proof demonstrates safety. Common sense should suggest that fishing near a former bombing range is potentially hazardous, unless a rigorous scientific study demonstrates otherwise.

When compared to abandoned privately owned hazardous sites, former Defense Department properties have features that should make it easier to return them to their original condition. Unlike many private waste sites that have been orphaned, there is no dispute over who is to blame, and thus no lengthy search for "potentially responsible parties." The Defense Department controls the cleanup process and clearly has the potential to act quickly. It possesses the exclusive expertise to understand what happened at the site in terms of operations, weapons deployment, and chemical release. It also has access to necessary funding for site evaluation and cleanup. Still missing, however, is necessary political will. Although the Defense Department budget is near $500 billion a year, less than half a percent of that amount has been spent on site restoration.

The Navy's sixty-year presence on Vieques not only contaminated large areas of the island, but also assured economic stagnation: compare the still quiet island of Vieques to thriving tourist economies on nearby St. Thomas and St. John. The high unemployment rate on Vieques is a result of the Navy's tenure, and like many other impoverished areas, is accompanied by a lack of health insurance and inaccessible medical care. Not only has the Navy left the island with a chemically polluted environment; it has left a poor population with long-lasting economic, health, and social hazards.

Since 1986 when the Superfund law was enacted, tens of thousands of government facilities and bases have been determined to need restoration, yet fewer than 1 percent of these have been fully restored. Cleanup expenses are so high that government has become content to extend restoration deadlines for listed sites until 2075. Meanwhile, those living near or within these facilities, that is, those who face the highest potential exposures, are limited in their ability to influence cleanup goals and deadlines. Access to information remains a significant issue: the problem is highly technical, and the military has not established the health-monitoring programs needed to examine possible associations between chemical exposures and health loss. And yet the government continues to transfer responsibility for these sites to other public and private institutions without completing full site restoration.

Can a nation prepare for war in a manner less destructive to the environment and less threatening to civilian health? Certainly it could, but doing so would require a far different culture than that demonstrated by the U.S. Defense Department in the twentieth century.

AIR POLLUTION

The surge in respiratory illnesses in the United States since 1980 has perplexed many scientists, but most agree that the change has been too sudden to be explained by genetic evolution and that environmental change could be the primary cause. Homes, schools, and other buildings built since the mid-1970s have better insulation and tighter windows and doors in order to conserve heat or reduce cooling costs, but these advances also reduce the exchange of indoor air with the usually cleaner outdoor air. Use of synthetic chemicals in building materials grew quickly during the same period, and eventually may prove to be the cause of the surge in illnesses. Plastics, for example, are often used in modern homes for siding, flooring, sewer and water pipes, electrical wiring, windows, doors, countertops, cabinets, wood finishes, paints, rugs, and furnishings, including bedding. Consumer products such as electronics, cosmetics, air fresheners, cleaners, and dryer sheets also emit mixtures of chemicals to indoor air. Children who spend more time indoors watching television, using computers, or playing video games increase their exposures to whatever is floating in the air.

Some outdoor pollutants such as diesel exhaust are more concentrated outside rather than indoors. The lure of diesel is its power; its downfall is its incomplete combustion, which produces a mixture of very dangerous substances. Recognized by scientists in the early 1970s to be mutagenic to bacteria, diesel exhaust is now associated with allergies, a variety of respiratory diseases including asthma, cardiac arrhythmias, heart attacks, diabetes, cancer, and premature death. The State of California declared diesel exhaust to be a "known human carcinogen" in 1990, and an "air toxic" in 1998. California and other scientists estimate that diesel emissions account for nearly 70 percent of the total risk from toxic air pollutants.

Diesel fuel is less refined than gasoline, and has more energy per gallon, but it burns less efficiently, causing it to emit a very complex mixture of chemicals. Beginning in 1971, early air quality regulation of diesel-related toxics focused on "total suspended particles," that is, particles of all sizes. The EPA effectively reduced the large-diameter carbon particles by nearly 90 percent (and decreased visible black smoke), but the required engine modifications also produced more of the smaller particles. By 1987, the agency had become convinced that particles smaller than 10 microns in diameter, PM-10, were more worrisome and set a limit for them. By 1997, the agency was recognizing that even smaller particles were still more threatening to health, so it set a new standard for PM-2.5, that is, those particles smaller than 2.5 microns.

This standard was challenged in court by the American Trucking Association soon after it was adopted and lower courts prevented its implementation. Not until the Supreme Court ruled in 2001 that the EPA could not consider costs when setting health-protective standards, as stated in the Clean Air Act, was the agency free to move ahead. Finally, in 2004, a full seventeen years following the last significant attempt to control particles, the EPA implemented the new PM-2.5 rule by declaring thirty-nine areas of the country beyond compliance. The glaciers are retreating faster than the EPA adopts new air quality rules (and in fact these two processes are related since carbon is very effective at absorbing solar radiation). More recent scientific reports demonstrate that the fight is far from over: nearly 90 percent of diesel particles are far smaller than the new 2.5 micron limit. Once again, the laws and regulations remain far behind the state of the science. These nanoparticles, less than a billionth of a meter in size, are larger in number and have a carbon surface that attracts some of the other gases in exhaust, including aldehydes, carbon monoxide, nitrogen oxides, and sulfur compounds, forming a toxic mixture. Moreover, their exceptionally small size means they more easily and deeply penetrate the lungs.

A rapid decline in hospital admissions for respiratory and cardiac distress in Atlanta during the Olympic Games in 2000 occurred when auto traffic was restricted in the city's center. In fact, the recent discovery that groups with preexisting illnesses are vulnerable to sudden and severe health loss following high pollution episodes is one of the most significant scientific findings in the field of environmental health in the past two decades. Dozens of peer-reviewed studies published during this period illustrate special vulnerability among those with illnesses including asthma, chronic obstructive pulmonary diseases, cardiovascular diseases, diabetes, and lung cancer. Nearly a quarter of the nation's population suffers from these illnesses. Those most at risk include children, the elderly, and those with compromised immune systems. The EPA estimated in 1999 that tens of thousands of premature deaths are caused annually by outdoor particulates that are smaller than PM-2.5. Although some scientists who are sponsored by the EPA are now acknowledging the seriousness of the threat posed by these ultrafine particles, no regulation limits human exposure.[8]

Another distinctive trait of the diesel engine is its longevity: a new long-haul diesel truck is now expected to last nearly thirty years, and new diesel cars, twenty years. Consequently, it will be difficult to improve air quality in the next generation by relying on improved engine design. Changes in fuel chemistry can have more immediate effects on emissions. For example, the removal of lead from gasoline quickly reduced blood lead levels in children who had inhaled the metal as airborne particles.

The environmental history of vehicle emissions is complicated by the wide range of vehicles, engines, and fuel types, which produce variable mixtures of particles and gases. In addition, the government's pollution data are generally averaged over extended periods and over large areas. By including less polluted times and places in the calculations, high pollution episodes are not discernable—the data show that the public is safe, even if bursts of high exposures may cause serious illness. The failure of federal officials to monitor individual or "personal" exposures has left tens of millions in the United States at serious risk of both acute and chronic illness. This situation is remarkably similar to the government's mismanagement of pesticide exposure. When pesticide residue levels are averaged across all foods, higher residues in a few foods become imperceptible. And when food intake is averaged across the entire U.S. population, the fact that children eat more of some foods is not discernable. But if only 1 percent of children consume foods with pesticide hazards, this means that 800,000 children are at risk. The government, by relying on averaging techniques to support its narrative of safety, is allowing a minority to be exposed to known toxic substances so that the majority of consumers can enjoy the pest reduction benefits that pesticides offer.

262 LESSONS LEARNED AND EMERGING THREATS

What matters most to your health is the quality of air where you spend your time, not where governments monitor pollution. Because we tend to live close to places where we burn fossil fuels—in homes and vehicles—the disparity in air pollution between your exposure and that predicted by federal monitors is likely to be large for carbon particles and hundreds of organic compounds in gaseous form. (Ozone, by contrast, tends to be more uniformly distributed than particulate matter.) Knowledge about the spatial variability of pollution grew from more intensive air monitoring—in a story remarkably similar to that of the Atomic Energy Commission's discovery that fallout settled unevenly around the world. Some elements of vehicle exhaust such as large-diameter particles do settle close to their source, while fine particles and gases may move regionally, and some others such as carbon dioxide circulate globally. The lesson here is that the only way to obtain a refined measurement of individual risk is to follow individuals through their daily routines. When we investigated the individual exposures of children riding diesel buses to and from school, for example, we discovered that they inhale air contaminated with nearly twenty times the number of fine particles than nearby outdoor air. The U.S. National Academy of Sciences has recognized the importance of this individualized approach for at least a decade, but the EPA has paid little attention.

Perhaps the most egregious defect in federal efforts to control air quality has been neglect of indoor environments, where people spend most of their time. Workplace bans have been most effective, as have protection of environments frequented by children. The most serious remaining oversight of indoor environments remains the home, which is perhaps being avoided due to privacy concerns. Several U.S. statutes—including federal pesticide, tobacco, consumer product, and toxic substance control laws—have attempted to influence indoor air quality. But these national laws have produced only weak regulations that rely on warning labels to educate consumers about dangerous exposures. Like other examples provided earlier, bans are far more effective than pollution limits, or restrictions by place or time that require intensive monitoring.

The EPA has been most effective at reducing emissions from large apparent sources such as mining, utility, chemical, and manufacturing companies, and least effective at managing and reducing emissions from highly decentralized sources such as vehicles. The rapid growth of U.S.

cities has concentrated the areas where energy is used—for transport, heating, and industrial activities—creating localized air pollution problems that often escape government detection. Highways have experienced increasing congestion as the U.S. population has continued its move from the city to the suburbs, and as people now generally drive farther to work each day. Although emissions per vehicle have dropped, the increasing numbers of cars and trucks have intensified our collective exposure to hazardous chemicals in exhaust, especially as we move slowly through traffic jams and construction sites.

The most compelling arguments for more restrictive outdoor federal air pollution stem from the now irrefutable facts that children's lung function decreases as traffic density increases near their residences (with asthma and bronchitis symptoms increasing the closer the home is to a highway), and that hospitalizations for asthma rise along with proximity to heavy traffic and especially, idling or traveling trucks. The growing consensus regarding the dangers of living, working, or playing near highway corridors is troubling for those hoping to reduce urban sprawl by locating new residential developments close to transit routes such as interchanges and hubs. Although such a shift could reduce a community's carbon emissions, it would also likely increase the number of those who breathe the particles and gases produced as they travel, consume electricity, and burn fossil fuels for heat.

CHAPTER SEVENTEEN

The Quiet Revolution in Plastics

THE NORTH PACIFIC Subtropical Gyre, a slow-moving current in the Pacific Ocean that swirls in an enormous clockwise spiral, has collected two large masses of plastic trash now twice the size of Texas, one between Japan and Hawaii and the other between Hawaii and California. In coastal regions, discarded plastics that are not recycled, land-filled, or incinerated often find their way into streams, rivers, and eventually the seas, where currents have quietly collected them in remote areas of the ocean. Plastic by one estimate now constitutes 90 percent of all trash floating in the world's oceans. Although many types of these resins tend not to degrade even after centuries, some will break into ever smaller pieces over time when exposed to sunlight. The mass of plastic particles in the two floating trash areas, known collectively as the Great Pacific Garbage Patch, is now six times the mass of plankton—small, abundant organisms consumed by many species of marine life. Many ocean animals mistake the waste for their natural foods, and predators can concentrate some plastic ingredients present in smaller prey. The largest predators tend to accumulate the highest concentrations of persistent chemicals.[1]

Since 1950, and with little notice, plastics have quickly and quietly entered the lives and bodies of most people on the planet. As of 2008, plastics composed nearly 70 percent of the synthetic chemical industry in the United States, where each year more than 100 billion pounds of

264

resins are formed into food and beverage packaging, electronics, building products, furnishings, electrical wiring, vehicles, toys, and medical devices. In 2007, the average American purchased more than 220 pounds of plastic, creating nearly $400 billion in sales; by comparison, during the same period total pesticide sales in the United States were approximately $11 billion.[2]

It is now impossible to avoid exposure to plastics. They surround and pervade our homes, bodies, foods, water supplies, and power delivery systems, and are used in a wide array of consumer products. Most individual homes constructed since 1985 are wrapped in plastic film such as Tyvek and their exterior shells are made from polyvinyl chloride (PVC) siding. Some modern homes receive water via PVC pipes, have their power distributed in plastic-coated electrical cables, and are blanketed in plastic insulation. Walls commonly are painted with plastic and epoxy compounds, countertops may be Formica or Corian, and wooden floors are coated with polyurethane finishes, polypropylene rugs, or polyvinyl chloride tiles.

The immersion begins early. Most infants spend twenty-four hours a day in plastic diapers and sleep in polyester pajamas on waterproof PVC mattresses treated with plastic flame retardants. Many rest beneath polyester fleece blankets in cribs normally surrounded by polyurethane foam padding. The nursery may have plasticized carpets, wallpaper, or furniture painted with epoxy resin paints. Plastic toys and teethers are likely to sit near the crib, along with baby wipes stored in plastic pop-top containers and a variety of creams and powders that are also packaged in plastic. Vinyl blinds or shades often cover windows, and fresh air may blow into the room through plastic screening.

Foods and beverages are normally packaged in plastic, including soda and soups with can linings coated with plastic epoxy resins, and milk bottles made from high-density polyethylene. Snack-sized food and beverage packages made from or lined with plastic are increasingly popular because they are so convenient, and as the package size diminishes, the use of plastic increases. Between 1997 and 2005, annual sales of small bottles of water—those holding less than one liter—increased from 4 billion to nearly 30 billion bottles.[3]

Nearly 80 percent of the world's toys are made in China and increasingly are formed from plastic resins, which are favored due to their malleability, durability, and low shipping weight. Tracing the chemical composition of these toys is nearly impossible given the complexity of supply chains, poor record keeping, and an absence of chemical regulation. Soft plastic toys, including dolls, rubber ducks, inflatable toys, balls, and baby-care items, are often made with polyvinyl chloride. The billions of video games, computers, MP3 players, cameras, and cell phones purchased each year in the United States use a wide variety of plastic resins. Although the useful life of electronic products is often less than three years, many will not degrade for hundreds of years unless burned. More than 100 million cell phones—which include not only several types of plastics but also hazardous elements including lead and mercury—are discarded every year in the United States alone. Planned and accidental obsolescence guarantees extraordinary waste.[4]

The plastic content of vehicles such as cars, trucks, buses, and aircraft has been increasing steadily as automakers strive to diminish weight and raise fuel efficiency. By 2008, new cars contained an average of nearly three hundred pounds of many different types of plastic, including electronics, air conduits, bumpers, airbags, wiring, dashboards, seats, door panels, flooring, and ceilings. Almost 7.5 million new vehicles are sold in the United States each year, and 2.5 billion pounds of their plastic components have little hope of being recycled if made from polyvinyl chloride or polycarbonate.

LABELING FAILURE

It has never been possible to know what chemicals are in the plastics that U.S. consumers encounter every day. Ingredients need not be labeled under federal law, and most manufacturers are unwilling or unable to disclose these contents or their sources. Some products are labeled to facilitate recycling but not to identify chemicals used in their manufacture. Indeed, often the only clue consumers have to the chemical identity of the plastics they use is the voluntary resin code designed to identify products that should and should not be recycled—but it offers little usable information. Category 3, for example, includes not only polyvinyl chloride made with phthalate softeners, but other resins and additives as well. And Category 7, "other," provides neither consumer information regarding ingredients nor much meaningful guidance for recycling. The absence of helpful federal labeling has led many states to mandate the use of some codes on certain types of bottles, but the result has been, at best, a confusing set of overlapping state standards with little monitoring or enforcement.[5]

Even the recycling information is generally ignored. The American Plastics Council estimated that only about 5 percent of all plastics manufactured are recycled (Figure 17.1). Recycling failure occurs for many reasons, including inconvenience and low redemption fees. It is also clear that the true costs of plastics, including the energy required to manufacture them, the environmental contamination caused by their disposal, the loss of health, and the recycling and eventual disposal costs, are not reflected in product prices. Collection, recycling, and redistribution charges for plastics were so high in 2006 that the price of virgin resin was 40 percent lower than that of recycled resin.[6]

Few Americans realize that most recycled polyethylene bottles are exported to China, now the world's largest destination for plastic waste. (U.S. export costs to China are inexpensive because the imbalance of trade often leaves empty space on container ships returning there.) Recycling and pollution control standards are very limited in China, however. One investigation tracking plastic waste from the United Kingdom to a Chinese village revealed that the plastic was melted without measures to protect workers from fumes, and wastes were poured directly into a river.[7]

Moreover, in the vast majority of cases, recycling plastic only extends its life by one generation—it is only a matter of time before the secondary product is discarded as waste in either landfills or incinerators, which each release hazardous chemicals into the environment. Even if polyethylene bottles used to contain water and sodas are recycled and used to make products such as polyethylene carpets or fleece jackets, these are rarely recycled again. And few plastic food containers are made into new food containers because to be recycled this way they are required to be coated with a layer of virgin plastic to prevent chemical contamination (although Food and Drug Administration regulations allow the layer to be only 1 millimeter in thickness, and the rule is rarely monitored or enforced).[8]

Most plastic is produced from natural gas and petroleum products, which have a high carbon content. The American Chemistry Council estimated in 2008 that the plastics industry accounts for 5 percent of the nation's consumption of natural gas and petroleum. This would mean the plastics industry consumes nearly 1.1 trillion cubic feet of natural gas and 400 million barrels of oil annually.[9] Calculating the energy used for

Symbol	Plastic Type	Amount Produced (Billions lbs.)	Percent Recycled	Product Examples
♻ 1	PETE Polyethylene Terephthalate	3.4	20%	Convenience-size beverage bottles, mouthwash bottles, boil-in-bag pouches
♻ 2	HDPE High Density Polyethylene	4.2	10%	Milk jugs, trash bags, ice cube trays, storage containers
♻ 3	PVC Polyvinyl Chloride (includes DEHP)	2.5	0%	Cooking oil bottles, meat packaging, baby bottle nipples
♻ 4	LDPE Low Density Polyethylene	10	2%	Produce bags, food wrap, bread bags, zip-lock bags, baby bottle liners
♻ 5	PP Polypropylene	5.1	5%	Yogurt containers, straws, margarine tubs, spice containers
♻ 6	PS Polystyrene	2.0	1.5%	Styrofoam cups, take-home boxes, egg cartons, meat trays
♻ 7	Other (includes Bisphenol A)	6.2	<1%	Polycarbonate baby bottles, 5-gallon water cooler bottles, meat trays, toddler fruit cups

FIGURE 17.1. Plastics: amount produced and percentage recycled. Tens of billions of pounds of plastics are either buried or burned in the United States each year, leading to air, soil, and water pollution. Among all types of plastic, only 5 percent has been recycled in recent years. Most residents of wealthier nations have detectable levels of many chemical ingredients of plastics in their bodies. *Sources:* Environment and Human Health, Inc., John Wargo, Mark Cullen, Hugh Taylor.

pumping, processing, transporting, and refrigerating bottled water, the Earth Policy Institute calculated the annual fossil fuel footprint of bottled water consumption in the United States alone to be the equivalent of more than 50 million barrels of oil.[10]

PLASTICS AND HUMAN HEALTH

What are the implications for human health of this surge in chemical exposures from plastics? In the late 1980s, Theo Colborn, a wildlife biologist who worked with the World Wildlife Fund for many years, began to compile and interpret the patchwork of scientific evidence collected since the 1950s suggesting that commercial chemicals and pollutants could mimic human hormones. Colborn's concern grew from her observations of reproductive failure among many different species of fish and wildlife. She convened a conference and generated a hypothesis that human exposure to "endocrine disrupting" chemicals might explain the rise of reproductive and developmental disorders recognized since 1985.[11]

More than a thousand chemicals are now suspected of affecting normal human hormonal activity. These include many pharmaceuticals, pesticides, plasticizers, solvents, metals, and flame retardants. Many of these compounds were identified through observations of so-called sentinel species: animals that experienced dramatic changes after being exposed and thus may harbinger similar trouble for humans. The insecticide DDT, for example, induced reproductive failure in many predatory birds including eagles, ospreys, falcons, and hawks, and some species of fish during the 1950s and 1960s. In later decades, fish swimming near paper mill sewage outfalls and thus exposed to dioxins exhibited estrogenic, androgenic, anti-androgenic, and anti-thyroid effects. Alligators swimming in waters contaminated by the insecticide dicofol developed reproductive abnormalities, including abnormally small phalluses, following a 1980 spill in Lake Apopka, Florida. Their egg survival rates declined and both males and females grew deformed sexual organs. (Alligators studied in nearby unpolluted lakes, meanwhile, exhibited none of these conditions.) And different species of birds suffered reproductive failures following exposures to DDT, polychlorinated biphenyls (PCBs), and polycyclic aromatic hydrocarbons (PAHs) in the North American Great Lakes region, as well as in Washington State's Puget Sound and the Baltic Sea in northern Europe.[12]

Our understanding of the potential of chemicals to mimic human hormones grew initially from the search for synthetic estrogens. Some synthetic estrogens elevate the risk of reproductive disorders. The drug diethylstilbestrol (DES), for example, is a synthetic estrogen formerly prescribed to prevent miscarriages, preterm birth, and other pregnancy problems. By 1953 published studies were demonstrating that DES prevented neither miscarriages nor preterm births. Many physicians, however, continued to prescribe the chemical until 1971, when it was reported to cause clear-cell adenocarcinoma, a rare form of vaginal cancer, among girls and young women who had been exposed to the drug while in their mother's womb.[13] Daughters of women who took DES developed other problems as well, including reduced fertility, premature births, miscarriages, and an elevated risk of breast cancer. Sons of women who took the drug were more likely to have problems with undescended or not fully descended testicles as well as hypospadias, the premature exit of the urethra before the end of the penis. Additional studies identified third-generation effects among DES-exposed mice, suggesting possible risks to grandchildren of DES-exposed mothers that are not yet clear. The DES history contains lessons and warnings about heightened human susceptibility to synthetic hormones during embryonic development, as well as the potential for adverse health effects in later generations.[14]

Normal growth and development among fetuses, infants, children, and adolescents is regulated in the body by a diverse set of hormones that promote or inhibit cell division. During 1987 two scientists at Tufts University Medical School continued research exploring chemicals that might inhibit cell reproduction. Anna Soto and Carlos Sonnenschein exposed live breast cancer cells to different concentrations of estrogen and then closely examined their rates of growth and reproduction. Normally they found that the rate of cell division depended predictably on the concentration of estrogen administered. During one of their experiments, however, they were surprised to find that all of their samples, even those untreated with estrogen, were dividing wildly. Suspicious that some laboratory contaminant was the cause, they investigated the air quality and the possible failure of serum filtration techniques used to remove estrogen. Nearly four months after continued experimentation, they switched the brand of their polystyrene test tubes from those manufactured by Corning. When cells contained in non-Corning tubes divided again at rates that depended on the dose of estrogen administered, they then believed that some compound in the Corning tubes had caused them to lose experimental control. When contacted, Corning admitted that the company had altered the ingredients of the tubes, but refused to identify the chemical change, claiming their right to trade secrecy. Soto and Sonnenshein worked for two additional years to discover on their own that the newer tubes were made with p-nonylphenol, an alkylphenol commonly added to polyvinyl chloride and polystyrene products to improve durability and chemical stability. When they injected the p-nonylphenol into female rats that had had their ovaries removed, their uterine growth increased just as it would if estrogen were present. Further research taught them that nonylphenols are also ingredients or degradation products of detergents, pesticides, and personal care products. European nations had banned the use of nonylphenol ethoxylates in commercial cleaning agents or agreed to end their use by 2000. But U.S. government regulators had largely neglected the chemical, and had not tested it for possible hormonal activity.[15]

Many additional scientists oriented their research to better understand the potential for industrial chemicals to mimic human hormones. Their interest coincided with a growing consensus within the NAS that children are often at greater risk than adults of damaging their health following exposures because of their rapidly growing but immature organ systems, hormone pathways, and metabolic systems. In addition, young children breathe more air, consume more food, and drink more water per pound of body weight than adults, and thus have greater exposure to any chemicals present in their environments. As mentioned earlier, in 1993 the NAS recognized in particular the susceptibility of the very young to pesticides, and in 1996 Congress adopted the Food Quality Protection Act. Recognizing that tens of thousands of pesticide products had not been tested for their potential effects on the human endocrine system, Congress required the EPA to develop an "endocrine disruptor screening program" to understand their risks. A similar screening requirement was added that year to the Safe Drinking Water Act amendments. Despite these legislative efforts, the EPA has made little progress in research or regulation in more than a decade, claiming the program is underfunded and that experts disagree over how best to conduct the tests.[16]

Chemical manufacturers have long criticized the use of animal evidence to infer human health risk. In 1999, however, the NAS considered the dangers posed by hormonally active chemicals and agreed that the health effects seen in animals do signal a threat to humans, especially when they are well correlated with an increasing prevalence of human illness. Many forms of human illness associated with abnormal hormonal activity have become more commonplace during the past several decades, including infertility, miscarriage, breast cancer, prostate cancer, obesity, and various neurological and neurobehavioral problems. More recently, numerous scientists have reported health abnormalities in laboratory animals similar to those reported in wildlife studies. The U.S. government has in fact long relied on animal data for sounding the alarm for chemical threats to human health. Since 1970, the EPA has relied almost exclusively on animal evidence as a surrogate for human studies to judge pesticide dangers, and many bans rest on evidence of harmful effects in animals together with known human exposures.[17]

THE BPA CONTROVERSY

Bisphenol-A (BPA), the primary component of hard and clear polycarbonate plastics, illustrates well the endocrine disruption problem. Each year several billion pounds of BPA are produced in the United States. The CDC has found, in results consistent with those found in other countries, that 95 percent of human urine samples tested have measurable BPA levels. BPA has also been detected in human serum, breast milk (in higher concentrations than in blood serum), maternal and fetal plasma, amniotic fluid, and placental tissue at birth.

Higher concentrations of BPA are found in children than adults, perhaps because they consume more food relative to their body weights than do adults. Concentrations may also vary by dietary patterns. Females and those with lower incomes had significantly higher concentrations than males and those who were better off financially. BPA also travels easily across the placenta, and levels in many pregnant women and their fetuses were similar to those found in animal studies to be toxic to the reproductive organs of male and female offspring.[18]

Government scientists believe that the primary source of human BPA exposure is foods, especially those that are canned. BPA-based epoxy resins are used to line the interior of cans that contain food and beverages and the chemical can migrate from the resins into the foods. On average, canned foods constitute 17 percent of the U.S. diet, but this varies;

in particular, some who depend on institutional cafeterias such as those in schools or hospitals may be more intensively exposed than those who predominantly eat fresh foods.[19]

Rapid growth in the bottled water industry has meant the greater use of polycarbonate refillable five-gallon water jugs. In 1997, the FDA found that BPA migrated from these containers into bottled water at room temperature and that concentrations increased over time. Another study reported that boiling water in polycarbonate bottles increased the rate of migration by up to 55-fold, suggesting that it would be wise to avoid filling polycarbonate baby bottles with boiling water to make infant formula from powders. The finding also raises questions about the potential leaching of the chemical from polycarbonate containers used to brew and store hot beverages such as coffee makers and pots.[20]

Some clear, hard plastic baby bottles, toddler drinking cups, and water bottles are made with plastics containing BPA. Manufacturers of polycarbonate baby and water bottles claim that their products are safe, despite the levels of BPA that may leach from their products. One study found that older worn bottles leach more BPA than new containers. I purchased a new polycarbonate baby bottle in 2008 and found an insert that warned against scratching the interior surface with an abrasive material. It also advised users to avoid the use of harsh detergents, not to clean the bottle in a dishwasher, and to discard the bottle if it appears worn. No reasons were provided for this advice, but it certainly reflects warnings in the scientific literature about practices that increase BPA migration from bottles to liquids.[21]

Many additional products may contain BPA but are more difficult to identify. Scientists have reported BPA detected in nonstick-coated cookware (United Kingdom), PVC stretch film used for food packaging (Spain), recycled paperboard food boxes (Japan), and clothing treated with fire retardants (Sweden). Tetrabromo-bisphenol-A, for example, is the most commonly used flame retardant and its concentrations have increased in human blood since the 1970s. By 2002 it was detected at concentrations 3.5 times higher in children under the age of four than in adults.[22]

Since 1995 numerous scientists have reported that BPA caused health effects in animals that were similar to some that are becoming more prevalent in humans. These conditions include breast and prostate cancer, declines in sperm counts, abnormal penile or urethra development in males, early sexual maturation in females, neurobehavioral problems, obesity and type 2 diabetes, and immune system disorders.[23]

The discovery of BPA's hormonal activity following exceptionally low-dose exposures was a surprise. Some scientists believe that there are two plausible mechanisms for BPA to disrupt normal endocrine function. The first and more commonly assumed process is that BPA binds with estrogen receptors in the cell's nucleus that regulate the transfer of genetic information to newly synthesized molecules of RNA—thereby interfering with the interaction of the receptors and estrogen. Another recently discovered possibility is that BPA may bind with estrogen receptors in cell membranes that regulate calcium transport and intracellular signaling. This second process has been reported at exceptionally low part per trillion doses, that is, at exposures nearly 1,000 times lower than the EPA's recommended acceptable limit (the agency does not regulate human exposure to the chemical).[24]

The kinds of effects found in animals exposed at doses well below those that EPA has claimed are safe will also sound familiar to those who take note of the latest headlines regarding human health problems. They include abnormalities of the reproductive tract in female rats; reduced daily sperm production and testis weight, and reduced anogenital distance in male mice; early onset of puberty in female mice; higher rates of meiotic failure during conception, specifically an increase in the number of eggs and embryos that are aneuploid (have an abnormal number of chromosomes) in mice; reduced sperm counts in both rat and mouse studies; mammary gland development, including neoplastic lesions in rats and mice; increased prostate size and precancerous prostatic lesions in rats; higher rates of prostate cancer in male rats; insulin resistance in adult mice; increased body weight in mice; altered immune function among exposed mice; impaired learning and memory among rats; and hyperactivity, increased aggression, and elimination of sex differences in behavior among exposed rats and mice.[25]

In 2007, the U.S. National Institutes of Health convened a panel of thirty-eight scientists to review the state of research on BPA-induced health effects and better understand the apparent adverse effects that followed administration of low doses to test animals. The panel issued a strong warning about the chemical's hazards: "There is chronic, low level exposure of virtually everyone in developed countries to BPA. The

:ientific literature on human and animal exposure to low publis\ in relation to *in vitro* mechanistic studies reveals that human dose BPA is within the range that is predicted to be biologically eyver 95 percent of people sampled. The wide range of adverse low doses of BPA in laboratory animals exposed both during nent and in adulthood is a great cause for concern with regard otential for similar adverse effects in humans."[26]

ne U.S. National Toxicology Program reviewed the panel's assess- .ent and issued these comments in April 2008: "The scientific evidence that supports a conclusion of some concern for exposures in fetuses, infants, and children comes from a number of laboratory animal studies reporting that 'low' level exposure to bisphenol A during development can cause changes in behavior and the brain, prostate gland, mammary gland, and the age at which females attain puberty. Because these effects in animals occur at bisphenol A exposure levels similar to those experienced by humans, the possibility that bisphenol A may alter human development cannot be dismissed."[27]

The American Chemistry Council, which advocates for the plastics industry, has criticized most scientific research that has reported an association between BPA or DEHP and adverse health effects. The council's complaints have included claims that sample sizes are too small; that animals are poor models for understanding hazards to humans; that discovering exposure does not prove that the chemical causes the observed health effects; that the experimental route of exposure in animal studies is different from that experienced by humans or from the route known to pose the greatest hazard; that doses administered in animal studies are normally far higher than those experienced by humans; that the mechanism of chemical action is poorly understood; that health effects among those exposed are not necessarily "adverse"; and that humans normally experience exposure to complex and ever changing mixtures of chemicals that might play some role in the etiology of disease.[28]

Industry advocates commonly level these kinds of criticisms against scientists who report an association between exposure to their products and health risks. And it is true that the design of the scientific research can always be improved. Research on plastics, however, now comprises a large and robust literature reporting adverse health effects in laboratory animals and wildlife at even low doses. Claims of associations between

BPA and hormonal activity in humans are strengthened by ʀ
that everyone is routinely exposed and by the rising incidence ᵤₛ
human diseases similar to those induced in animals once dosed w
chemical. Two competing narratives—one forwarded by indepen.
scientists and the other promoted by industry representatives—have
lied on the same evidence and delayed government action to protect t,
health of citizens through bans or restrictions.

DEHP EXPOSURE

The history of another plastic resin, DEHP, is remarkably similar to that
of BPA. DEHP has long been used to soften polyvinyl chloride (PVC).
Each year several hundred million pounds of DEHP are produced in the
United States, most of which are added to PVC to provide varying degrees
of flexibility. DEHP does not bind chemically to PVC; thus it can easily
separate from it and move through the environment or be absorbed by
plants, animals, and humans. The enormous volume of PVC in consumer
and industrial products, its persistence, and its routine disposal help to
explain why human exposure to DEHP is so common.

DEHP has been found in human blood, seminal fluid, amniotic fluid,
breast milk, and saliva. Younger children normally have higher concentrations of DEHP than adults, possibly due to their higher food consumption
per pound of body weight, their tendency to put plastic toys and pacifiers
in their mouths, and perhaps closer contact with PVC flooring products.
The DEHP dose experienced by nursery school children was found to be
about twice as high as that of adults, and the highest exposures were reported for children six months to four years old. A study measuring levels
of phthalates in personal air samples collected from pregnant women in
New York and Poland found DEHP in 100 percent of both air and urine
samples. Women of childbearing age were reported to have significantly
higher phthalate exposures than other adults, and fetal and maternal
exposure are closely related. A German study of DEHP found that nearly
one-third of the men and women exceeded the EPA's recommended exposure limit—a level that many independent scientists believe is not health
protective, because it far exceeds levels reported to induce adverse effects
in some animal studies. A Taiwanese study found that 85 percent of the
study participants exceeded the EPA limit.[29]

The major source of exposure to DEHP for most children is food. The

FDA allows the use of DEHP in can coatings, adhesive paper manufacturing, single and repeated use containers, cellophane, and as a metal foil lubricant. Processing equipment, including plastic tubing, surface coatings, and gaskets used in the food industry, may also contain DEHP. The chemical is considered an indirect additive in packaged foods due to its use in some plastic wraps, heat seal coatings for metal foils, closure seals for containers, and printing inks for food wrappers and containers. It is known to migrate into food from plastics during processing and storage. In fact, the chemical or its metabolites has been reported in many foods common in a child's diet, including milk, cheese, meat, margarine, eggs, cereal products, baby food, infant formula, and fish. The presence of phthalates in infant formula and breast milk in the United States, Canada, Germany, Finland, and Denmark is well documented, as is its inclusion in baby food. DEHP can bind to fat, cross the placenta to expose a fetus, and pass into breast milk.[30]

Drinking water can also be a source of the compound. The Natural Resources Defense Council detected DEHP in several types of bottled water it tested, at levels just below the EPA tap water limit. The government has not established a bottled water standard. Inexplicably, while tap water is regulated by the EPA, bottled water is regulated as a food by the FDA, which historically has expressed little if any concern over food and beverage packaging materials migrating into the human diet.[31]

DEHP is one of six phthalates that are common indoor air pollutants due to extensive PVC use in building materials, furnishings, and consumer products. Several studies have reported indoor air concentrations exceeding levels outdoors. It has been used in furniture upholstery, mattresses, wall coverings, and vinyl flooring—as well as children's mattresses or waterproof mattress covers. Children living in homes that are heavily fragranced may be exposed to DEHP in higher concentrations: Consumers Union tested eight common fragrances for phthalates, some advertised to be free of the compounds, and all contained DEHP. Fragrances are most common in cosmetics, air fresheners, and candles, but they are also added to some foods, toys, and a wide variety of other consumer products such as plug-in air fresheners. The Cosmetic, Toiletry and Fragrance Association has claimed that its member cosmetic companies do not use DEHP; perhaps the chemical leached into the fragrance from plastic containers.[32]

In the United States, DEHP was the most common plasticizer used in soft PVC products intended for children until the early 1980s, when U.S. manufacturers and the Consumer Product Safety Commission (CPSC) reached a voluntary agreement to remove it from toys intended for mouthing such as bottle nipples, teethers, pacifiers, and rattles. But DEHP is still found in toys designed for young children in the United States and abroad: the overwhelming majority of toys purchased in the United States are made in China, a nation that has no restrictions on the use of DEHP.[33]

The chemical is also the most commonly used phthalate in medical devices such as tubing and intravenous bags. Newborns and infants undergoing some medical procedures may have 100 to 1,000 times the exposure experienced by the general population. Many hospitals are switching to plastics that do not contain DEHP, especially in neonatal infant care units, due to concerns about the health of this very vulnerable population.[34]

Several toxicological studies undertaken in the 1970s noted that prenatal exposure of rats to DEHP resulted in higher rates of skeletal malformations and cleft palates, as well as fewer live fetuses at birth. Other researchers in the 1980s reported that prenatal, suckling, and adult rats exposed to DEHP experienced reduced liver function.[35] By 1999, scientists were reporting that DEHP alters sexual differentiation in male rats, producing malformations of the reproductive tract.[36] Numerous subsequent studies on animals reported an association between DEHP exposure and both diminished testicular function and impaired androgen-dependent developmental processes. Others noted that male rodents exposed to DEHP before or shortly after birth exhibit a variety of developmental and reproductive abnormalities, including undescended testicles, reduced anogenital distance, hypospadias, female-like areolas or nipples, and other anatomical differences in infant male rats, as well as decreased sperm production and testosterone levels.[37]

By 2003, associations between DEHP and problems in the human body were also being documented. Researchers from Harvard and the CDC, for example, reported that men with low sperm counts and impaired sperm quality generally had higher phthalate levels, with the highest phthalate concentrations found in those men with the lowest sperm counts. The same investigators in 2007 found that DNA damage to human sperm was associated with DEHP metabolites at levels similar to those reported for the U.S. population at large.[38]

Expanding on results from the animal studies, researchers in 2005 reported a relationship between maternal exposure to phthalates during pregnancy and the distance between the penis and anus in young boys. The mother's prenatal urinary concentrations of several phthalates were measured and compared to the anogenital distance of their infant sons. A significantly shorter distance was reported in boys exposed to higher levels of phthalates during pregnancy. Shorter distances were also associated with an increased proportion of boys with incompletely descended testicles and small genital size.[39] A scientific panel convened by the U.S. National Toxicology Programs, upon reviewing studies like these, concluded that exposure to DEHP can present a risk to the development of the reproductive tract of male infants.[40]

Additional studies report an association between DEHP and respiratory illness, including asthma. Others suggest that DEHP may contribute to allergic reactions in people. Phthalates have been measured in residential air and are common components of air and house dust. A Swedish study found a positive association between allergic asthma in children and DEHP in house dust, noting that DEHP in house dust correlated with the amount of PVC in flooring. Another study related the risk of asthma to the presence of plastic wall materials that can contain 40 percent DEHP. And yet another study suggests that the development of lung problems in the first two years of life may be linked to exposure to plastic interior surfaces. This finding is consistent with other epidemiologic studies involving children living in Norway, Finland, Sweden, and Russia.[41]

A NEED FOR GREATER REGULATION

Given this growing consensus that BPA and DEHP are associated with a variety of adverse health effects, and that nearly everyone is routinely exposed, how have plastics escaped serious attention by the U.S. Congress and regulatory agencies? The answer may seem familiar given these governmental agencies' history concerning the threats of radionuclides, toxic landscapes, pesticides, and air pollution: a lack of premarket testing of health and environmental effects has given many policymakers and consumers an unfounded impression of safety.

The EPA, FDA, Consumer Product Safety Commission, and Occupational Safety and Health Administration all claim they have insufficient information to set environmental standards that would assure health and environmental protection. Each agency, however, has failed to demand essential data from plastics manufacturers, and each has neglected to set enforceable standards to protect the health of children or anyone else. In part, these failures can be blamed on the inadequate Toxic Substances Control Act, which requires manufacturers to submit information they already possess only if EPA requests it. The agency may demand premarket testing, but it faces the burden to demonstrate that new chemicals pose a significant risk before it may require manufacturers to submit additional data. Consequently, the overwhelming majority of chemicals in the marketplace have unknown effects on the environment and on human health.[42]

Despite neglect of the issue by U.S. regulatory agencies, several government research institutes have played important roles in defining the seriousness and extent of the plastics problem. The National Institute for Environmental Health Sciences funded many of the first studies that reported the diversity of health effects described earlier. Similarly, the CDC developed some of the earliest reports of widespread human exposures to chemicals in plastic. And, as noted earlier, the National Toxicology Program has expressed increasing concern over the extent of exposure and the growing literature on health effects.

The primary source of exposure to most people in the United States is food and water. The safety of the U.S. food supply is the responsibility of several agencies, including the FDA, EPA, and USDA. Of these, the FDA has the authority to regulate or ban plastic resins used in food packaging as indirect food additives. But it approved BPA for use in food and beverage containers in the 1960s, claiming the chemical was "generally regarded as safe." The agency reaffirmed this finding in 2007—even though PVC packaging containing DEHP has been restricted in a number of European nations. European restrictions on DEHP in children's products are also tighter than those in the United States. The European Union restricts the use of DEHP in toys and childcare products and has drafted legislation to restrict its use in medical devices. Initially the European Union prohibited DEHP and five other phthalates from soft PVC toys intended to be placed in the mouth by children under the age of three, noting specifically concern about testicular damage. It then broadened the ban to include other childcare articles, including any product intended to aid the sleep, relaxation, hygiene, or feeding of children, as well as any item likely or intended to be mouthed by children.[43]

Two decades ago, the Consumer Product Safety Commission and U.S. toy manufacturers came to a voluntary agreement to restrict concentrations of DEHP in children's pacifiers, teethers, and rattles to 3 percent. But even if this rule were followed to the letter, it still would not protect American children from violations of the limit by foreign manufacturers: enforcing such a concentration limit for all imported products would require testing of millions of shipments to identify violations.

The FDA has not restricted DEHP use in medical equipment, nor has it required that medical equipment containing the chemical be labeled. In response to a plea by the American Medical Association that the FDA require labeling of all medical products containing the compound, the FDA instead published a public health notification in 2002 recommending that plastics containing DEHP not be used for medical procedures that might expose newborn boys, women pregnant with boys, and boys entering puberty.[44]

The United States has no national plastics management policy. Instead it has woven a patchwork quilt of ineffective laws with administrative responsibility fractured among at least four regulatory agencies: the EPA, FDA, Consumer Product Safety Commission, and Occupational Safety and Health Administration. None of these bureaus have demanded premarket testing of plastic ingredients, none have required warnings on plastic products, and none have limited production, environmental release, or human exposure. As a result, the entire U.S. population continues to be exposed to hormonally active chemicals from plastics without their knowledge or consent.

Despite production volumes that have long exceeded hundreds of millions of pounds per year, neither the plastics industry nor the government have devised any rational recycling plan for these products; instead they have cast a blind eye on the mountains of plastic wastes that now clog landfills, incinerators, and the oceans. Consumers who separate their trash often assume plastics will be recycled, while in fact the overwhelming majority is burned or buried, producing a toxic brew that moves through the environment and comes to rest in our bodies.

The history of plastics illustrates the failure of both government and manufacturers to control the chemical industry. The manufacturers have been eager to innovate and sell more chemicals and products, while the government has condoned industry behavior by its regulatory neglect. What should be done? The nation needs a comprehensive plastics policy, just as we have a national pesticide policy, designed to protect both environmental quality and human health. Key elements of a national policy should include tough government regulations that demand premarket testing and prohibit chemicals that do not quickly degrade into harmless compounds. Plastics that do not meet these standards should be heavily taxed and phased out of production. Federal redemption fees for products containing plastics should be set at levels tied to chemical persistence, toxicity, and production volume. These fees should be high enough that consumers have a strong incentive to recycle. Consumers need to be educated about the content and effects of the resins to make responsible choices in the marketplace, so we need mandatory labeling of plastic ingredients and origin. Finally, the chemical industry itself needs to replace persistent and hazardous chemicals with those that are proven to be safe. Green chemistry will hopefully define the future of the industry, but before that occurs, manufacturers should take responsibility for cleaning up residues from the more than one trillion pounds of plastic wastes that they have produced over the past fifty years.

CHAPTER EIGHTEEN

Green Intelligence

I BEGAN THIS BOOK wondering whether government will ever have the capacity and will to protect society from dangerous levels of pollution and commercial chemicals. My primary conclusion is that environmental risks are usually poorly understood by society at large, and neglected by states, corporations, and individuals.

The central problem is widespread public misunderstanding of the presence and danger of chemicals in everyday environments. Innocent acceptance of chemical mixtures into our lives has led to preventable exposures and serious loss of health. The histories of strontium-90, iodine-131, DDT, chlorpyrifos, mercury, diesel exhaust, and plastic resins provide different examples of what has grown to be a pervasive problem. Many will suffer additional damage from poorly tested and carelessly managed chemicals markets unless the laws that govern them are changed in fundamental ways. The last third of the twentieth century produced the first wave of environmental law, reflecting widespread dissatisfaction with many of the problems previously described. The failure of many of these statutes demands a second wave, one designed to create an environmentally intelligent society that is self-conscious of the subtle ways that humanity is transforming the chemistry of the planet and our bodies. Dozens of statutes require fundamental change, but instead of addressing each law individually, we should take a more expansive approach. What principles should become the bedrock of all environmental law, so that we may best understand and manage serious risks? The answers to this question fall into two categories: the gathering of intelligence, and the management of risk.

PRINCIPLES FOR INTELLIGENCE GATHERING

Historically, successful management of dangers to the environment and human health have occurred only after intensive intelligence gathering efforts identified the magnitude and distribution of each problem. The cases described in previous chapters demonstrate that the following types of knowledge are necessary to minimize environmental threats to human health.

*Track the Sources and Movement of Hazardous Chemicals
    and Technologies*

It is essential to know where dangerous chemicals are released into the environment and to track their movement. The task was relatively simple for atmospheric weapons tests; each explosion was clearly identifiable and intensively studied. Vehicle manufacturers, too, are easily identifiable, as are the vehicles themselves. Hazardous landscapes, however, are often disguised. And pesticides pose the greatest challenge because they are marketed as more than 20,000 separate products and released both indoors and outdoors.

The 1976 Resource Conservation and Recovery Act provides one model to account for hazardous chemicals from "cradle to grave"—that is, from production to disposal. This type of identification and tracking system would surely increase the cost of products, but it would also reduce the costs associated with restoring environmental damage and managing health loss among those who unknowingly encounter them. If required, it would also provide an incentive to simplify and reduce the complexity of product ingredients, and encourage the use of ingredients that readily decompose into nontoxic compounds.

Pay Attention to Persistence and Mobility

Chemical persistence and mobility are the primary traits that lead to chronic human exposure. Compounds with high persistence and mobility are especially deserving of strict regulation or bans. The long lifespans of radionuclides, chlorinated pesticides, metals, PCBs, and some solvents make it possible for them to accumulate as they move through the environment and enter food chains, where they can end up settling in human tissues. Nuclear waste provides perhaps the best example of a very long-lasting and mobile compound; it requires a management plan spanning tens of thousands of years.

The Persistent Organic Pollutants Treaty, signed in December 2000 by 122 countries, provides one example of an attempt to identify and phase out chemicals that do not break down. But the problem of persistence must be considered along with an understanding of chemical potency. Methylmercury remains in the human body only between 44 and 180 days, but it still can be extremely hazardous and build up in the body if mercury-containing fish are a regular part of the diet. Bisphenol-A, too, has a half-life of less than a week in the human body and so may seem to pose a low threat, but its detection in nearly everyone examined, as well as its ability to mimic estrogen at exceptionally low doses, make it worthy of strict regulation or a ban.

Chemical persistence often leads to its mobility, explaining why DDT is still in global food chains despite having been banned by most nations nearly three decades ago. Lead is indestructible and paints that contain it have caused nearly a million U.S. children a year to experience unsafe blood levels. And once a building is demolished the debris may be buried or burned, releasing the element into the ground or atmosphere. Chlorinated solvents are often less stable but still are capable of building up in foods consumed by humans. Other chemicals such as some organophosphate insecticides can break down into even more hazardous by-products. For all of these reasons it is crucial to understand chemical persistence and mobility before releasing new compounds into the marketplace.

Determine Where the Chemicals Come to Rest
Understanding persistence and mobility will be meaningless unless we test our environment for the presence of chemical residues. Our perception of environmental purity and contamination is shaped by the vigilance and precision of environmental surveillance. No one suspected that milk might accumulate residues of DDT, strontium-90, lead, dioxins, and PCBs, so for years after these compounds' release into the environment, no one tested milk or food for their presence.

Our misunderstanding of how compounds persist and move through the atmosphere, rain, soils, plants, animals, and eventually humans via the food supply would not have happened if environmental testing were more thorough, or if information had been shared across agencies and fields of study. All four of the histories detailed here included a recognition that mammals' milk concentrated residues. Had independent scientists been given this initial data—and the opportunity to pursue further testing and related questions—human milk may well have been tested, and other means of transmission considered, prior to the introduction of many of these toxins. After all, when the story of human breast milk contamination by radionuclides did break in the 1950s, the public outcry spurred changes that guide regulatory policies even today.

Develop a Thorough Understanding of Exposure
Assume that you know where chemicals are released, their half-lives, tendency to move, and where they are likely to be found in the environment and human tissues—all categories of knowledge just described. None of this information will explain how humans are likely to be exposed, unless you consider very specific patterns of human behavior. For example, no one understood that children's blood lead levels could be predicted by knowing the age and condition of housing, together with the tendency of toddlers to mouth windowsills with flaking leaded paints. For nearly half a century, the EPA neglected to consider how variability in the human diet would lead to predictable patterns of pesticide exposure—just as the AEC neglected to consider that milk intake by children would foretell levels of strontium-90 in their bones. Nor did the government understand that a diet rich in swordfish and tuna would lead to high levels of mercury, detectable in human hair. Today, too, few people understand that as their commuting times increase in urban settings, so does exposure to fine particulates and other components of vehicle exhaust. Before this intelligence existed regulators held simplistic images of uniform behavior and exposure often based on limited data. The ultimate purpose of this book is to suggest how exposure to similar dangers can best be avoided, though unless human behavior is studied with considerable care, neither you nor the government would know how health protection can be assured.

Improve Toxicity Testing

It is clear from recent U.S. and European Union analyses that most chemicals in the marketplace are inadequately tested for possible toxic effects on human health—leaving us without the basic information we need to make regulatory decisions. The EU REACH testing program is an important step toward producing the necessary intelligence to protect environmental quality and health, and importantly, it shifts the burden of proof for safety to the manufacturers themselves. REACH, which loosely stands for the "registration, evaluation, authorisation, and restriction of chemical substances," will require toxicity testing by manufacturers of more than 30,000 compounds.

But implementation of REACH will not alone solve the problem. Toxicity testing must also expand and become more nuanced so that we can understand the delayed effects of acute and chronic exposures. In particular, it seems important to look for possible connections between exposures to pesticides, solvents, metals, and pharmaceuticals and those health disorders that have recently become much more prevalent, including neurological decline among the elderly, childhood developmental disorders, cancers of the reproductive tract (breast and prostate), diabetes, and immune disorders. Although human studies are incomplete, the carcinogenic, neurotoxic, and endocrine-disrupting potential of these compounds has become increasingly apparent in animal and other laboratory tests.

Account for Variations in Susceptibility

Many late-twentieth-century advances in environmental regulation grew from a realization that children were more exposed and susceptible to environmental hazards than adults and so needed special protection under the law. Whether it was the discovery that radionuclides from nuclear weapons testing concentrated in the bones of fetuses and children, that lead from gas and paints induced neurological deficits in children, that mercury emitted from coal-fired power plants and concentrated in fish threatened normal fetal development, or that vehicle emissions cause and exacerbate respiratory illness in the young, the unique susceptibility of children shaped environmental policy in dramatic ways.

Our understanding of these variations must grow to include other at-risk populations. Susceptibility to environmentally induced illness may also exist among the elderly, those with specific genetic traits, and those with underlying illness or disability. More than 20 million Americans have asthma diagnosed by a physician and vehicle emissions are known to exacerbate the intensity and duration of attacks. Mortality increases among those with cardiovascular and respiratory disease following temporary increases in air pollution, and those at risk tend to be middle-aged or elderly. And those of Puerto Rican descent living in the United States have higher asthma rates than do other U.S. ethnic minorities or whites.

The concept of susceptibility should also include the ability to manage health loss. Those who are poor, illiterate, undereducated, and uninsured are less likely to seek routine medical care; instead their illnesses often are left undiagnosed until they become much more costly to treat.

Today, the best way to investigate differences in susceptibility among various populations is to track long-term health trends. A very promising model for this approach is the U.S. National Children's Study. Created by the National Children's Health Act of 2000, the study is designed to follow the health, environmental quality, and behavior of 100,000 children from birth to age twenty-one.[1] The Women's Health Study, too, was designed to understand patterns of vascular diseases and cancer among nearly 28,000 female health professionals, and is producing crucial insights into environmental, behavioral, and genetic factors that contribute to health and illness among women over the age of forty-five.[2]

PRINCIPLES FOR MANAGING RISK

Knowledge of the risks associated with technologies and products normally lags far behind recognition of their benefits. The national security benefits of weapons testing and military training were immediately apparent to government and military officials, while the hazards of nuclear fallout took years to understand. Similarly the benefits of pesticides, which protect crops, reduce vector-borne diseases, and preserve many consumer products, were more readily obvious than their risks.

As we have seen, there has been little incentive to speed up the discovery of risks so that policymakers and consumers can weigh the costs and benefits before using a product or technology. New chemicals and technologies are normally produced, traded, and incorporated into products sold internationally before they are tested to understand their persistence, environmental behavior, and effects on ecological and human health. In addition, consumer products diffuse into international markets at a pace now dictated by electronic transfers of money, goods, and services on the Internet. Protecting the public's health in this context will require changes in both priorities and policies.

Recast the Issue of Intellectual Property Rights
Many U.S. laws protect as trade secrets or confidential business information a wide range of knowledge that could help inform users about the potential dangers of using the product (for example, product ingredients, chemical releases to the environment, contamination levels, and toxicity data). Trade secrecy law, defamation law, and confidentiality provisions of various statutes have thus limited the public's ability to recognize important health hazards in products like building materials, paints, food additives, and genetically modified substances. Although the EPA reports some chemical releases to the environment as required by several statutes, gross releases are difficult to translate into probable levels of exposure and risk, and many chemicals and polluters are exempt from the reporting requirements.[3]

Recommendation: Knowledge of risk should be public, not private, property. We need to amend current laws to establish the public's right to know about any hazardous substance—chemical, biological, and physical—in air, water, food, consumer products, workplaces, schools, recreation areas, and other places, especially those that are frequented by children and women of childbearing age. When individuals report adverse health effects associated with consumer products to manufacturers, this information should immediately be reported to the government and be publicly available.

Improve Surveillance
Without information on where and how chemicals are produced and released, how they move through the environment, and their toxicity, regulators can never know where hazards lie and will be unable to ensure that exposures are safe. Past monitoring efforts have been incomplete, and in particular have produced an insufficient understanding of how exposures are likely to differ across space, time, and among demographic groups—defined by age, income, and ethnicity. Current health surveillance efforts

290 LESSONS LEARNED AND EMERGING THREATS

are also insufficient to detect important childhood illnesses and disabilities that may be associated with environmental hazards.

Recommendation: Laws should be amended to ensure that information regarding environmental hazards is extensive and publically available. In addition, considerable progress should be made immediately in improving surveillance efforts already authorized by laws governing the quality of air, food, water, soils, and consumer products. Surveillance should occur at the scale and frequency necessary to understand and manage children's exposure to hazards. Monitoring should be conducted in schools, play areas, vehicles, and occupational settings to better understand children's exposure to environmental health hazards. National registries should also be developed or expanded for major childhood diseases such as cancer, birth defects, respiratory diseases, and neurological and behavioral disorders including autism. The National Children's Study will provide invaluable information on patterns of exposure, the prevalence of various illnesses, and the relative potency of diverse hazards.

Provide Adequate Fair Warning

Current product labeling practices provide the public with little understanding and warning of consumer product hazards. Product benefits are normally prominent, while hazards are rarely labeled with clarity. Even if hazards are labeled, explanations are often in technical language printed in small type, which can keep consumers from fully grasping the risks. Ingredients often remain unspecified, and percentages or other useful comparisons are rarely provided. Many manufacturers assume that hazards will be manageable if printed directions are followed. Labeling, however, will never be an effective means of managing risks when children, those with limited literacy, or those with inadequate technical competence have access to these products.

Recommendation: Standardize product warnings to alert and educate consumers about product hazards. Fair warning demands unambiguous symbols that are easily recognized and understood.

Set Health-Protective Standards

Most environmental laws do not demand that pollution and contamination standards or limits be set to ensure protection of human health. The Food Quality Protection Act is the one clear exception, demanding that standards be set to assure "a reasonable certainty of no harm." Other statutes permit technological feasibility, economic costs of compliance, and uncertain information to be considered and balanced against health threats as government agencies make choices to limit pollution, contamination, or risk.

Recommendation: Establish health protection as a uniform standard for environmental laws. Amend laws to require that contamination and pollution limits protect the health of children, infants, pregnant women, and other susceptible groups—a change that would require the reevaluation of thousands of existing standards not yet set to protect health. Strict deadlines should be set to achieve health protective standards for air, water, food, soil, and consumer products.

Regulate Chemical Mixtures
Human exposure to mixtures of hazardous chemicals is largely unstudied and unregulated. No laws now limit the cumulative cancer, neurological, or respiratory health risks that may be posed by mixtures of chemical compounds moving through the environment; instead these threats are permitted to accumulate chemical-by-chemical from contaminated air, water, food, soils, and consumer products.

Recommendation: Ensure that exposures to chemical mixtures, especially children's exposures, are managed to prevent additive or synergistic effects. An important and immediate first step would be to reduce the use and release of chemical mixtures at sites where children spend most of their time: homes, schools, recreational areas, and vehicles such as cars and school buses.

Inform the Public about Landscape Hazards
The public is rarely informed of the location of facilities where hazardous compounds are or were developed, tested, or stored—or where accidental contamination has occurred; in fact, lands and industrial sites that pose significant threats to human health are often disguised. These areas may include solid and hazardous waste repositories, underground storage tanks, former agricultural lands, contaminated aquifers, industrial sites, military facilities, and transport and power corridors.

Recommendation: Centralize current and historical information on facilities, lands, and aquifers that have a history of contamination by creating a national registry of sites that are known to be hazardous. Establish clear criteria for which sites should be listed and how the information should be presented, and include all known sources of emissions on the registry. If followed, these recommendations could guide parental choices concerning where to live, send children to school, and play. This knowledge also would improve current residents' potential to influence those who own or manage hazardous properties and encourage real estate market values to more truly reflect environmental quality and contamination.

Improve Transparency

The EPA possesses knowledge about environmental hazards that it chooses to keep from the public. Such secrecy interferes with the public's ability to participate effectively in decisions about exposures, especially when these decisions rest on detailed scientific or technical information. In 2002 the EPA began removing important information files from its websites and closed some of its libraries. Although the agency claimed that it had to make these changes due to budgetary cuts, the effect has been to impede public access even more and to further centralize the agency's decision-making authority.

Recommendation: Congress should require the EPA to demonstrate how its restriction of access to this essential information serves the public interest. Some information will need to be kept secret—for example, if its disclosure would pose obvious and direct threats to national security, reveal exclusively held patent information, or harm someone's economic or physical security. But in no case should the government withhold knowledge of hazardous conditions from those exposed.

Use Precaution When Data Are Uncertain

Few laws demand the use of precautionary policy, or safety factors, when setting allowable emissions or contamination limits, or levels of acceptable exposure. Two exceptions include the Food Quality Protection Act and the Clean Air Act, which both require that safety margins protect against underestimates of health threats when information is uncertain. The history of toxicity testing, however, suggests that government scientists and regulators have underestimated both chemical toxicity and children's exposure. Once fully tested, few chemicals are found to be less toxic than originally believed.

Recommendation: Regulations that limit exposures should assure the protection of fetuses, infants, and children with an adequate margin of safety. The burden of proving safety should rest with those who control the technology. Until credible evidence of safety is produced, cautious assumptions should be applied to set standards for allowable concentrations in air, water, food, consumer products, and those environments in which children live and play—homes, schools, recreational facilities, and vehicles.

Encourage the Development of Healthier Alternatives
The field of "green chemistry" seeks to design less-hazardous compounds for industrial and commercial applications. These efforts have grown in part from manufacturers' attempts to design less persistent and safer pesticides and pharmaceuticals, a trend that began early in the twentieth century. The search for chemicals with no or low hazard is essential to create a more sustainable environment, but the search will likely be gradual and lead to innovations that could reduce exposures over decades.[4] Recent innovations are promising, but they are not likely to reduce our exposure to the thousands of chemicals we normally encounter in outdoor and indoor environments until far in the future. In addition, great care needs to be taken to avoid introducing yet another incompletely tested technology that may pose unanticipated risk. An example with enormous implications for human exposure to chemicals is "nanotechnology," or the design of ultrafine molecules that may, for example, improve the efficiency of drug delivery, reduce friction, and increase water repellency. Improvement of functionality, comfort, or aesthetics may be important, but history teaches us that the extraordinarily small size of these molecules should make us worry about their movement and distribution through the environment, as well as human tissues. In other words, the pursuit of green chemistry is not the only solution we need to encourage to reduce our exposure to chemical hazards. Our society needs near-term protective strategies now.

Recommendation: Public and private organizations should invest heavily in the search for safer compounds and technologies that would reduce human exposure to dangerous substances. Congress should empower the EPA to manage the introduction of new chemicals with great care to ensure that older, riskier technologies are replaced by those that are safer. And nanotechnology should be regulated to ensure that exposure to dangerous molecules does not increase.

For too long our society has been left to fend for itself to judge the safety of outdoor and indoor environments. Corporations churn out thousands of new products and chemicals each year, advertising their functionality, convenience, efficiency, aesthetic and health-promoting effects, durability, and reasonable cost. Rarely do we hear about their ingredients, hazards, or eventual fate as waste. Public discourse, too, is dominated by the private sector, which has effectively shaped our values, our consumer behavior, and how we assess risk. The effect has been the creation of more chemicals, pollution, and waste than ever before in human history, which in turn has accelerated a chemical transformation of the planet and our bodies, as well as a growing incidence of human illness associated with degraded environments.

Staying the course would mean continuing our exposure to poorly evaluated chemical mixtures long into the future. Effective protection will require far tougher regulations, informed and proactive consumers, and recognition by corporations that environmental responsibility is profitable. All of these changes need to be guided strategically by the types of green intelligence just described that would make chemical dangers much easier to recognize, contain, and avoid. Without this intelligence, the global chemical experiment on public health will remain wildly out of control.

THE LEGAL REFORMS outlined earlier describe how our nation could provide greater protection for all children and their families. Progress on these initiatives, however, is likely to be incremental. Meanwhile, there are ways you can reduce your personal exposure to some dangers by strategically changing what you buy, how you use it, and how you live your daily life.

Your exposure to contaminants depends not only on the presence of hazardous chemicals in your environment, but also on your behavior. Consuming large quantities of just a few types of food, indiscriminately using pesticides, commuting in congested traffic, and driving a diesel vehicle are just a few examples of activities that could put you in contact with a predictable mixture of risky compounds. Your best personal defense begins with an assessment of your daily habits and an inventory of the environments you frequent, that is, your home, a school or workplace, recreational areas, and any vehicles you drive or ride in.

Most people encounter at least some environmental contaminants in their diet, water, air, and consumer products; the advice that follows, then, focuses on how you might refine your behavior in these areas to reduce your exposure and risks. These broad guidelines and suggestions are meant to apply primarily to residential settings; however, they could be adapted easily for educational, occupational, recreational, and transport environments where children spend their time.[1]

DIET

- Breastfeed your babies. Each of the histories presented earlier demonstrates the accumulation of a different contaminant in human breast milk. Yet the American Academy of Pediatrics recommends that breastfeeding provides the healthiest milk for babies. There is also some evidence that breastfeeding reduces the risks of asthma and cancer, and may increase the pace of intellectual development. The nutritional, psychological, and immune system benefits of breastfeeding are well established.
- Drink eight glasses of water a day to promote a healthy body-weight, to ensure hydration, and to help flush hazardous chemicals from your body.
- Eat low on the food chain and include many fresh fruits and vegetables. Consuming a diet rich in fruits, vegetables, and grains provides many nutritional benefits, and offers protection from other types of health problems, including colon cancer and cardiovascular disease.
- Avoid animal fats. Reduce your intake of the persistent chemicals that may accumulate in the fat of fish and animals by steering clear of these fats. Drink nonfat milk; eat lean meat, poultry, and fish; and trim away fat from meats before cooking.
- What foods do your children eat most? Try to buy these from a certified organic grower. Purchase basic and raw organic foods rather than those that are processed and include multiple ingredients. Remember that children who consumed an organic diet had lower residue levels of numerous pesticides in their urine than those eating conventional foods. The National Organic Standard is not perfect, but certified products will reduce your toxic chemical body burden.
- Try limiting your food purchases to items with fewer than five ingredients. If you cannot pronounce any ingredient, or do not understand it, avoid it. Be skeptical of the word "natural" in food labeling, because it does not guarantee that the product is free from synthetic chemicals.
- Encourage your supermarket manager to stock foods that are certified to be pesticide-free. As more people buy these foods, prices will continue to fall as they have during the past decade.

- Be sure to wash your fresh fruits and vegetables before eating them. Check to see if fruits and vegetables are coated in wax, which may encase pesticide residues. Peeling the waxed skin can limit your exposure.
- Enjoy fish, but learn which species are likely to contain methyl-mercury (generally the larger predatory fish), and especially avoid these if you are a woman of childbearing age. Pregnant or nursing mothers and young children should strictly limit or avoid shark, swordfish, king mackerel, tuna, and tilefish. Shrimp, scallops, salmon, and tilapia, by contrast, have demonstrated the lowest levels of mercury when tested by the FDA.

DRINKING WATER

- Filter your water to reduce levels of any metals or organic pollut-ants, which are a special risk to children and pregnant women. Many tap filters are both inexpensive and effective. If you use a filter, follow replacement guidelines carefully to avoid excessive bacterial buildup.
- Avoid bottled water if possible and remember that urban supplies are the best-tested sources of drinking water in the nation. If your water is supplied by a public system, your local water supplier is obligated to provide a list of the chemicals it tests for in your water, the results, and a description of any water treatment it employs.
- Avoid using hot water from the tap for cooking or drinking. Lead leaches more easily from hot water lines, so be certain never to mix infant formula with hot tap water.
- Let cold water run for several minutes in the morning before consuming it. Depending in part on its acidity or alkalinity, water that sits in pipes for several hours may accumulate chemicals that can leach from pipes and fixtures.
- Test your water for lead. Testing is especially important for apart-ment dwellers, because flushing (letting the water run before consuming it) may not be as effective in high-rise buildings with lead-soldered central piping. Older buildings are more likely to have water lines and fixtures containing lead than those built after 1980.

- If your water is drawn from a private well, and you suspect contamination, have your water tested for pesticides, metals, and other possible contaminants by a state-certified laboratory. If hazardous chemicals are detected, contact your local health agency to understand whether the concentrations pose a health threat.
- If you rely on a neighborhood or community well, try to agree with your neighbors to avoid using pesticides and fertilizers in the vicinity of the supply source—generally within one-quarter mile (although pesticides have migrated farther both in surface- and groundwater).
- Consider the land-use history of your home. If you have a well and live next to farmland, on former farmland, or near industrial, commercial, or institutional facilities, you may be at risk. The major contaminants of concern for children are metals, fertilizers, pesticides, fuels, and solvents.

AIR QUALITY

- In most locations, outdoor air is cleaner than indoor air. Opening windows and doors will often improve indoor air quality.
- If you live near a highway, industrial or manufacturing facility, airport, power plant, bus depot, incinerator, or an agricultural area, you may experience higher outdoor than indoor pollution. Learn about the times when emissions are likely to be highest and avoid ventilation during these periods.
- Do not permit smoking in your home, vehicles, or office.
- Ventilate while cooking. If you have a gas stove, use an overhead range hood that vents outdoors when you cook. Remember that burning food often produces carcinogenic substances on the food's surface, and releases high concentrations of low-diameter particulates to the air.
- Inspect furnaces, gas water heaters, and clothes dryers regularly to ensure proper ventilation. Never use a kerosene heater without outdoor ventilation.
- Woodstoves and fireplaces can be important sources of indoor air pollution if not well ventilated. If you use a wood stove or boiler, be certain it is well vented, and preferably is fitted with a catalytic

converter. Avoid using these heating devices during still and humid days when chimney smoke will settle to the ground and increase in concentration. If you smell smoke, you have a problem.

- Clean air conditioners, humidifiers, and heat exchangers semi-annually to avoid a buildup of mold and bacteria.
- Test your home to determine if radon is a problem. If it is, consult your local health officials about ventilation techniques.
- Be certain to have fire alarms and carbon monoxide detectors installed and maintain their batteries.
- To reduce vehicle exhaust in your home, do not keep your car running while in a garage, and avoid backing into attached garages. In addition, try not to store fuels in your garage because fumes typically evaporate from most containers. This is more important if bedrooms are situated overhead.
- Limit the amount of carpet used in the home, because it harbors chemical residues, molds, and animal dander, and is difficult to clean well. Hard, smooth surfaces are usually preferable for those with respiratory illnesses such as asthma.
- When purchasing new carpeting, if possible, find a place to unroll it and air it out for at least several days before installing it.
- Provide maximum ventilation when painting or using solvents or strong cleaning solutions. Use water-based (low or zero VOC) paints, wood finishes, and sealants to avoid breathing indoor fumes following application.
- Minimize use of air fresheners, fragrances, and deodorizers.
- Try to schedule your painting, floor refinishing, and other renovations during seasons when you can keep windows open to ventilate fumes and dusts.
- Pregnant women and young children should avoid exposure to home renovation and construction areas.
- Be certain to store fuels, automotive supplies, and solvents in childproof containers—well out of reach of little hands.
- If you or your children have asthma or other respiratory problems, pay attention to daily air-quality forecasts for ozone and particulate levels. Avoid intensive exercise outdoors during periods of high air pollution. Ozone levels usually peak between

midday and evening. Also avoid intensive exercise next to heavily traveled highways.

- If you live adjacent to farmland, golf courses, or recreational areas or have neighbors who routinely apply pesticides, learn about the timing of applications. Be certain to keep children and pets inside, and close your windows and doors.

- Think carefully about where you live, and learn the land-use history of your home, school, workplace, and recreational area with air quality in mind. Have toxic chemicals or waste ever been stored nearby? Is the property former agricultural land? Which industries were located in your area, and are there today? If possible, avoid settling on or very near former industrial and commercial properties.

CONSUMER PRODUCTS

- Avoid using pesticides, especially indoors. Residues may persist long after homes are sprayed. If you have a nuisance insect problem, try to solve it physically rather than chemically. Check screening of doors and windows to prevent insect entry. Try to minimize food crumbs, spills, and scraps that attract insects on floors, counters, and cabinets. Check houseplants for bugs and wash them off instead of spraying plants with pesticides.

- Use "low-VOC" or "no-VOC" paints. Read the label on your paint can. One gallon of oil-based paint may contain five pounds of volatile organic compounds that will be released into the air after the paint is applied. Water-based paints contain approximately a third of this amount. Even better, "no-VOC" paints are now available and commonly used in hospitals and nursing homes.

- Use natural cleansers. They commonly perform as effectively as harsh detergents and solvents. Try cleaning using mild detergents and water before employing chemical and antibacterial products.

- Use nontoxic toys and art supplies. Young children should be allowed to use only those art supplies that have a nontoxic label. The Art and Creative Materials Institute (ACMI) certifies that art supplies do not contain materials in sufficient quantities to be toxic or injurious to humans or to cause acute or chronic health problems.

- Mercury-containing products, such as thermometers and compact fluorescent light bulbs, should be handled and disposed of with great care, preferably at hazardous material collection sites. Spilled mercury should never be vacuumed or touched.
- Solid wood furnishings are preferable to particle and fiberboard, which contain adhesives and preservatives that may give off fumes indoors. Some particle board contains formaldehyde, a chemical that irritates the respiratory tract of some, can cause asthma, and is classified as "reasonably anticipated to be a human carcinogen" by the National Toxicology Program.

Changing your chemical environment will not be easy given the current lack of reliable information about specific hazards. Still, self-education, vigilance, and common sense can lead you to reduce your exposure to toxic substances as you live your daily life. Moreover, the choices you make will have a ripple effect. When enough shoppers begin to select safer goods, manufacturers will respond to this demand by creating products that are less harmful to human health and the environment. When enough people make choices about where they live, what they buy, and how they travel with environmental issues in mind, businesses and governments will adapt. Hopefully, too, the process of contemplating your exposures to hazardous chemicals will encourage you to press politically for more openness and public debate within those government agencies responsible for keeping us all from harm. By thinking historically, ecologically, causally, critically, and imaginatively, we all can develop the green intelligence necessary to navigate more safely among the often-imperceptible chemical dangers in our environment.

ACKNOWLEDGMENTS

THIS BOOK could not have been completed without the help of numerous colleagues, students, friends, family, and dozens of others who were willing to answer my inquiries and offer guidance. I especially wish to thank those at Yale University Press who have supported this project, including Jean Thomson Black, Jenya Weinreb, Tina Weiner, Sarah Clark, Margaret Otzel, and Julie Carlson.

I also want to acknowledge the many academic and public health doctors who provided inspiration, including Michelle Bell, John Blanton, Barry Boyd, David Brown, Mark Cullen, Durland Fish, Richard Jackson, Philip Landrigan, Robert LaCamera, Brian Leaderer, Herbert Needleman, Fredrica Perrera, Ellen Silbergeld, Lynn Goldman, Theodore Holford, Hugh Taylor, and Joseph Warshaw.

There were many colleagues at Yale who were always ready to listen and provide alternative interpretations of uncertain information. Special thanks in this regard go to William Burch, Garry Brewer, Paul Anastas, Gus Speth, James Scott, Karl Turekian, Jeffrey Park, Ron Smith, Daniel Kevles, and William Seagraves. Several others who have had a strong influence on my thinking include Sheila Jasanoff at Harvard, and Daniel Krewski.

I benefited greatly from the Ph.D. students who have worked with me. In particular, Matthew Auer, Eva Cuadrado, Carlos Gonzales, Jeffrey

Albert, Judith Pongsiri, Achim Halpaap, Sanjay Baliga, Philip Johnson, Laura Robb, and Sarah Smiley Smith challenged my thinking and made the book stronger. Yale College students who helped with research and provided important insights include Marina Spitkovskya, Laura Hess, and Thomas Santoro.

Others who have regularly provided criticism and advice include Wendy Gordon, Michael Lerner, Charlotte Brody, Nancy and Jim Chuda, Tessa Hill, Elizabeth Sword, Nancy Alderman, Ken Cook, Richard Wiles, Frances Beinecke, John Peterson Myers, and Timothy and Claire Barnett.

I am indebted to my friends and colleagues at Environment and Human Health, Inc., including Nancy Alderman, Susan Addiss, Barry Boyd, David Brown, Mark Cullen, Robert LaCamera, Hugh Taylor, Linda Wargo, and Jane Bradley. The chapters on air quality began as EHHI reports that I coauthored with Mark Cullen, David Brown, Nancy Alderman, and Linda Wargo. The chapter on plastics both summarizes and expands on a report that I coauthored with Hugh Taylor, Linda Wargo, Mark Cullen, and Nancy Alderman.

Charts and tables for the book were expertly prepared by Jane Bradley of Capservices, Inc. Maps were prepared with the exceptional help of Abraham Parrish, director of Yale's Sterling Map Library, who devised a historical geographic information system that led to many of the insights discussed in the nuclear weapons and military testing chapters. Stacey Maples, also of the Sterling Map Library, developed final maps with great skill and patience.

Colleagues in Puerto Rico who helped me to understand the history of the U.S. Navy's relationship with the territory and the environmental and health effects of military operations there include Carmen Ortiz-Roque, Biff Browning, Radamez Tirado, Roberto Rabin, Myrna Pagan, Zaida Torrez, Carlos Ventura, and Carlos Zenón.

Finally, I am deeply grateful for the indulgence, patience, and expert research advice from my wife, Linda Wargo, and our children, Adam, Kate, and Ellie.

The following abbreviations have been used in the notes:

ACHRE   U.S. Department of Energy, Advisory Committee on Human
                    Radiation Experiments
AEC   Atomic Energy Commission
AFSWP   Armed Forces Special Weapons Project
ATSDR   Agency for Toxic Substances and Disease Registry
CDC   Centers for Disease Prevention and Control
CFR   Code of Federal Regulations
DDRS   Declassified Documents Reference System
DOD   U.S. Department of Defense
DOE   U.S. Department of Energy
EPA   Environmental Protection Agency
FDA   Food and Drug Administration
GAO   Government Accountability Office
NCI   National Cancer Institute
NIEHS   National Institute of Environmental Health Sciences
NRC   Nuclear Regulatory Commission

PROLOGUE

1. K. Sexton, L. L. Needham, and J. L. Pirkle, "Human Biomonitoring of
   Environmental Chemicals," *American Scientist* 92 (2004): 38–45; see also
   http://www.cdc.gov/exposurereport.
2. U.S. Bureau of the Census, "Demographic Trends in the Twentieth
   Century," 2002, http://www.census.gov/prod/2002pubs/censr-4.pdf.
3. The U.S. Environmental Protection Agency estimates that 5,400 chemicals

are produced at 10,000 pounds per year, and nearly 30,000 are produced in quantities exceeding 2,200 pounds per year. See EPA, "High Production Volume (HPV) Challenge: Six Basic Tests Are Necessary for a Minimum Understanding of a Chemical's Hazard—Acute Toxicity; Chronic Toxicity; Developmental and Reproductive Toxicity; Mutagenicity; Ecotoxicity; and Environmental Fate," December 2008, http://www.epa.gov/HPV. See also EPA, Office of Pollution, Prevention and Toxics, "Chemical Hazards Data Availability Study: What Do We Really Know about the Safety of High Production Volume Chemicals?" April 1998, http://www.epa.gov/HPV/pubs/general/hazchem.htm.

4. EPA, "High Production Volume (HPV) Challenge"; GAO, *Chemical Regulation: Options Exist to Improve EPA's Ability to Assess Heath Risks and Manage Its Chemical Review Program,* June 2005, http://www.gao.gov/new.items/d05458.pdf. The European Union is facing a similar conundrum but in 2006 responded with more aggressive demands for chemical testing. It adopted a law known as REACH (Registration, Evaluation, Authorization and Restriction of Chemicals) that requires toxicity and environmental fate testing for 1,500 of the 100,000 chemicals used or traded among member nations. REACH established the European Chemicals Agency, and its primary purpose is to restrict chemicals of greatest concern, including 1,500 compounds produced at high volume believed to pose the most significant threat to health. Criteria for inclusion on the priority list remain highly debated. For example, chemicals that are hormonally active and mimic or block estrogens or androgens have been excluded from consideration until 2012. The REACH program should produce data over the next several decades to encourage substitution of less risky for more risky chemicals, meaning that real risk reduction will still be a generation away, and only for a small subset of chemicals already in the marketplace. See European Commission, *REACH in Brief,* October 2007, http://ec.europa.eu/environment/chemicals/reach/pdf/2007_02_reach_in_brief.pdf. The United States has no similar effort under way, with the exception of an "endocrine disruptor screening program" that produced little regulatory action more than a decade after its requirement by the Food Quality Protection Act of 1996.

5. James Gustave Speth, *The Bridge at the Edge of the World: Capitalism, the Environment, and Crossing from Crisis to Sustainability* (New Haven: Yale University Press, 2008).

CHAPTER 1. PERFECTING THE ART OF TERROR

1. Jack Niedenthal, *For the Good of Mankind: A History of the People of Bikini and Their Islands,* 2d ed. (Majuro, M.H.: Bravo Press, 2001); see also Ruth Levy Guyer, "Radioactivity and Rights Clashes at Bikini Atoll," *American Journal of Public Health* 91, no. 9 (2001): 1371–1376.

2. Jonathan Weisgall, *Operations Crossroad: The Atomic Tests at Bikini Atoll* (Annapolis, Md.: Naval Institute Press, 1994), 415.

3. Stafford Warren, "The Role of Radiology in the Development of the Atomic Bomb," in Kenneth D. Allen, ed., *Radiology in World War II* (Washington, D.C.: U.S. Surgeon General's Office, 1966); see also AEC, "Staff Report on Bikini Atoll Environmental Levels of Plutonium," DDRS, DOE/EH-0471, 1970, https://www.osti.gov/opennet/advancedsearch.jsp.

4. Oklahoma Geological Survey Observatory, "Catalog of Known and Putative Nuclear Explosions from Unclassified Sources," http://nuclearweapon archive.org/library/catalog; see also http://nuclearweaponarchive.org/usa/tests/index.html.

5. Lee DuBridge, "What about the Atomic Tests?" *Bulletin of the Atomic Scientists* (May 15, 1946): 7, 16; see also Lloyd J. Graybar, "The 1946 Atomic Bomb Tests: Atomic Diplomacy or Bureaucratic Infighting?" *Journal of American History* 72, no. 4 (March 1986): 888–907.

6. Barton Hacker, *Elements of Controversy* (Berkeley: University of California Press, 1994), 57.

7. AEC, "Radioactive Fallout Hazards from Surface Bursts of Very High Yield Nuclear Weapons," AFSWP technical analysis report no. 507, May 1954, declassified December 13, 2000, National Security Archive, George Washington University, http://www.gwu.edu/~nsarchiv/NSAEBB/NSAEBB94/.

8. Samuel Glasstone and Phillip J. Dolan, eds., *The Effects of Nuclear Weapons*, 3d ed. (Washington, D.C.: DOD and DOE, 1977), 437.

9. AEC, "Radioactive Fall-Out Hazards." A 10-megaton device was detonated on Enewetak Atoll in November 1952, but only "crosswind and upwind fallout data were obtained."

10. Vitaly I. Khalturin et al., "A Review of Nuclear Testing by the Soviet Union at Novaya Zemlya, 1955–1990," *Science and Global Security* 13 (2005): 1–42; see also William Burr and Hector L. Montford, eds., *The Making of the Limited Test Ban Treaty, 1958–1963*, National Security Archive, George Washington University, 2003, http://www.gwu.edu/~nsarchiv/NSAEBB/NSAEBB94/#48. P. L. Podvig, ed., *Russian Strategic Nuclear Forces* (Cambridge: MIT Press, 2001), 620.

11. Defense Nuclear Agency, "Analysis of Radiation Exposure—Service Personnel on Rongerik Atoll," (Washington, D.C.: DNA 001-85-C-0101, July 1987), http://www.dtra.mil/documents/rd/DNATR86120.pdf.

12. R. A. Conard et al., *BNL 908 T-371: Medical Survey of the People of Rongelap and Utirik Islands Nine and Ten Years after Exposure to Fallout Radiation* (Springfield, Va.: U.S. Dept. of Commerce, National Bureau of Standards, Clearinghouse for Federal Scientific and Technical Information, 1965). See also E.P. Cronkite, *BNL 384 T-71: Twelve-Month Postexposure Survey on Marshallese Exposed to Fallout Radiation* (Upton, N.Y.: Brookhaven National Laboratory, 1955).

13. G. Johnson, "Micronesia: America's Strategic Trust," *Bulletin of the Atomic Scientists* (February 1979): 51.
14. "Report on H-bomb Hidden Three Months," *New York Times*, March 25, 1955.
15. Ibid.
16. U.S. Congress, Joint Committee on Atomic Energy, *The Nature of Radioactive Fallout and Its Effects on Man: Summary Analysis of Hearings May 27–29 and June 3–7, 1957*, 85th Cong. (Washington, D.C.: U.S. Government Publishing Office, 1957).
17. J. Belcher, "Bikini Islanders Lose Again to Radiation," *Los Angeles Times*, July 23, 1978.
18. Ibid.
19. W. L. Robison, W. A. Phillips, and C. S. Colsher, "Dose Assessment at Bikini Atoll," June 8, 1977, Lawrence Livermore Laboratories, prepared for U.S. Energy Research and Development Administration, http://www.hss .doe.gov/healthsafety/env_docs/200108.pdf.
20. U.S. Congress, "Effect of Radiation on Human Health," Hearings before the Select Committee on Health and the Environment of the Committee on Interstate and Foreign Commerce, House of Representatives, 96th Cong., 2d sess., vol. 1.: January and February 1978, serial no. 95-179, and vol. 2: July 1978, serial no. 95-180 (Washington, D.C.: U.S. Government Printing Office, 1979).
21. Johnson, "Micronesia," 14.
22. "Assuming that they had all been there since 1970 and received the average estimated integrated total dose of 2.6 rems for the period, based upon known radiation-induced risk data, one would expect only 0.005 total cases of leukemia to develop in that population as a result of their radiation exposure. The need for further medical examinations is not indicated based on possible radiation effects associated with such low doses." R. Conrad, *Radiological Status of the Bikini People: A Summary Report* (Upton, N.Y.: Brookhaven National Laboratory, 1978).
23. J. Weisgall. "The Nuclear Nomads of Bikini," *Foreign Policy* 39 (Summer 1980): 74–98.
24. The Defense Nuclear Agency established cleanup levels: 40 picograms of plutonium 238/240 per gram of soil (pg/g) were allowed for residential islands, 80 pg/g on agricultural islands, and 160 pg/g on food-gathering islands.
25. "Enewetak Islanders Plan Return," *Honolulu Star Bulletin*, September 26, 1979; "Problems Unresolved on Enewetak Atoll," *Honolulu Star Bulletin*, March 26, 1980.
26. G. Johnson, "Paradise Lost," *Bulletin of the Atomic Scientists* (December 1980), quoting Rosalie Bertell.
27. Ibid., 28.
28. Additionally, four underwater atomic explosions sent millions of additional tons of radioactive seawater, fish, coral, sand, and plant life miles into the sky. Three atmospheric tests were conducted in the South Atlantic Ocean.

29. W. Pincus, "Bikinians Must Quit Island for at Least Thirty Years, Hill Told," *Washington Post,* May 23, 1978; see also Congressional Research Service, Report to Congress, "Republic of the Marshall Islands Changed, Circumstances Petition to Congress," March 14, 2005, http://digital.library .unt.edu/govdocs/crs/data/2005/meta-crs-7588.tkl.

CHAPTER 2. THE STRONTIUM-90 ODYSSEY

1. Opening comments of Dr. John Bugher, AEC, Division of Biology and Medicine, in "Meeting Transcript: Biophysics Conference," January 18, 1955, National Archives, Washington, D.C., 1.

2. "Worldwide Effects of Atomic Weapons," Project Sunshine, R-251-AEC (amended), Project Technical Information Services, Department of Commerce: AECU-3488, August 6, 1953, http://www.osti.gov/energycitations/ product.biblio.jsp?osti_id=4346140. This study, conducted at the RAND Corporation in Santa Monica, California, in the summer of 1953, launched Sunshine Project Technical Information Services, which was part of the Department of Commerce.

3. "Wild Bill of the Atom: Willard Libby," *New York Times,* June 8, 1957.

4. Comments of W. Libby, AEC, Division of Biology and Medicine, in "Meeting Transcript: Biophysics Conference," January 18, 1955, National Archives, Washington, D.C., 8.

5. R. A. Dudley to Gertrude Steel, October 16, 1953, National Security Archive, George Washington University, http://www.gwu.edu/~nsarchiv/ radiation/dir/mstreet/commeet/meet11/brief11/tab_i/br11i1c.txt.

6. AEC, Division of Biology and Medicine, "Meeting Transcript, Biophysics Conference," January 18, 1955, National Archives, Washington, D.C., 2; National Security Archive, George Washington University, http://www .gwu.edu/~nsarchiv/radiation/dir/mstreet/commeet/meet15/brief15/tab_d/ br15d2a.txt.

7. Bugher opening comments, "Meeting Transcript: Biophysics Conference," 190–191. "I don't think that in other areas where one wants human material collected you have ever been successful unless you had someone locally responsible who got paid by the sample. Over the years in the Rockefeller Foundation in the yellow fever studies, we got tens of thousands of liver specimens all through Latin America by that scheme. Of course, we lost a few agents, too, who got shot or knifed, but not very many actually."

8. E. B. Lewis, "Aspects of Somatic Effects of Fallout Radiation," statement in U.S. Congress, *Fallout from Nuclear Weapons Tests,* Hearings before the Special Subcommittee on Radiation of the Joint Committee on Atomic Energy, 86th Congress, 1st sess., May 5, 6, 7, and 8, 1959 (Washington, D.C.: U.S. Government Printing Office, 1959), vol. 2, 1553.

9. L. J. Kulp, W. R. Eckelmann, and A. R. Schulert, "Strontium-90 in Man," *Science* 125 (1957): 219–225.

10. Responsibility for monitoring the fallout was then transferred to the U.S.

Public Health Service until 1970, when the Environmental Protection
Agency took over the role.

11. J. L. Kulp and A. R. Schulert, *Strontium-90 in Man and His Environment*
(Palisades, N.Y.: Geochemical Laboratory, Lamont Geological Observatory,
May 1962).

12. Kulp, Eckelmann, and Schulert, "Strontium-90 in Man," 219–225.

13. E. Gamerekian, "A-Fallout Piling Up in U.S. Soil," *Washington Post,*
March 3, 1959.

14. "Atom Blast Sweeps 250 miles," *New York Times,* July 16, 1957; Health and
Human Services, "Transcript of Meeting on Statistical Considerations on
Field Studies on Thyroid Diseases in School Children in Utah-Arizona,"
ACHRE no. HHS-022395-A, December 3, 1965 (Washington, D.C.: U.S.
Department of Energy, Office of Human Radiation Experiments, 1995). See
also H. M. Parker, "Radiation Exposure from Environmental Hazards,"
United Nations International Conference on the Peaceful Uses of Atomic
Energy, August 1955, reprinted in Ronald L. Kathren, Raymond W. Baal-
man, and William J. Bair, eds., *Herbert M. Parker: Publications and Other
Contributions to Radiological and Health Physics* (Richland, Wash.: Battelle
Press, 1986), 494–499.

15. E. Gamarekian, "Serious Fallout Cases Uncovered in Middle West,"
Washington Post, June 7, 1959.

16. E. W. Pfeiffer, "Some Aspects of Radioactive Fallout in North Dakota,"
North Dakota Quarterly (Autumn 1958): 93–99.

17. U.S. Congress, *Fallout from Nuclear Weapons Tests,* Hearings before the
Special Subcommittee on Radiation of the Joint Committee on Atomic
Energy, 86th Cong., 1st sess., May 5, 6, 7, and 8, 1959 (Washington, D.C.:
U.S. Government Printing Office, 1960).

18. "Lack of Data Noted on Fallout Effect," *New York Times,* November 1, 1956.

19. Edwin Diamond, "AEC Fallout Estimates Upset by New Evidence,"
Washington Post and *Times Herald,* May 23, 1957.

20. U.S. Congress, *The Nature of Radioactive Fallout and Its Effect on Man,*
Hearings of the Joint Committee on Atomic Energy, 85th Cong. (Washington, D.C.: U.S. Government Printing Office, May–June 1957 and May 1959),
4 vols.

21. L. J. Kulp, "Worldwide Distribution of Strontium-90 and Its Uptake in
Man," *Bulletin of the Swiss Academy of Medical Sciences* 14 (1958): 419–433;
Kulp, Eckelmann, and Schulert, "Stronium-90 in Man," 219–225.

22. U.S. Congress, Hearings before the Special Subcommittee on Radiation of
the Joint Committee on Atomic Energy, 86th Cong., 1st sess.: (a) May 1957
hearings on fallout; (b) May 1959 hearings on fallout; (c) June 1959 hearings on the biological and environmental effects of nuclear war (Washington, D.C.: U.S. Government Printing Office, 1957–1960).

23. Comments by W. Libby in "Meeting Transcript: Biophysics Conference."

24. A. R. Shulert, "Strontium-90 in Alaska," *Science* 136 (April 1962): 146–148;

also in Kulp and Schulert, *Strontium-90 in Man and His Environment*, vol. 1: *Summary*.

25. L. J. Kulp, A. R. Shulert, and E. J. Hodges, "Strontium 90 in Man III: The Annual Increase of This Isotope and Its Pattern of Worldwide Distribution in Man Are Defined," *Science* 129, no. 3358 (1959): 1249–1255.

26. U.S. Congress, Joint Committee on Atomic Energy, *Report* (August 24, 1959).

27. White House, Memorandum of Conference with the President, June 24, 1959, DDRS CDROM Id: 1978070100292.

28. Kulp, Shulert, and Hodges, "Strontium 90 in Man III."

29. Kulp and Shulert, *Stronium-90 in Man and His Environment*, vol. 1: *Summary*.

30. M. Welsh et al., "Penicillin Blood and Milk Concentrations in the Normal Cow Following Parenteral Administration," *Science* 108 (1948): 185–187.

31. AEC, Division of Biology and Medicine, *Report on Project Gabriel*, July 1954, 18; http://www.gwu.edu/~nsarchiv/radiation/dir/mstreet/commeet/meet11/brief11/tab_i/br11i1a.txt.

32. Ibid., 12.

33. L. K. Bustad et al., "I-131 in Milk and Thyroid of Dairy Cattle Following a Single Consumption Event and Prolonged Daily Administration," *Health Physics* 9 (July 25, 1963).

34. GAO, *Nuclear Health and Safety: Examples of Post World War II Radiation Releases at US Nuclear Sites*, GAO/RCED94–51FS (Washington, D.C.: GAO, November 24, 1993), note 6.

35. Michele Gerber, *On the Home Front: The Cold War Legacy of the Hanford Nuclear Site* (Lincoln: University of Nebraska Press, 1992), 280 n. 95. See also E. Mart, "Hanford's Air Monitoring Program from 1945–1955: A Compilation of Notes," Memo to the File, citing S. Cantril and J. Healy, "Iodine Metabolism with Reference to Iodine-131," October 22, 1945, http://www.cdc.gov/NIOSH/ocas/pdfs/sec/hanford/sec50part2c.pdf.

36. R.A. Dudley to Gertrude Steel, October 16, 1953.

37. Comments by W. Libby in "Meeting Transcript: Biophysics Conference," 12–13.

38. "Stevenson Sees Cover-up on Bomb," *New York Times*, November 3, 1956.

39. W. F. Libby, "An Atoms for Peace Proposal," Letter to H.M. Kalckar, National Institutes of Health, June 10, 1957; http://www.gwu.edu/~nsarchiv/radiation/dir/mstreet/commeet/meet15/brief15/tab_d/br15d2c.txt.

40. Ibid.

41. Consumers Union, "The Milk All of Us Drink—and Fallout," *Consumer Reports*, March 1959.

42. U.S. Congress, *Fallout from Nuclear Weapons Tests*, esp. table 7: AEC HASL Human Milk Samples from Los Angeles and San Francisco, 2124–2125.

43. U.S. Congress, Joint Committee on Atomic Energy, *The Nature of Radioactive Fallout and Its Effects on Man: Summary Analysis of Hearings May 27–*

29, and June 3–7. See esp. 168, figure 4, demonstrating global pattern and concentrations of fallout from Pacific tests.

44. S. L. Simon et al., "Transfer of 131I into Human Breast Milk and Transfer Coefficients for Radiological Dose Assessments," *Health Physics* 82, no. 6 (2002): 796–806.

45. S. A. Lough, "Fallout and Our Food Supply," *Food Drug and Cosmetic Law Journal* (February 1962).

46. "Test Clears Milk of Strontium 90," *New York Times*, December 20, 1961.

47. Tom Wicker, "President Toasts Milk with Milk," *New York Times*, January 24, 1962.

48. "Expert Minimizes Strontium Peril," *New York Times*, November 16, 1956.

CHAPTER 3. EXPERIMENTS ON HUMANS

1. Quoted in Michele Gerber, *On the Home Front: The Cold War Legacy of the Hanford Nuclear Site,* 3d ed. (Lincoln: University of Nebraska Press, 2007), 25.

2. Michele Gerber, "Our Nuclear Legacy: The Hanford Engineer Works Comes to the Columbia Basin," *Columbia* 7, no. 3 (Fall 1993): 24–27, 34–38.

3. W. D. Norwood to Robert Stone, September 4, 1944, DOE Scientific and Technical Information HW-7-589, http://www.osti.gov/bridge/product .biblio.jsp?query_id=0&page=0&osti_id=10134859.

4. Leslie R. Groves, *Now It Can Be Told: The Story of the Manhattan Project* (New York: Harper, 1962), 50.

5. ACHRE, *Final Report, 1995* (Washington, D.C.: U.S. Government Printing Office, 1995), chapter 11, 516, http://dewey.tis.eh.doe.gov/ohre/roadmap/ achre/chap11_2.html.

6. Ibid.

7. Gerber, *On the Home Front.*

8. Ibid.

9. This section summarizes material presented in ACHRE, *Final Report, 1995,* chapter 11.

10. J. Healy, "Oral History of Jack Healy," November 28, 1994, in DOE, *Human Radiation Studies: Remembering the Early Years,* http://www.orau.org/ptp/ Library/oralhistories/healy.pdf.

11. Ibid.

12. Barton C. Hacker, *The Dragon's Tail: Radiation Safety in the Manhattan Project, 1942–1946* (Berkeley: University of California Press, 1987), 22–23.

13. Gerber, "Our Nuclear Legacy."

14. ACHRE, *Final Report, 1995,* chapter 7, and "Briefing Book," vol. 13, tab E, http://www.tis.eh.doe.gov/ohre/roadmap/achre/chap7_4.html; Samuel Crane, "Irradiation of the Nasopharynx," *Annals of Otology, Rhinology, and Laryngology* 55 (1946): 779–788; H. L. Holmes and J. D. Harris, "Aerotitis Media in Submariners," *Annals of Otology, Rhinology, and Laryngology* 55 (1946): 347–371.

15. ACHRE, *Final Report, 1995*, chapter 7, and part 4: Findings.

16. Task Force on Human Subject Research, to Philip Campbell, Commissioner, Commonwealth of Massachusetts Executive Office of Health and Human Services, Department of Mental Retardation, April 1994, in Task Force on Human Subject Research, *A Report on the Use of Radioactive Materials in Human Subject Research That Involved Residents of State-Operated Facilities within the Commonwealth of Massachusetts from 1943 to 1973* (ACHRE No. MASS-072194-A), 1, National Security Archive, George Washington University, http://www.gwu.edu/~nsarchiv/radiation/dir/mstreet/commeet/meet4/brief4.gfr/tab_e/br4e1.txt.

17. Clemens E. Benda, Director of Research, the Walter E. Fernald State School, to Parent, 28 May 1953, as cited in Task Force on Human Subject Research, *Report on the Use of Radioactive Materials*, appendix B, document 23.

18. Robert S. Harris, Professor of Biochemistry and Nutrition, Massachusetts Institute of Technology, to Clemens E. Benda, May 1, 1953, in Task Force on Human Subject Research, *Report on the Use of Radioactive Materials*, appendix B, document 21, 1.

19. Task Force on Human Subject Research, *Report on the Use of Radioactive Materials*.

20. C. E. Nurnberger and A. Lipscomb, "Transmission of Radioiodine (Iodine-131) to Infants through Human Maternal Milk," *Journal of the American Medical Association* 150 (1952): 398–400.

21. ACHRE, *Final Report, 1995*, chapter 7, 320.

22. R. G. Cuddihy, "Hazard to Man from Iodine-131 in the Environment," *Health Physics* 12 (1966): 1021–1025.

23. E. E. Martmer, K. E. Corrigan, H. P. Charbeneau, and A. Sosin, "Study of the Uptake of Iodine (Iodine-131) by the Thyroid of Premature Infants," *AMA American Journal of Diseases of Children* 17 (1955): 503–505.

24. R. T. Morrison, J. A. Birkbeck, T. C. Evans, and J. I. Routh, "Radioiodine Uptake Studies of Newborn Infants," *Journal of Nuclear Medicine* 4 (1963): 162–166.

25. AEC, Division of Biology and Medicine, *Monthly Status and Progress Report* (Iowa City: University of Iowa, June 1953), in DOE Archives, record group 326, Division of Biology and Medicine, box 3363, folder 23.

26. R. M. Hagstrom, S. R. Glasser, A. B. Brill, and R. M. Heyssel, "Long-Term Effects of Radioactive Iron Administered during Human Pregnancy," *American Journal of Epidemiology* 90, no. 1 (1969): 1–10; P. F. Hahn et al., "Iron Metabolism in Human Pregnancy as Studied with the Radioactive Isotope Fe59," *American Journal of Obstetrics and Gynecology* 61 (1951): 477–486; W. Pribilla, T. H. Bothwell, and C. A. Finch, "Iron Transport to the Fetus in Man," in *Symposium on Iron in Clinical Medicine*, eds. R. O. Wallerstein and S. R. Mettier (Berkeley: University of California Press, 1958), 58–64.

27. R. D. Lloyd, W. S. Zundel, C. W. Wagner, and R. C. Pendleton, "Short Cesium-137 Half-Time in Patients with Muscular Dystrophy," *Nature* 220 (1968): 1029–1031.
28. ACHRE, *Final Report, 1995,* chapter 7, 333.
29. Ibid., chapter 7.
30. Ibid.
31. National Research Council, *The Biological Effects of Atomic Radiation: A Report to the Public from a Study by the National Academy of Sciences* (Washington, D.C.: National Academies of Science Press, 1956).
32. National Security Council, "Conclusions from AEC Paper," March 21, 1957, DDRS, http://galenet.galegroup.com/servlet/DDRS?locID=29002, CDROM Ide: 1991090102728.
33. B. Commoner, in U.S. Congress, *Fallout from Nuclear Weapons Tests,* 2170.
34. Consumers Union, "The Milk All of Us Drink."
35. Barry Commoner, U.S. Congress, Joint Committee on Atomic Energy, Hearings of the Special Subcommittee on Radiation, "Fallout from Nuclear Weapons Tests," May 5, 6, 7, 8, and 9, 1959, 2170.
36. NCI, *Estimated Exposures and Thyroid Doses Received by the American People from Iodine-131 in Fallout Following Nevada Atmospheric Nuclear Bomb Tests* (Bethesda, Md.: NCI, 1997); National Research Council, Institute of Medicine, *Exposure of the American People to Iodine-131 from the Nevada Nuclear-Bomb Tests: Review of the National Cancer Institute Report and Public Health Implications* (Washington, D.C.: National Academy Press, 1999).
37. Ibid.; United Nations Scientific Committee on the Effects of Atomic Radiation, *Sources and Effects of Atomic Radiation: Sources and Effects of Ionizing Radiation* (New York: United Nations, 2000).
38. CDC and NCI, *Report on the Feasibility of a Study of the Health Consequences to the American Population from Nuclear Weapons Tests Conducted by the United States and Other Nations,* May 2005, appendices E, F, G, and H, http://www.cdc.gov/nceh/radiation/fallout/default.htm.
39. National Research Council, Institute on Medicine, *Exposure of the American People to Iodine-131.*
40. AEC, *Annual Report to the National Security Council, Fiscal Year 1955,* DDRS, CDROM Id: 1993030100996.
41. "Thirteen at Yale Favor H-bomb Test Ban," *New York Times,* November 4, 1956.
42. W. E. Nervik, M. I. Kalkstein, and W. F. Libby, "Purification of Milk for Calcium and Strontium with DOWEX-50 W Resin," University of California Radiation Laboratory, report 2674, available through the U.S. Office of Science and Technical Information, Oak Ridge, Tenn., OSTI ID: 4371966.
43. White House, Memorandum of Conference with the President, June 24, 1959, DDRS, CDROM Id: 1978070100292; "Expert Minimizes Strontium Peril," *New York Times,* November 16, 1956.
44. White House, "Informal WH meeting with Eisenhower on Briefing for

Macmillan Talks," Memorandum of Conference with the President, March 17, 1959, DDRS, CK3100309924.

45. White House, "Conference with President Dwight D. Eisenhower Regarding: Talking Points for British Prime Minister Harold Macmillan's U.S. Visit; German Reunification; Treaty between the U.S.S.R. and West Germany; U.S. Relations with West Germany; U.S. Relations with Great Britain; Nuclear Weapons Testing; Egypt's Anti-Communist Stance," Memorandum of Conference with the President, March 17, 1959, Department of State, DDRS, CK3100149716.

46. White House, Memorandum of Conference with the President, Memorandum of Kistiakowsky, Persons, Goodpaster's March 30, 1960 Meeting with President Eisenhower Relative to a Plan for the Coordinated Conduct of Research in the Seismic Field in Connection with Nuclear Testing, March 30, 1960 (declassified 1996), DDRS, CK3100085315.

47. U.S. Department of State, "Treaty Banning Nuclear Weapon Tests in the Atmosphere, in Outer Space and Under Water," August 5, 1963, http://www.state.gov/t/ac/trt/4797.htm.

CHAPTER 4. NUCLEAR ACCIDENTS

1. L. Arnold, *Windscale 1957: Anatomy of a Nuclear Accident*, 2d ed. (Basingstoke, Eng.: Palgrave Macmillan, 1995).

2. N. Stewart and R. Crooks, "Long-Range Travel of the Radioactive Cloud from the Accident at Windscale," *Nature* 182 (1958): 627–628.

3. Atomic Energy Office, *Accident at Windscale No. 1 Pile on 10th October, 1957,* (London: Her Majesty's Stationery Office, 1957), reprinted in *Nature,* 180 (1957): 1043.

4. P. Ortmeyer and A. Makhijani, "Let Them Drink Milk," *Bulletin of the Atomic Scientists* (November–December 1997).

5. K. Love, "Radioactive Area in England Irked," *New York Times,* October 31, 1957.

6. Ibid.

7. J. Lelyveld, "On Coast of England, Jobs vs. Atoms," *New York Times,* April 6, 1986.

8. R. Wakeford. "The Windscale Reactor Accident—50 Years On," *Journal of Radiological Protection* 27 (2007): 211–215.

9. S. Diamond, "Reactor Fallout Is Said to Match Past World Total," *New York Times,* September 23, 1986.

10. S. Diamond, "Nine U.S. Reactors Said to Share Characteristics with One in Ukraine," *New York Times,* May 3, 1986.

11. D. Marples, "The Chernobyl Disaster: Its Effect on Belarus and Ukraine," in James K. Mitchell, ed., *The Long Road to Recovery: Community Responses to Industrial Disaster* (Tokyo: United Nations University Press, 1996).

12. John Greenwald, "Deadly Meltdown," *Time,* May 12, 1986.

13. P. Boffey, "Aides Say Radioactivity Has Arrived in U.S.," *New York Times,* May 6, 1986.

14. Ronald Sullivan, "Fallout Found in U.S. Is Said to Pose No Risk," *New York Times,* May 13, 1986.

15. F. Barringer, "From Children of Chernobyl, Stories of Flight and of Fears," *New York Times,* June 4, 1986.

16. F. Clines, "Once Again, Chernobyl Takes a Toll," *New York Times,* September 30, 1989.

17. D. Marples, "Chernobyl: Five Years Later," report prepared for Commission on Security and Cooperation in Europe, Washington, D.C., April 26, 1991.

18. S. Diamond, "Moscow Now Sees Chernobyl Peril Lasting for Years," *New York Times,* August 22, 1986.

19. F. Clines, "A New Arena for Soviet Nationalism: Chernobyl," *New York Times,* December 30, 1990.

20. M. Brown. "Nuclear Disaster: More Contamination, Fallout Spreads," *New York Times,* May 6, 1986.

21. F. Clines, "Chernobyl Cloud Keeps Welsh Lamb Off the Table," *New York Times,* July 3, 1986.

22. S. Schmemann, "Chernobyl and the Europeans: Radiation and Doubts Linger," *New York Times,* June 12, 1988.

23. J. Tagliabue, "Keeping Tainted Food Off Third World Shelves," *New York Times,* February 8, 1987.

24. E. J. Dionne, "Europe Bars Food from Soviet Bloc," *New York Times,* May 13, 1986.

25. F. Clines, "Chernobyl Shakes Reindeer Culture of Lapps," *New York Times,* September 14, 1986.

26. F. Barringer, "Nuclear Disaster: What Press Is Saying," *New York Times,* May 3, 1986.

27. K. Schneider, "Hint of Crop Damage Roils U.S. Markets," *New York Times,* May 2, 1986.

28. ACHRE, *Final Report, 1995,* chapter 11, table 1.

29. F. Clines, "A New Arena for Soviet Nationalism: Chernobyl," *New York Times,* December 30, 1990.

30. S. Diamond, "Moscow Now Sees Chernobyl's Peril Lasting for Years," *New York Times,* August 22, 1986.

31. R. Gale, "Guidebook to Armageddon," *New York Times,* April 27, 1987.

32. P. Lewis, "Atomic Power Safety Steps Approved," *New York Times,* September 26, 1986.

CHAPTER 5. SOWING SEEDS OF PROTEST

1. Radames Tirado, interviews with the author, March and July 2004.

2. Katherine McCaffrey, *Military Power and Popular Protest: The U.S. Navy in Vieques, Puerto Rico* (New Brunswick, N.J.: Rutgers University Press, 2002).

3. César Ayala, *American Sugar Kingdom: The Plantation Economy of the Spanish Caribbean, 1898–1934* (Chapel Hill: University of North Carolina Press, 2001).

4. Ibid.

5. McCaffrey, *Military Power and Popular Protest.*

6. Ramón Bosque-Pérez and José Javier Colón Morera, *Puerto Rico under Colonial Rule: Political Persecution and the Quest for Human Rights* (Albany: State University of New York Press, 2005), esp. chapter 10: César Ayala and Viviana Carro, "Expropriation and Displacement of Civilians in Vieques, 1940–1950." See also U.S. Department of Navy, *Continued Use of the Atlantic Fleet Weapons Training Facility Inner Range (Vieques): Draft Environmental Impact Statement* (Arlington, Va.: TAMS Consultants, 1979).

7. César Ayala Casás and José Bolivar Fresneda, "The Cold War and the Second Appropriations of the Navy in Vieques," *CENTRO* 17, no. 1 (Spring 2006).

8. Ibid.

9. Associated Press, "Takeover Memories Alive in Vieques," *Puerto Rico Herald,* July 29, 2001.

10. *Weinberger v. Romero-Barcelo,* 456 U.S. 305 (1982).

11. Ibid. at 707.

12. *Romero-Barcelo v. Brown,* 643 F.2d 835 (1981). "Whether or not the Navy's activities in fact harm the coastal waters, it has an absolute statutory obligation to stop any discharges of pollutants until the permit procedure has been followed and the Administrator of the Environmental Protection Agency, upon review of the evidence, has granted a permit." The appellate court suggested that the Navy could request the President to exempt it from the requirements of the Federal Water Pollution Control Act.

13. Ivan Roman, "Vieques' Anger at Navy Dates to 1940s," *Orlando Sentinel,* July 16, 2001.

14. Carlos Zenón, interview with author, March 12, 2003.

15. TAMS, "Draft Environmental Impact Statement: Continued Use of the Atlantic Fleet Weapons Training Facility Inner Range (Vieques)," TAMS Consultants, Arlington, Va., December 1979.

16. Ibid.

17. Ibid.

18. Carlos Zenón interview.

19. Ibid.

20. Ibid. See also McCaffrey, *Military Power and Popular Protest.*

21. José Adams, interview with author, March 10, 2003.

22. J. Rice, "Causation of Nervous System Tumors in Children: Insights from Traditional and Genetically Engineered Animal Models," *Toxicology and Applied Pharmacology* 199 (2004): 175–191.

23. C. A. Stiller and D. M. Parkin, "International Variations in the Incidence of Neuroblastoma," *International Journal of Cancer* 52 (1992): 538–543. See

also L. Lau et al., "Factors Influencing Survival in Children with Recurrent Neuroblastoma," *Journal of Pediatric Hematology/Oncology* 26 (2004): 227–232.

24. Puerto Rico Department of Health, *Health Survey of Vieques Residents* (San Juan: Puerto Rico Department of Health, 1998); J. Ruillan, "Puerto Rican Secretary of Health," *Associated Press*, May 11, 2003; Cruz María Nazario, John Lindsay-Poland, and Deborah Santana, "Vieques Issue Brief: Health on Vieques, a Crisis and Its Causes," Task Force on Latin America and the Caribbean, June 2002, http://www.forusa.org/programs/puertorico/viequesupdate0602.html.

25. Zaidy Torres, director of Allianca Salude de Mulher, interview with the author, March 15, 2004.

CHAPTER 6. RAVAGING LANDSCAPES AND SEACOASTS

1. Globalsecurity.org, "Atlantic Fleet Weapons Training Facility (AFWTF)," available at http://www.globalsecurity.org/military/facility/afwtf.htm (accessed January 2009).

2. Public Hearings, *Vieques Lands Transfer Act of 1994*, H.R.3831, 103d Cong., 2d sess. (October 4, 1994); *Reserve Forces Facilities Authorization Act*, Public Law 93–166, *U.S. Statutes at Large* 87 (1974): 685.

3. Ibid.

4. TAMS, "Draft Environmental Impact Statement: Continued Use of the Atlantic Fleet Weapons Training Facility Inner Range (Vieques)," TAMS Consultants, Arlington, Va., December 1979.

5. ATSDR, "Public Health Assessment: Air Pathway Evaluation Isla de Vieques Bombing Range, Vieques, Puerto Rico," 2003, http://www.atsdr.cdc.gov/HAC/PHA/vieques4/vbr_toc.html.

6. Ibid. Two types of explosives were commonly used at Vieques, each with different by-products from the explosion reaction. One of these types was derived from organic nitrated compounds (i.e., only carbon, hydrogen, oxygen, and nitrogen). Examples include TNT, RDX, HMX, tetryl, Explosive D, Composition B (RDX and TNT), Octol (HMX and TNT), and Composition A-3 (RDX and wax). Carbon dioxide, nitrogen, and carbon monoxide are the primary by-products resulting from this type of explosive. The other explosive contains aluminum in addition to organic nitrated compounds. Examples include Tritonal (TNT and aluminum), H-6 (TNT, RDX, and aluminum), and Torpex (TNT, RDX, and aluminum). The by-products from a bomb made with this explosive include all the chemicals listed for the first type of explosive as well as acetylene, ethylene, phosphine, and aluminum oxide.

7. ATSDR, "Focused Public Health Assessment: Drinking Water Supplies and Groundwater Pathway Evaluation, Isla de Vieques Bombing Range Vieques, Puerto Rico," October 16, 2001, appendix F, http://www.atsdr.cdc.gov/HAC/PHA/vieques/vie_toc.html.

8. ATSDR, "Public Health Assessment: Air Pathway Evaluation."

9. U.S. Department of Defense, "Disclosure of Information on Project 112 to the Department of Veterans Affairs" 2003 Report to Congress, prepared pursuant to section 709(e) of the *National Defense Authorization Act for Fiscal Year 2003*, Public Law 107–314, *U.S. Statutes at Large* 116 (2002): 2458, http://armedservices.house.gov/comdocs/reports/2003exereports/03-08-12disclosure.pdf.

10. National Security Decision Memorandum (NSDM) 35, U.S. Policy on Chemical Warfare Program and Bacteriological/Biological Research Program, from National Security Advisor Henry A. Kissinger to the Vice President, the Secretary of State, the Secretary of Defense, etc., November 25, 1969, http://www.gwu.edu/~nsarchiv/NSAEBB/NSAEBB58/#doc8.

11. U.S. General Accounting Office, "Environmental Protection, DOD Management Issues Related to Chaff," GAO-NSIAD 98–219, 1998, http://www.fas.org/man/gao/nsiad-98-219.htm.

12. César Ayala Casás and José Bolívar Fresneda, "The Cold War and the Second Expropriations of the Navy in Vieques," *CENTRO* 18, no. 1 (Spring 2006): 7.

13. R. Oppenheimer to President Harry Truman, May 3, 1946, box 178 ("JSC Evaluation Board" folder), Oppenheimer Papers, Library of Congress, cited in J. Weisgall, *Operation Crossroads* (Annapolis, Md.: Naval Institute Press, 1994).

14. Defense Threat Reduction Agency, "Operation Crossroads Factsheet," 2002, http://www.dtra.mil/news/fact/nw_crossroads.html.

15. L. Mullenneaux, *Ni Una Bomba Mas: Vieques vs. U.S. Navy* (New York: Penington Press, 2000).

16. Phil Williams, "UGA Ecologist and Coral Reef Expert Part of Team that Discovered Bombs, Sunken Ships off Disputed Puerto Rican Island," University of Georgia press release, December 13, 1999.

CHAPTER 7. MERCURY

1. C. Ortiz-Roque and Y. López-Rivera, "Mercury Contamination in Reproductive Age Women on a Caribbean Island: Vieques," *Journal of Epidemiology and Community Health* 58 (2004): 756–757.

2. EPA, "Mercury Study Report to Congress: Whitepaper and Volume II: An Inventory of Anthropogenic Mercury Emission in the U.S." (Washington, D.C.: Office of Air Quality Planning and Standards and Office of Research and Development, 1997), http://www.epa.gov/oar/mercury.html (updated September 25, 2002).

3. United Nations Environment Program, *Pollution and Prevention Abatement Handbook* (Washington, D.C.: World Bank Publications, 1999).

4. P. F. Schuster et al., "Atmospheric Mercury Deposition during the Last 270 Years: A Glacial Ice Core Record of Natural and Anthropogenic Sources," *Environmental Science and Technology* 36, no. 11 (2002): 2303–2310.

5. Mercury occurs in three forms: elemental (Hg0), inorganic salts (Hg+1 and Hg+2), and organic forms (methylmercury, ethylmercury, and phenyl-mercury). The behavior of mercury in the environment and its threat to human health varies considerably by form or species. Most mercury released to the atmosphere is gaseous and elemental (Hg0), remaining in circulation for up to a year and traveling long distances. Gaseous divalent mercury (Hg+2) easily dissolves in water and is more rapidly transported to the earth's surface attached to rain, snow, and fog; it can also attach to fine particles floating in the atmosphere that eventually fall to earth. For these reasons divalent mercury has a far shorter atmospheric residency time. Other atmospheric pollutants influence the transformation of elemental to divalent mercury. Ozone increases the rate of conversion, thereby promoting its more rapid deposition. Reactions with sulfite ions, also emitted by burning fossil fuels, may reverse the reaction, promoting longer atmospheric residence. Mercury in "particulate" form interacts with soot and other large particles that are more quickly deposited on the earth's surface.

6. Northeast States for Coordinated Air Use Management (NESCAUM), *Northeast States and Eastern Canadian Provinces Mercury Study: A Framework for Action* (Boston: NESCAUM, 1998).

7. EPA, "Mercury Study Report to Congress"; National Research Council, *Toxicological Effects of Methylmercury* (Washington, D.C.: National Academy Press, 2000).

8. National Research Council, *Toxicological Effects of Methylmercury;* see also "Meeting of the Minds on Mercury," *Environmental Health Perspectives* 107, no. 1 (1999), http://ehpnet1.niehs.nih.gov/docs/1999/107–1/forum.html.

9. National Research Council, *Toxicological Effects of Methylmercury;* N. Sorenson et al., "Prenatal Methylmercury Exposure as a Cardiovascular Risk Factor at Seven Years of Age," *Epidemiology* 10 (1999): 370–375.

10. E. Silbergeld and V. Weaver, "Exposure to Metals: Are We Protecting the Workers?" *Occupational and Environmental Medicine* 64, no. 3 (2007): 141–142; E. K. Silbergeld, I. A. Silva, and J. F. Nyland, "Mercury and Autoimmunity: Implications for Occupational and Environmental Health," *Toxicology and Applied Pharmacology* 207, no. 2 (suppl) (2005): 282–292; E. M. Yokoo et al., "Low Level Methylmercury Exposure Affects Neuropsychological Function in Adults," *Environmental Health* 2, no. 1 (2003): 8; P. Mendola, S. G. Selevan, S. Gutter, and D. Rice, "Environmental Factors Associated with a Spectrum of Neurodevelopmental Deficits," *Mental Retardation and Developmental Disabilities Research Reviews* 8, no. 3 (2002): 188–197.

11. P. Grandjean et al., "Cognitive Deficit in Seven-Year-Old Children with Prenatal Exposure to Methylmercury," *Neurotoxicology and Teratology* 19 (1997): 417–428.

12. P. Davidson et al., "Effects of Prenatal and Postnatal Methylmercury Exposure from Fish Consumption of Neurodevelopment: Outcomes at Sixty-six Months of Age in the Seychelles Child Development Study," *Journal of the*

American Medical Association 280 (1998): 701–707; P. W. Davidson et al., "Neurodevelopmental Outcomes of Seychellois Children from the Pilot Cohort at 108 Months Following Prenatal Exposure to Methylmercury from a Maternal Fish Diet," *Environmental Research* 84, no. 1 (2000): 1–11.

13. National Research Council, *Toxicological Effects of Methylmercury*. In an additional study on contending with contradictory data, Jacobson writes: "A power analysis, conducted by computing standardized regression coefficients for the three studies, indicated that many of the Faroe Island findings were so subtle that the power to detect them in the Seychelles study, despite its large sample size, was only about 50 percent. Because prospective epidemiological studies are often hampered by limited control over confounding and other factors, including unmeasured between cohort differences in genetic vulnerability and nutritional adequacy, inferences about toxicity often depend heavily on a qualitative assessment of the weight of the evidence from multiple studies." See J. L. Jacobson, "Contending with Contradictory Data in a Risk Assessment Context: The Case of Methylmercury," *Neurotoxicology* 22, no. 5 (2001): 667–675. Responding to the NAS panel's comments, the authors of the Seychelles Islands report state: "There continues to be an association among both prenatal and postnatal exposure and the Preschool Language Scale Total Score, as well as an association in males only among postnatal exposure and scores on the Woodcock-Johnson Applied Problems and the Bender Gestalt drawing and copying errors. . . . These associations continue to suggest beneficial effects with increasing mercury levels that may reflect dietary benefits of fish consumption. In a population exposed to MeHg from consumption of ocean fish, we continue to find no evidence of adverse effects." P. W. Davidson et al., "Methylmercury and Neurodevelopment: Reanalysis of the Seychelles Child Development Study Outcomes at Sixty-six Months of Age," *Journal of the American Medical Association* 285, no. 10 (2001): 1291–1293.

14. P. Grandjean et al., "Delayed Evoked Potentials in Children Exposed to Methylmercury from Seafood," *Neurotoxicology and Teratology* 21, no. 4 (1999): 343–348; see also "More Evidence of Mercury Effects in Children," *Environmental Health Perspectives* 107, no. 11 (1999): A554–555.

15. Grandjean et al., "Delayed Evoked Potentials in Children"; see also "More Evidence of Mercury Effects in Children"; "Meeting of the Minds on Mercury," A12; P. Grandjean et al., "Methylmercury Neurotoxicity in Amazonian Children Downstream from Gold-mining," *Environmental Health Perspectives* 107, no. 7 (1999): 587–591; and P. Weihe et al., "Neurobehavioral Performance of Inuit Children with Increased Prenatal Exposure to Methylmercury," *International Journal of Circumpolar Health* 61, no. 1 (2002): 41–49.

16. See "Meeting of the Minds on Mercury," A12.

17. Grandjean et al., "Methylmercury Neurotoxicity in Amazonian Children."

18. Weihe et al., "Neurobehavioral Performance of Inuit Children."

322 NOTES TO PAGES 111–116

19. Sorensen et al., "Prenatal Methylmercury Exposure as a Cardiovascular Risk Factor," 370–375.

20. N. Frery et al., "Gold-Mining Activities and Mercury Contamination of Native Amerindian Communities in French Guiana: Key Role of Fish in Dietary Uptake," *Environmental Health Perspectives* 109, no. 5 (2001): 449–456.

21. CDC, "Blood and Hair Mercury Levels in Young Children and Women of Childbearing Age: United States, 1999," *Morbidity and Mortality Weekly Report* 50, no. 8 (2001): 140–143.

22. "Mercury in Seafood Linked to Infertility," Reuters, reuters.com, September 23, 2002.

23. J. Salonen et al., "Mercury Accumulation and Accelerated Progression of Carotid Atherosclerosis: A Population-Based Prospective Four-Year Follow-up Study in Men in Eastern Finland," *Atherosclerosis* 148 (2000): 265–273.

24. K. Yoshizawa et al., "Mercury and the Risk of Coronary Heart Disease in Men," *New England Journal of Medicine* 347, no. 22 (2002): 1755–1760. See also T. Rissanen et al., "Fish Oil–Derived Fatty Acids, Docosahexaenoic Acid and Docosapentaenoic Acid, and the Risk of Acute Coronary Events: The Kuopio Ischaemic Heart Disease Risk Factor Study," *Circulation* 102 (2000): 2677–2679; M. Ahlqwist et al., "Serum Mercury Concentration in Relation to Survival, Symptoms, and Diseases: Results from the Prospective Population Study of Women in Gothenburg, Sweden," *Acta Odontologica Scandinvica* 57 (1999): 168–174.

25. EPA, Kathryn Mahaffey, "Development of Methylmercury Reference Dose," http://www.masgc.org/mercury/abstracts.html (accessed July 22, 2003).

26. EPA, "Mercury Study Report to Congress."

27. CDC, "Blood and Hair Mercury Levels"; K. Mahaffey, EPA, communication with the author, July 23, 2003; D. Hattis and K. Silver, "Human Inter-individual Variability: A Major Source of Uncertainty in Assessing Risks for Noncancer Health Effects," *Risk Analysis* 14, no.4 (1994): 421–431.

28. National Research Council, *Toxicological Effects of Methylmercury.*

29. J. M. Hightower and D. Moore, "Mercury Levels in High-End Consumers of Fish," *Environmental Health Perspectives* 111, no. 4 (2003): 604–608. Care should be taken not to infer that these are average findings, because the sample was not random. Those sampled had sought out the help of a physician, complaining of nonspecific symptoms that had no other clear cause, leading the physicians to screen blood for metals.

30. EPA, "Fish Consumption Advisories," http://www.epa.gov/hg/advisories .htm (updated August 29, 2008).

31. FDA, "Safe and Sanitary Processing and Importing of Fish and Fishery Products," rule 21 CFR, part 123 (Washington, D.C.: FDA, 1995). Each firm must subscribe to the Hazard Analysis and Critical Control Point

(HAACP) system, also used by the U.S. Department of Agriculture (USDA) to identify and control contamination of beef and poultry products.

32. GAO, "Federal Oversight of Seafood Does Not Sufficiently Protect Consumers," Report to the Committee on Agriculture, Nutrition, and Forestry, U.S. Senate GAO-01–204, 2001, http://www.gao.gov/new.items/d01204.pdf.

33. J. Houlihan and R. Wiles, *Focus Pocus: Internal Documents Reveal that the FDA Is Failing in Its Public Health Obligation to Warn Pregnant Women about Mercury in Tuna* (Washington, D.C.: Environmental Working Group, 2002).

34. GAO, "Federal Oversight of Seafood Does Not Sufficiently Protect Consumers," 20. Different nations have consumer warning systems similar to that of FDA. Health Canada has a guideline of 0.5 ppm of total mercury content—more protective than the EPA guideline of 1 ppm for methylmercury alone. Canada also recommends no more than one meal per week of swordfish, shark, or tuna, and not more than one meal per month for women of child-bearing age and children. Britain is less protective, advising women who are or intend to become pregnant or breastfeeding to restrict their intake to two medium-sized cans of tuna or one tuna steak per week. Given average levels of mercury detected in tuna by FDA (0.17 mcg/gram of fish), this recommendation would expose British women to levels of mercury 11.5 times higher than the maximum advised by the National Academy of Sciences and the EPA. If the tuna contains mercury at the 1 ppm FDA advisory limit, the British advisory would result in a dose 68 times higher than the NAS suggested.

35. *Science and Health: The Mercury Story,* broadcast January 21, 2005, http://www.pbs.org/now/science/mercuryinfish.html.

36. S. Schober et al., "Blood Mercury Levels in U.S. Children and Women of Childbearing Age, 1999–2000," *Journal of the American Medical Association* 289 (2003): 1667–1674.

37. National Research Council, *Toxicological Effects of Methylmercury.*

38. FDA and EPA, "FDA and EPA Announce the Revised Consumer Advisory on Methylmercury in Fish," March 19, 2004, http://www.fda.gov/bbs/topics/news/2004/NEW01038.html.

39. *The People of the State of California v. Tri-Union Seafoods, LLC, et al.,* case no. CGC-04-432394 (San Francisco Sup. Ct., filed June 21, 2004). In 2005, Lester Crawford, commissioner of the FDA, announced that California's suit to comply with the Proposition 65 labeling requirements conflicted directly with the requirements of the FFDCA, and is therefore preempted by the federal law. The District court found in favor of the seafood company, but the U.S. Court of Appeals for the Third Circuit reversed the lower court's opinion, finding that the FDA had not adopted a regulation that would preempt California's action. See U.S. Court of Appeals for the Third

Circuit, *Deborah Fellner v. Tri-Union Seafoods,* no. 07-1238, August 19, 2008.

40. *Food Quality Protection Act of 1996, U.S. Code* (2000), title 21, section 346a.

41. Chrissie Cole, "U.S. Agencies at Odds over Consumer Advice on Fish," Environmental Working Group, Washington, D.C., December 15, 2008, http://www.ewg.org/node/27463.

42. ATSDR, "Public Health Assessment: Fish and Shellfish Evaluation Isla Vieques Bombing Range Vieques, Puerto Rico," June 27, 2003, 2–3, http://www.atsdr.cdc.gov/hac/pha/viequesfish/viequespr-toc.html.

43. National Research Council, *Toxicological Effects of Methylmercury.*

44. ATSDR, "Public Health Assessment: Fish and Shellfish Evaluation Isla Vieques Bombing Range Vieques, Puerto Rico." In this report, ATSDR compares exposure estimates to its recommended limit of 0.3 micrograms/ kilogram of body weight/day. If exposures exceed the limit, ATSDR places an asterisk next to the estimate, and the accompanying note states: "Estimated exposure exceeds health guideline."

45. Ibid., 2–3.

46. Ibid., 28.

CHAPTER 8. WASTELAND OR WILDERNESS?

1. *Defense Base Closure and Realignment Act of 1990,* Public Law 101–510, http://www.defenselink.mil/brac/docs/legis03.pdf; *Floyd D. Spence National Defense Authorization Act for Fiscal Year 2001,* Public Law 106–398, http://www7.nationalacademies.org/ocga/Laws/PL106_398.asp, as amended by section 1049 of the *National Defense Authorization Act for Fiscal Year 2002,* Public Law 107–107, http://www7.nationalacademies.org/ocga/laws/PL107_107.asp. This statute required the Navy to transfer federal lands to the Department of the Interior to be managed by the U.S. Fish and Wildlife Service as a National Wildlife Refuge.

2. Leatherbacks may grow to 1,500 pounds and live for fifty years or more. Leatherbacks can spend the vast majority of their lives in the deep ocean. Little is known about their migratory patterns, although some turtles have been fitted with transmitters and tracked by satellites. One was found to travel nearly 7,500 miles in one year. See British Broadcasting Corporation, http://www.hswri.org/leatherback_journeys.htm.

3. The Department of the Navy transferred property on the eastern end of the island to the administrative jurisdiction of the Department of Interior under the *Floyd D. Spence National Defense Authorization Act for Fiscal Year 2001,* as amended by section 1049 of the *National Defense Authorization Act for Fiscal Year 2002.*

4. U.S. Congress, *The Wilderness Act of 1964,* Public Law 88–577, 88th Cong., 2d sess. (September 3, 1964).

5. Each state and U.S. territory has the opportunity to list one site without a

NOTES TO PAGES 128–139

relative ranking procedure normally used to set cleanup priorities. Puerto Rico chose the Vieques/Culebra complex.

6. Congress reauthorized and amended Comprehensive Environmental Response, Compensation, and Liability Act (CERCLA), commonly known as Superfund, in 1986 when it created the Defense Environmental Restoration Program that manages site restoration of defense facilities. Program goals include "having an approved cleanup process in place or cleanup complete at 100 percent of all such properties by the end of fiscal year 2014," http://www.gao.gov/new.items/d01557.pdf.

7. Oscar Diaz, interview with the author, July 24, 2006.

8. Center for Public Environmental Oversight, "Toxic Ranges," October 2002, http://www.cpeo.org/pubs/pub.html.

9. Aaron Wildavsky, "Richer Is Safer," *Public Interest* 60 (Summer 1980): 23–39.

10. Department of the Navy, "Base Realignment and Closure Office Management Program," http://www.bracpmo.navy.mil.

11. U.S. Fish and Wildlife Service, "Alaska Maritime National Wildlife Refuge," http://alaskamaritime.fws.gov.

12. CIA, *The 2006 World Factbook,* https://www.cia.gov/cia/publications/factbook/geos/mq.html.

13. GAO, "Federal Facilities: Consistent Relative Risk Evaluations Needed for Prioritizing Cleanups," GAO/RCED-96–150, June 1996, http://www.gao.gov/archive/1996/rc96150.pdf; Defense Base Closure and Realignment Commission, "2005 Defense Base Closure and Realignment Commission Report," September 8, 2005, http://www.brac.gov/docs/final/Volume1BRACReport.pdf.

14. GAO, "Environmental Contamination: Cleanup Actions at Formerly Used Defense Sites," GAO-01-557, July 2001, http://www.gao.gov/gao-01-1012sp. The GAO estimated that two hundred additional training areas may be similarly contaminated.

15. Center for Public Environmental Oversight, "Center for $90 Million Club: A List of Military Facilities Exceeding $90 Million in Cleanup Costs," 2001, http://www.cpeo.org/pubs/club/90mil.htm.

16. S. Breyer, *Breaking the Vicious Circle: Toward Effective Risk Regulation* (Cambridge, Mass.: Harvard University Press, 1995).

CHAPTER 9. CAPITALIZING ON INNOCENCE

1. The facts of this chapter are derived from court findings, government documents, and news reports. See *Dow Chemical Co. v. Ebling,* 723 N.E.2d 881 (Ind. Ct. App. 2000); 753 N.E.2d 633 (Ind. S.Ct. 2001), Interlocutory Appeal no. 22S05-0008-CV-481. See also J. Wargo, *Our Children's Toxic Legacy* (New Haven: Yale University Press, 1998). The quotations are printed in the news reports cited in this chapter's notes.

2. K. Hensel, *The Toxic Effect*, WishTV, 2008, transcript available at http://www.wishtv.com/.

3. Ibid.

4. J. Morris, "The Stuff in the Backyard Shed," *U.S. News and World Report*, November 8, 1999.

5. Wargo, *Our Children's Toxic Legacy*.

6. R. H. Hill et al., "Pesticide Residues in Urine of Adults Living in the U.S.," *Environmental Research* 71 (1995): 99–108. This article reports on research conducted in 1994 in the National Health and Nutrition Examination (NHANES) III study.

7. EPA, "Interim Reregistration Eligibility Decision for Chlorpyrifos," 2006, http://www.epa.gov/oppsrrd1/REDs/chlorpyrifos_ired.pdf.

8. Ibid., 17 and table 7.

9. Charles R. Kratzer, Celia Zamora, and Donna L. Knifong, "Diazinon and Chlorpyrifos Loads in the San Joaquin River Basin, California, January and February 2000," U.S. Geological Survey, 2000, http://pubs.usgs.gov/wri/wri02-4103/.

10. EPA, "Drinking Water Assessment of Chlorpyrifos," memorandum from Michael Barrett to Steven Knitzner, November 13, 1998, 8, in author's possession.

11. EPA, *Drinking Water Assessment of Chlorpyrifos* (Washington, D.C.: Office of Prevention, Pesticides and Toxic Substances, 1998).

12. *National Drinking Water Regulations, 2008*, in *U.S. Code*, title 42, chapter 6a, subchapter 12, part A, sec. 300f, http://www.law.cornell.edu/uscode/search/index.html.

13. EPA, *Drinking Water Assessment of Chlorpyrifos*.

14. R. B. Leidy, C. G. Wright, and H. E. Dupree, "Chlorpyrifos in the Air and Soil of Houses Eight Years after Its Application for Termite Control," *Bulletin of Environmental Contamination and Toxicology* 52 (1994): 131–134.

15. Morris, "The Stuff in the Backyard Shed."

16. EPA, *Drinking Water Assessment of Chlorpyrifos*.

17. Carol Browner, "Dursban Announcement," June 8, 2000, EPA Press Release, http://www.epa.gov/history/topics/legal/03.htm.

18. T. Slotkin, "Developmental Cholinotoxicants: Nicotine and Chlorpyrifos," *Environmental Health Perspectives* 107 (1999): 71–80.

19. M. Bjørling-Poulsen, H. R. Andersen, and P. Grandjean, "Potential Developmental Neurotoxicity of Pesticides Used in Europe," *Environmental Health* 22, no. 7 (October 2008); 50.

20. *Safe Drinking Water Act, as Amended through Public Law 107–377*, December 31, 2002, *U.S. Code* 42 (2002), sec. 300f et seq. For updated information about the act, see http://epw.senate.gov/envlaws/envlaws.htm.

21. J. Blondell to D. Smegal, memorandum, "Chlorpyrifos Incident Review Data," April 20, 2000, http://www.epa.gov/pesticides/foia/reviews/059101/059101-429-20-04-2000.pdf. See also R. W. Whitmore, J. E.

Kelly, and P. L. Reading, *National Home and Garden Pesticide Use Survey Final Report,* vol. 1 (Research Triangle Park, N.C.: Research Triangle Institute, 1992).

22. Wargo, *Our Children's Toxic Legacy,* chapters 8 and 9.

23. EPA, "Chlorpyrifos Revised Risk Assessment and Agreement with Registrants, 2000," http://www.epa.gov/pesticides/op/chlorpyrifos/agreement .pdf.

24. *Dow Chemical Co. v. Ebling,* 723 N.E.2d 881 (Ind. Ct. App. 2000).

25. These include Syngenta, Makhetshim-Agan, Drexel, Prentis, Gowan, and Aventis.

26. W. E. Robbins, T. L. Hopkins, and G. W. Eddy, "Metabolism and Excretion of Phosphorus-32 Labeled Diazinon in a Cow," *Journal of Agricultural and Food Chemistry* 5 (1957): 509–513.

27. J. C. Derbyshire and R. T. Murphy, "Diazinon Residues in Treated Silage and Milk of Cows Fed Powdered Diazinon," *Journal of Agricultural and Food Chemistry* 10 (1962): 384–386.

28. EPA, "Quantitative Usage Analysis for Diazinon," 1999, http://www.epa .gov/oppsrrd1/op/diazinon/usage.pdf.

29. P. L. Soren et al., "Dearomatized White Spirit Inhalation Exposure Causes Long-Lasting Neurophysiological Changes in Rats," *Neurotoxicology and Teratology* 18 (1996): 67–76. See also L. M. Pedersen and K-H Cohr, "Biochemical Pattern in Experimental Exposure of Humans to White Spirits, I: The Effects of a Six-Hour Single Dose," *Acta Pharmacologica et Toxicologica* 55 (1984): 317–324; L. M. Pedersen and K-H Cohr, "Biochemical Pattern in Experimental Exposure of Humans to White Spirits, II: The Effects of Repetitive Exposures," *Acta Pharmacologica et Toxicologica* 55 (1984): 325–330.

30. ATSDR, "Stoddard Solvent Toxicity, Biological Fate," Case Studies in Environmental Medicine, 2002, http://www.atsdr.cdc.gov/HEC/CSEM/ stoddard/biologic_fate.html; ATSDR, "Public Health Statement for Stoddard Solvent," June 1995, http://www.atsdr.cdc.gov/toxprofiles/phs79.htm.

31. Massachusetts Department of Environmental Protection, Office of Research and Standards, "Updated Petroleum Hydrocarbon Fraction Toxicity Values for the VPH/EPH/APH Methodology," May 2002, http://www .mass.gov/dep/cleanup/laws/policies.htm.

32. *Dow Chemical Co. v. Ebling,* 723 N.E.2d 881 (Ind. Ct. App. 2000).

33. Wargo, *Our Children's Toxic Legacy.*

34. Ibid.

35. EPA, "Organophosphate Pesticides: Revised OP Cumulative Risk Assessment," 2002, http://www.epa.gov/pesticides/cumulative/rra-op.

36. EPA, "Diazinon Summary," December 5, 2000, http://www.epa.gov/ oppsrrd1/op/diazinon/summary.htm.

37. T. L. Litovitz et al., "1996 Annual Report of the American Association of Poison Control Centers Toxic Exposure Surveillance System," *American Journal of Emergency Medicine* 15 (1997): 494.

38. J. Blondell to D. Smegal, memorandum, "Chlorpyrifos Incident Review Data," April 20, 2000, http://www.epa.gov/pesticides/foia/reviews/059101/059101–429–20–04–2000.pdf. See also Whitmore, Kelly, and Reading, *National Home and Garden Pesticide Use Survey Final Report*, vol. 1.

39. EPA, "Dursban 2E Label," 2005, http://oaspub.epa.gov/pestlabl/ppls.home.

40. Browner, "Dursban Announcement."

CHAPTER 10. WITHOUT WARNING

1. J. Morris, "The Stuff in the Backyard Shed," *U.S. News and World Report*, November 8, 1999.

2. Philip Shabecoff and Alice Shabecoff, *Poisoned Profits: The Toxic Assault on Our Children* (New York: Random House, 2008), 139–140.

3. K. Hensel, *The Toxic Effect*, WishTV, 2008, transcript available at http://www.wishtv.com.

4. For human health loss and exposure compensation, see *Radiation Exposure Compensation Act*, Public Law 101-426, *U.S. Statutes at Large* 104 920 (1990): 920; for property loss compensation, see *Alaska Native Claims Settlement Act*, Public Law 92-203, *U.S. Statutes at Large* 85 (1971): 688.

5. American Law Institute, *Restatement of the Law, Third, Torts: Products Liability*, reporters J. A. Henderson, Jr., and Aaron D. Twerski (Washington, D.C.: American Law Institute, 1998).

6. See *Jones v. George*, 61 Tex. 345 (1884). This case involved a claim against the seller of "Royall's Patent or Receipt for Killing Worms," a product that failed to accomplish the control of cotton worms. Cited in Brief Amicus Curiae for the U.S. in Support of Plaintiffs-Appellants, *Etcheverry, et al., v. Tri-Ag Service, Inc., Bayer Corp., et al.* California Supreme Court (2000).

7. American Law Institute, *Restatement of the Law, Third, Torts: Products Liability*, http://www.ali.org/ali_old/promo6081.htm.

8. Manufacturers place "express warrantees" at the end of most labels. A common warrantee includes the following claim: "(Company X) warrants that this product conforms to the chemical description on the label and is reasonably fit for the purposes stated on the label when used in strict accordance with the directions, subject to the inherent risks set forth below. . . . It is impossible to eliminate all risks associated with use of this product. Plant injury, lack of performance, or other unintended consequences may result because of such factors as use of the product contrary to label instructions (including conditions noted on the label, such as unfavorable temperatures, soil conditions, etc.), abnormal conditions (such as excessive rainfall, drought, tornadoes, hurricanes), presence of other materials, the manner of application, or other factors, all of which are beyond the control of (the company) or seller. All such risks are assumed by the buyer."

9. This failure to deliver information to those most likely to be exposed is sometimes termed a "downstream dissemination failure." *Dow Chem. Co.*

v. Ebling, 753 N.E.2d 633 (Ind. 2001); *Dow Chemical Co. v. Ebling* 723 N.E.2d 881 (Ind. Ct. App. 2000).

10. *Dow Chemical Co. v. Ebling,* 723 N.E.2d 881 (Ind. Ct. App. 2000).

11. U.S. Congress, *1991 National Literacy Act,* Public Law 102–73, 102d Cong., 1st sess. (1991). Congress defined literacy as "an individual's ability to read, write, and speak in English, and compute and solve problems at levels of proficiency necessary to function on the job and in society, to achieve one's goals, and develop one's knowledge and potential."

12. Ibid., http://www.nifl.gov/nifl/facts/reading_facts.html#sadults.

13. American Foundation for the Blind, "Key Definitions of Statistical Terms," 2008, http://www.afb.org/section.asp?SectionID=15&DocumentID=1280.

14. New York Attorney General, "Dow AgroSciences LLC Notice of Proposed Litigation," April 2, 2003, http://www.oag.state.ny.us/media_center/2003/apr/apr02a_03.html, and http://www.oag.state.ny.us/press/2003/dec/dec15a_03html.

15. EPA, "EPA Fines DowElanco for Failure to Report Pesticide Health Effects," EPA Press Release, May 2, 1995, archived through http://www.epa.gov. See also *Dow Chemical Co. v. Ebling,* 723 N.E.2d 881 (Ind. Ct. App. 2000).

16. EPA, "Interim Reregistration Eligibility Decision for Chlorpyrifos," July 31, 2006, EPA 738-R-01-007, http://www.epa.gov/oppsrrd1/REDs/chlorpyrifos_ired.pdf.

17. EPA, Pesticide Product Label System, "Dursban LO Label" and "Dursban 2E Label," 2008, http://oaspub.epa.gov/pestlabl/ppls.home.

18. DowAgroSciences, "Dursban 4E Label," EPA registration no. 62719–11, http://oaspub.epa.gov/pestlabl/ppls.home; see also ATSDR, "Agencies Are Sharing Strategies for More Effective Response to Methyl Parathion Health Threat," May 2003, http://www.atsdr.cdc.gov/HEC/HSPH/hsph73-1.html.

19. See advertisements by Dow as early as January 1986 in the journal *Pest Control* promoting lower environmental concentrations and lower toxicity: "Dow Makes a Quantum Leap in Pest Control," *Pest Control* 54, no. 1 (January 1986): 37–40; "High-Tech with 'Time Release' Products," *Pest Control* 54, no. 7 (July 1986): 50–57; and "The Ultimate Survivor Is No Match for the Ultimate Insecticide," *Pest Control* 55, no. 1 (January 1987): 36–37.

20. EPA, *Label Review Manual,* chapter 19: "The Consumer Labeling Initiative and Pesticide Labels," http://www.epa.gov/oppfead1/labeling/lrm/chap-19.htm; "Labeling Devices for Pesticides and Devices," 40 *CFR* 156.10 (1999); see also *Cigarette Labeling and Advertising Act of 1965,* Public Law 89–92 (1965), http://uscode.house.gov/download/pls/15C36.txt; and *Public Health Cigarette Smoking Act of 1969,* Public Law 91–222 (1970), http://tobaccodocuments.org/atc/71066088.html.

21. *Dow Chemical Co. v. Ebling,* 723 N.E.2d 881 (Ind. Ct. App. 2000).

22. Morris, "The Stuff in the Backyard Shed."

23. Although the language of FIFRA on preemption does not specifically

prohibit state tort claims, many courts have relied on a U.S. Supreme Court ruling in *Cipollone v. Leggett,* 505 U.S. 504 (1992) concluding that civil actions claiming a failure to warn "are pre-empted to the extent that they rely on a state-law "requirement or prohibition . . . with respect to . . . advertising or promotion."

24. *Dow Chemical Co. v. Ebling,* 723 N.E.2d 881 (Ind. Ct. App. 2000).

25. "Liability for injury under Indiana's Product Liability Act is premised on the claim that the product in question is in "a defective condition unreasonably dangerous." *Ind. Code,* sec. 34-20-2-1, http://www.in.gov/legislative/ic/code/title34/ar20/ch2.html.

26. EPA, "EPA Fines DowElanco for Failure to Report Pesticide Health Effects."

27. U.S. Constitution, Article 6, clause 2, available at http://caselaw.lp.findlaw.com/data/constitution/article06.

28. *Federal Insecticide, Fungicide and Rodenticide Act, U.S. Code* 7 (2005), sec. 136v : "(a) In general: A state may regulate the sale or use of any federally registered pesticide or device in the State, but only if and to the extent the regulation does not permit any sale or use prohibited by this subchapter. (b) Uniformity: Such state shall not impose or continue in effect any requirements for labeling or packaging in addition to or different from those required under this Act."

29. *Bates et al. v. Dow AgroSciences LLC,* 544 U.S. 431 (2005).

30. The Clean Air Act, Section 209b, permits California to adopt vehicle emission restrictions more restrictive than the federal government. Section 177 of the same statute authorizes states to adopt California's standards if their air quality is not in compliance with 58 *CFR* 4166 (1993), http://books.nap.edu/openbook.php?record_id=11586&page=331.

CHAPTER 11. THE DDT DILEMMA

1. FDA, "Pesticide Program Residue Monitoring 2003," May 2005, http://www.cfsan.fda.gov/~dms/pes03rep.html.

2. World Health Organization, "The Global Malaria Action Plan for a Malaria Free World," 2008, http://www.rbm.who.int/gmap/toc.html.

3. These statistics are rough estimates, because mortality and morbidity are presumed to be worst in areas were medical care is least available, and records are rarely or poorly kept. The estimates are also imprecise due to the difficulty of disentangling the relative roles of malaria versus malnourishment, intestinal disease, and acute respiratory infections in inducing death. WHO's malaria death estimates have ranged between 1 and 2 million people per year for much of the past decade.

4. "DDT Cost Cut 40 percent since July," *New York Times,* December 29, 1944.

5. DDT replaced pyrethrum (a naturally derived pesticide from chrysanthe-

mum flowers), in the spray containers, formerly manufactured by Japan before the war.

6. *Regulations for the Enforcement of the Federal Insecticide, Fungicide and Rodenticide Act,* 47 CFR 6493, October 2, 1947.

7. L. S. Henderson, "DDT in the Home," *Yearbook on Agriculture* (Washington, D.C.: U.S. Department of Agriculture, 1947), 643–647.

8. G. Woodard et al., "Accumulation of DDT in the Body Fat and Its Appearance in the Milk of Dogs," *Science* 102 (1945): 177–178.

9. Horace Telford and James Guthrie, "Transmission of the Toxicity of DDT through the Milk of White Rats and Goats," *Science* 102 (1945): 647.

10. "DDT," *New York Times,* March 10, 1946.

11. K. Noren and D. Meironyte, "Certain Organochlorine and Organobromine Contaminants in Swedish Human Milk in Perspective of Past Twenty to Thirty Years," *Chemosphere* 40 (2000): 1111–1123.

12. "Farmers Warned on DDT: Expert Says It Appears in Milk, Meat after Crop Dusting," *New York Times,* May 25, 1947.

13. "Oversolution of DDT Is Fatal to Wildlife," *New York Times,* April 16, 1949.

14. M. M. Ellis, B. A. Westfall, and M. D. Ellis, "Toxicity of DDT to Goldfish and Frogs," *Science* 100 (1944): 477.

15. William Laurence, "Food Gain Sighted in Insect Control," *New York Times,* April 3, 1952.

16. "Pre-Testing Urged for Pest Poisons," *New York Times,* September 9, 1949.

17. U.S. Congress, *Federal Insecticide Act of 1910, U.S. Statutes at Large* 36 (1910): 331.

18. T. R. Dunlap, *DDT: Scientists, Citizens, and Public Policy* (Princeton, N.J.: Princeton University Press, 1981), 46; and J. C. Whorton, *Before "Silent Spring"* (Princeton, N.J.: Princeton University Press, 1974), 186.

19. Douglas Hurt, "The Poison Squad," *Ohio Historical Society Timelime* 2 (1985): 64.

20. "Tolerances and Exemptions from Tolerances for Pesticide Chemicals in or on Raw Agricultural Commodities," 21 *CFR* 120.101 (1955).

21. "Food Additives: DDT," 21 *CFR* 121.226–227 (1963).

22. "Tolerances for Residues of DDT," 25 *CFR* 12289 (1960). Establishes tolerances "in the fat of meat from cattle, hogs, and sheep."

23. "Food Additives Permitted in Food for Human Consumption: DDT, DDD, and DDE in Milk and Manufactured Dairy Products," 32 *CFR* 4059 (1967). The producers in 1965 signing the request included the Geigy Chemical Corporation, Diamond Alkali Co., Allied Chemical Corporation, Montrose Chemical Corporation, and Olin Mathieson Chemical Corporation.

24. "Benefits and Dangers of DDT," *New York Times,* September 2, 1945.

25. "Regulations for the Enforcement of Federal Insecticide, Fungicide and Rodenticide Act," 49 *CFR* 3885 (1949).

26. "DDT Residues in Apple Pomace," 21 *CFR* 120.147b (1967).

27. Food additive tolerances for processed foods (residues that concentrate

during processing) appear in 21 *CFR*, part 193. Feed additive tolerances for pesticides used on crops that become animal feeds are published in 21 *CFR*, part 561. See the FDA's searchable database at http://www.accessdata .fda.gov/scripts/cdrh/cfdocs/cfcfr/cfrsearch.cfm.

28. R. E. Hodgson and W. L. Sweetman, "What to Feed a Cow," *Yearbook on Agriculture* (Washington, D.C.: U.S. Department of Agriculture, 1946).

29. Wargo, *Our Children's Toxic Legacy*.

30. "Food Man Depicts Fight on Pesticides," *New York Times*, February 1, 1952.

31. *Food Quality Protection Act of 1996*, U.S. Public Law 104–170, http://www .epa.gov/pesticides/regulating/laws/fqpa.

32. E. P. Laug, "Occurrence of DDT in Human Fat and Milk," *Archives of Industrial Hygiene and Occupational Medicine* 3 (1951): 245; G. W. Pearce et al., "Examination of Human Fat for Presence of DDT," *Science* 116 (1952): 254.

33. World Wildlife Fund, *Chemical Trespass: A Toxic Legacy* (London: World Wildlife Fund, 1999).

34. P. J. Landrigan et al., "Chemical Contaminants in Breast Milk and Their Impacts on Children's Health: An Overview," *Environmental Health Perspectives* 110 (2002): 313–315.

35. EPA, "Dicofol Registration Eligibility Document," epa.gov/oppsrrd1/REDs/ factsheets/0021fact.pdf, 1998.

36. "Bromoxynil, Diclofop-methyl, Dicofol, Diquat, Etridiazole, et al., Proposed Tolerance Actions," 72 *CFR* 147.41913–41931 (August 1, 2007).

37. "Citing Children, EPA Is Limiting Use of Pesticides," *New York Times*, August 3, 1999.

38. ATSDR, "National Alert: Illegal Use of Methyl Parathion Insecticide," 1996, http://www.atsdr.cdc.gov/alerts/961213.html; ATSDR, "Agencies Are Sharing Strategies for More Effective Response to Methyl Parathion Health Threat," May 2003, http://www.atsdr.cdc.gov/HEC/HSPH/hsph73-1.html.

39. World Health Organization, "WHO Gives Indoor Use of DDT a Clean Bill of Health for Controlling Malaria," 2006, http://www.who.int/media centre/news/releases/2006/pr50/en/index.html.

CHAPTER 12. WHAT IS ACCEPTABLE RISK?

1. Banned compounds include DDT, aldrin, dieldrin, chlordane, hepatchlor, 2,4,5-T, HCB, toxaphene, DBCP, EDB, and Alar.

2. J. Wargo, *Our Children's Toxic Legacy* (New Haven: Yale University Press, 1998).

3. Compensation is required by the Fifth Amendment to the U.S. Constitution: "nor shall private property be taken for public use without just compensation."

4. Pesticides are classified as "carcinogens" primarily based on animal studies. See National Resource Council, *Regulating Pesticides in Food: The Delaney Paradox* (Washington, D.C: National Academy Press, 1987).

5. National Research Council, *Pesticides in the Diets of Infants and Children* (Washington, D.C.: National Academies Press, 1993).

6. "Tolerances and Exemptions for Pesticide Chemical Residues," *U.S. Code* 21 (2000): sec. 346a.

7. The administrator must find tolerances to be safe, meaning that there is a reasonable certainty that no harm will result from aggregate exposure to the pesticide chemical residue, including all anticipated dietary exposures and all other exposures for which there is reliable information. See ibid.

8. The Food Quality Protection Act's requirement to consider a common mechanism of toxicity states: "In establishing, modifying, leaving in effect, or revoking a tolerance or exemption for a pesticide chemical residue, the Administrator shall assess the risk of the pesticide chemical residue based on . . . (III) available information concerning the cumulative effects on infants and children of such residues and other substances that have a common mechanism of toxicity," ibid.

9. Federal Food, Drug and Cosmetic Act, Section 408 (b)(2)(C). The FFDCA was amended by the Food Quality Protection Act of 1996. The agency may replace the tenfold additional safety factor only if it finds reliable data that demonstrate children would be sufficiently protected from increased exposures that could result from reducing the margin of safety. See EPA, "Determination of the Appropriate FQPA Safety Factor(s) in Tolerance Assessment," 2002, http://www.epa.gov/oppfead1/trac/science/determ.pdf.

10. S. Makris et al., "A Retrospective Analysis of Twelve Developmental Neurotoxicity Studies Submitted to the U.S. EPA Office of Prevention, Pesticides and Toxic Substances (OPPTS)," presentation to the FIFRA Scientific Advisory Panel, December 9, 1998, http://www.epa.gov/scipoly/sap/1998/december/neuro.pdf.

11. T. Slotkin, "Developmental Cholinotoxicants: Nicotine and Chlorpyrifos," *Environmental Health Perspectives* 107 (1999): 71–81; EPA, "Chlorpyrifos Reregistration," http://www.epa.gov/pesticides/reregistration/chlorpyrifos/. This website is the portal for EPA's risk assessments for chlorpyrifos.

12. National Research Council, *Pesticides in the Diets of Infants and Children* (Washington, D.C.: National Research Council, 1993).

13. R. R. Dietert et al., "Workshop to Identify Critical Windows of Exposure for Children's Health: Immune and Respiratory Systems Work Group Summary," *Environmental Health Perspectives* 108 (2000): 383–490. See also S. D. Holladay and R. J. Smialowicz, "Development of the Murine and Human Immune System: Differential Effects of Immunotoxicants Depend on Time of Exposure," *Environmental Health Perspectives* 108 (2000): 463–473; S. D. Holladay and M. I. Luster, "Developmental Immunotoxicology," in C. Kimmel and J. Buelke-Sam, eds., *Developmental Toxicology*, 2d ed. (New York: Raven Press, 1994), 93–118.

14. EPA, *Endocrine Disruptor Screening and Testing Advisory Committee: Final Report*, 1998, http://www.epa.gov/endo/pubs/edspoverview/finalrpt.htm.

15. S. Olin and B. Sonawane, "Workshop to Develop a Framework for Assessing Risks to Children from Exposure to Environmental Agents," *Environmental Health Perspectives* 111 (2003).

16. These limits do not mean that concentrations are normally found in milk or animal feeds, but that the chemicals are approved for use on crops fed to animals. Instead tolerances are limits of pesticide residues or metabolites that may appear in cow's milk. Animal feed tolerances are often lower than milk tolerances, because many chemicals may concentrate in animal and milk fats.

17. U.S. Department of Agriculture, Food Safety and Inspection Service (FSIS), "The FSIS National Residue Program," 1999, http://www.fsis.usda .gov/ophs/blue99/sec6tab.pdf. FSIS authority is derived from the Federal Food Drug and Cosmetic Act, which requires that residues of pesticides, drugs, and other contaminants be restricted to tolerance or action levels. If these levels are exceeded, then the food is declared "adulterated." Slaughterhouses are responsible to ensure that adulterated products do not enter the commerce under the Federal Meat Inspection Act, the Poultry Products Inspection Act, and the Egg Products Inspection Act. Products regulated by FSIS include horses, bulls, beef cows, dairy cows, heifers, steers, veal, heavy calves, sheep, lambs, goats, hogs, boars/stags, sows, young chickens, mature chickens, turkeys, ducks, geese, rabbits, and egg products.

18. Food Safety and Inspection Service methods also were incapable of detecting six different carbamate insecticide residues during the same period.

19. EPA, Office of Water, "Drinking Water Standards and Health Advisories," 2008, http://www.epa.gov/ost/drinking/standards/dwstandards.pdf.

20. U.S. Geological Survey, "Pesticides in Groundwater: U.S. Geological Survey Fact Sheet FS-244–95," http://www water.usgs.gov/nawqa/pnsp/pubs/fs244-95.

21. EPA, Exposure Factors Handbook, 2007, http://cfpub.epa.gov/ncea/cfm/recordisplay.cfm?deid=20563; EPA, *Child-Specific Exposure Factors Handbook*, 2002, http://fn.cfs.purdue.edu/fsq/WhatsNew/KidEPA.pdf.

22. R. M. Whyatt et al., "Within- and Between-Home Variability in Indoor-Air Insecticide Levels during Pregnancy among an Inner-City Cohort from New York City," *Environmental Health Perspectives* 115, no. 3 (2007): 383–389; M. K. Williams et al., "An Intervention to Reduce Residential Insecticide Exposure during Pregnancy among an Inner-City Cohort," *Environmental Health Perspectives* 114 (2006): 1684–1689.

23. "There is insufficient use information and exposure data to assess exposure resulting from use in vehicles (i.e. planes, trains, automobiles, buses, boats) and other current label uses." EPA, "Chlorpyrifos Revised Risk Assessments," 2000, http:// www.epa.gov/pesticides/op/chlorpyrifos.htm.

24. R. Boyle, "Flying Fever," *Audubon* (June 2000).

25. Laura Mansnerus, "Pesticide Spraying in City Is Illegal," *New York Times,* September 21, 2000.

26. Norimitsu Onishi, "Risk of Mosquito-Borne Virus Changes the Routines of Many," *New York Times,* September 16, 1996.

27. Tracey Wilson, "Judge Stops Pesticide Spray near School," *Los Angeles Times,* May 26, 2001.

28. These include chlorfenvinphos, ethion, ethyl parathion, fenamiphos, fenthion, fonofos, isazophos, isofenphos, mevinphos, monocrotophos, phosphamidon, sulfotepp, and sulprofos.

29. T. Colborn, "A Case for Revisiting the Safety of Pesticides: A Closer Look at Neurodevelopment," *Environmental Health Perspectives* 114 (2006): 10–17.

30. EPA, *Organophosphate Pesticides: Revised Cumulative Risk Assessment, 2002,* http://www.epa.gov/pesticides/cumulative/rra-op; EPA, *Cumulative Risk Assessment: 2006 Update,* http://www.epa.gov/pesticides/cumulative/2006 -op/index.htm.

31. Ibid.

32. EPA, "Triazine Cumulative Risk Assessment," March 28, 2006, http:// www.epa.gov/oppsrrd1/cumulative/triazine_fs.htm.

33. EPA, "Chromated Copper Arsenate Revised Risk Assessments," 2003, http://www.epa.gov/oppad001/reregistration/cca.

CHAPTER 13. AIRBORNE MENACE

1. Oak Ridge National Laboratory, *Transportation Energy Data Book,* 27th ed., 2008, http://www-cta.ornl.gov/data/index.shtml, 1–17.

2. This chapter summarizes and builds on research reported in 2006 by Environment and Human Health, Inc.: see J. Wargo, M. Cullen, H. Taylor, L. Wargo, N. Alderman, and S. Addiss, *Plastics That May Be Harmful to Children and Reproductive Health* (North Haven, Conn.: Environment and Human Health, Inc., 2006), http://www.ehhi.org/reports/plastics/ehhi; see also Interagency Forum on Child and Family Statistics, *America's Children: Key National Indicators of Well-Being, 2000,* http://www.childstats.gov/ ac2000/toc.asp. These estimates are based on sampling strategies that vary among pollutants, and are often derived from fixed sampling sites that produce estimates averaged over days or longer periods of time.

3. See American Academy of Pediatrics Committee on Environmental Health, "Ambient Air Pollution: Health Hazards to Children," *Pediatrics* 114 (2004): 1699–1707.

4. National Library of Medicine, National Institutes of Health, "Tox Town: Chemical Ozone," July 13, 2005, http://toxtown.nlm.nih.gov/text_version/ chemical.php?name=ozone.

5. M. L. Bell, F. Dominici, and J. M. Samet, "A Meta-Analysis of Time-Series Studies of Ozone and Mortality with Comparison to the National Morbidity, Mortality, and Air Pollution Study," *Epidemiology* 16 (2005): 436–445; M. L. Bell et al., "Ozone and Short-Term Mortality in Ninety-five U.S.

Urban Communities, 1987–2000," *Journal of the American Medical Association* 292 (2004): 2372–2378.

6. EPA, Office of Air Quality Planning and Standards, "Air Quality Index: A Guide to Air Quality and Your Health," 2007, http://www.epa.gov/airnow/aqibroch/aqi.html#9.

7. Joel Schwartz and Lucas M. Neas, "Fine Particles Are More Strongly Associated Than Coarse Particles with Acute Respiratory Health Effects in Schoolchildren," *Epidemiology* 11 (2000): 6–10.

8. P. S. Gilmour et al., "The Procoagulant Potential of Environmental Particles (PM10)," *Journal of Occupational and Environmental Medicine* 62 (2005): 164–171.

9. A. C. A. Pope et al., "Lung Cancer, Cardiopulmonary Mortality, and Long-Term Exposure to Fine Particulate Air Pollution," *Journal of the American Medical Association* 287 (2002): 1132–1141.

10. M. Svartengren et al., "Short-Term Exposure to Air Pollution in a Road Tunnel Enhances the Asthmatic Response to Allergen," *European Respiratory Journal* 15 (2000): 716–724.

11. EPA, "General Air Quality Information," 2008, http://yosemite.epa.gov/r10/airpage.nsf.

12. Oak Ridge National Laboratory, *Transportation Energy Data Book*, 27th ed., 2008, http://cta.ornl.gov.

13. U.S. Census Bureau, "Americans Now Spend over a Hundred Hours a Year Commuting," 2005, http://usgovinfo.about.com/od/censusandstatistics/a/commutetimes.htm.

14. D. Schrank. and T. Lomax, "Urban Mobility Report," Texas Transportation Institute, 2004, http://mobility.tamu.edu.

15. Oak Ridge National Laboratory, *Transportation Energy Data Book*, 27th ed., 1–17.

16. U.S. Department of Transportation, Bureau of Transportation Statistics, "Transportation Statistics Annual Report," 2001, http://www.bts.gov/publications/transportation_statistics_annual_report/2001/html/chapter_04_figure_01_069.html.

17. Energy Information Administration, "September 2005 Monthly Energy Review," September 28, 2005, http://tonto.eia.doe.gov/merquery/mer_data.asp?table=T01.09.

18. DOE, "Analysis of Technology Options to Reduce the Fuel Consumption of Idling Trucks," Center for Transportation Research, Argonne National Laboratory, Argonne, Ill., June 2000, http://www.transportation.anl.gov/pdfs/TA/15.pdf.

19. Green Car Congress, "Redesign of Toyota Vitz (Echo) with Stop-Start System Hits Japanese Market," February 10, 2005, http://www.greencarcongress.com/2005/02/redesign_of_toy.html.

20. Masayoshi Minato, "Japanese Gov't to Promote 'Idling Stop' Drive Campaign," *Kyodo News*, April 7, 2005.

21. R. J. Delfino et al., "Symptoms in Pediatric Asthmatics and Air Pollution: Differences in Effects by Symptom Severity, Anti-Inflammatory Medication Use and Particulate Averaging Time," *Environmental Health Perspectives* 106 (1998): 751–761.

22. J. A. Wiley et al., "Study of Children's Activity Patterns: Final Report," report no. A733-149, California Air Resources Board, Sacramento, 1991.

23. R. A. Etzel et al., "Acute Pulmonary Hemorrhage in Infants Associated with Exposure to Stachybotrys Atra and Other Fungi," *Archives of Pediatric and Adolescent Medicine* 152 (1998): 757–762; Committee on Environmental Health, American Academy of Pediatrics, "Toxic Effects of Indoor Molds," *Pediatrics* 101 (1998): 712–714.

24. J. J. Quackenboss, M. Krzyzanowski, and M. D. Lebowitz, "Exposure Assessment Approaches to Evaluate Respiratory Health Effects of Particulate Matter and Nitrogen Dioxide," *Journal of Exposure Analysis and Environmental Epidemiology* 1 (1991): 83–107; V. Hasselblad, D. M. Eddy, and D. J. Kotchner, "Synthesis of Environmental Evidence: Nitrogen Dioxide Epidemiology Studies," *Journal of Air Waste Management Association* 42 (1992): 662; M. H. Garrett et al., "Respiratory Symptoms in Children and Indoor Exposure to Nitrogen Dioxide and Gas Stoves," *American Journal of Respiratory and Critical Care Medicine* 158 (1998): 891–895; B. P. Lanphear et al., "Residential Exposures Associated with Asthma in U.S. Children," *Pediatrics* 107 (2001): 505–511; B. J. Smith et al., "Health Effects of Daily Indoor Nitrogen Dioxide Exposure in People with Asthma," *European Respiratory Journal* 16 (2000): 879–885; J. M. Samet and M. J. Utell, "The Risk of Nitrogen Dioxide: What Have We Learned from Epidemiological Studies?" *Toxicology and Industrial Health* 6 (1990): 247–262.

25. Institute of Medicine, National Academy of Science, *Clearing the Air: Asthma and Indoor Air Exposures* (Washington, D.C.: National Academy Press, 1999).

26. American Lung Association, "Adolescent Smoking Statistics," November 2003, http://www.lungusa.org/press/tobacco/not_stats.html.

27. Mark A. Schuster, Todd Franke, and Cung B. Pham, "Smoking Patterns of Household Members and Visitors in Homes with Children in the United States," *Archives of Pediatric Adolescent Medicine* 156 (2002): 1094–1100.

28. CDC, "More U.S. Households Adopting Smoke-free Home Rules," 2007, http://www.cdc.gov/media/pressrel/2007/r070524.htm.

29. R. A. Etzel, "Active and Passive Smoking: Hazards for Children," *Central European Journal of Public Health* 5 (1997): 54–56; J. R. DiFranza and R. A. Lew, "Morbidity and Mortality in Children Associated with the Use of Tobacco Products by Other People," *Pediatrics* 97 (1996): 560–568; D. Mannino et al., "Environmental Tobacco Smoke Exposure and Health Effects in Children: Results from the 1991 National Health Survey," *Tobacco Control* 5 (1996): 13–18; S. L. Gortmaker et al., "Parental Smoking and the Risk of Childhood Asthma," *American Journal of Public Health* 72 (1982): 574–579;

American Academy of Pediatrics, Committee on Environmental Hazards, "Involuntary Smoking: A Hazard to Children," *Pediatrics* 77 (1986): 755–757.

30. R. Ehrlich et al., "Childhood Asthma and Passive Smoking: Urinary Cotinine as a Biomarker of Exposure," *American Review of Respiratory Disease* 145 (1992): 594–599.

31. F. D. Gilliland et al., "Maternal Smoking during Pregnancy: Environmental Tobacco Smoke Exposure and Childhood Lung Function," *Thorax* 55 (2000): 271–276.

32. Committee on Environmental Health, American Academy of Pediatrics, *Handbook of Pediatric Environmental Health*, ed. Ruth Etzel, 1999, https://www.nfaap.org/netforum/eweb.

33. EPA, "The Inside Story on Indoor Air Quality," 2001, http://www.epa.gov/iaq/voc.html; A. J. Harrison, Jr., "An Analysis of the Health Effects, Economic Consequences, and Legal Implications of Human Exposure to Indoor Air Pollutants," *South Dakota Law Review* 37, no. 289 (1991–1992): 292–293.

34. D. Norback et al., "Asthmatic Symptoms and Volatile Organic Compounds, Formaldehyde, and Carbon Dioxide in Dwellings," *Occupational and Environmental Medicine* 52 (1995): 388–395; M. L. Burr, "Indoor Air Pollution and the Respiratory Health of Children," *Pediatric Pulmonology* 18 (suppl.) (1999): 3–5.

35. M. Krzyzanowski, J. J. Quackenboss, and M. D. Lebowitz, "Chronic Respiratory Effects of Indoor Formaldehyde Exposure," *Environmental Research* 52 (1990): 117–125; G. Smedje, D. Norback, and C. Edling, "Asthma among Secondary Schoolchildren in Relation to the School Environment," *Clinical Experimental Allergy* 27 (1997): 1270–1278; T. Godish, "Aldehydes," in J. Spengler, J. Samet, and J. McCarthy, *Indoor Air Quality Handbook* (New York: McGraw-Hill, 2001). See also Federal Emergency Management Agency, "Formaldehyde and Travel Trailers," July 20, 2007, http://www.fema.gov/news/newsrelease.fema?id=36730.

36. G. Wieslander et al., "Asthma and the Indoor Environment: The Significance of Emission of Formaldehyde and Volatile Organic Compounds from Newly Painted Indoor Surfaces," *International Archives Occupational Environmental Health* 69 (1997): 115–124; J. R. Beach et al., "The Effects on Asthmatics of Exposure to a Conventional Water-Based and a Volatile Organic Compound-Free Paint," *European Respiratory Journal* 10 (1997): 563–566.

37. C. Shim and M. H. Williams, Jr., "Effect of Odors in Asthma," *American Journal of Medicine* 80 (1986): 18–22.

38. D. Shusterman, "The Health Significance of Environmental Odor Pollution: Revisited," *Journal of Environmental Medicine* 1, no. 1 (1999): 249–258.

39. State of California, Air Resources Board, "Hazardous Ozone Generating

'Air Purifiers,'" September 2007, http://www.arb.ca.gov/research/indoor/ozone.htm.

40. GAO, "Air Pollution: Hazards of Indoor Radon Could Pose a National Problem," 1986, cited in A. Reitz, "The Legal Control of Indoor Air Pollution," *Boston College Environmental Affairs Law Review* 25 (1998): 247; EPA, personal communication to B. Feingold, October 23, 2006; *Toxic Substances Control Act (TSCA)*, Public Law 94–469, *U.S. Statutes at Large* 90 (1976): 2003.

CHAPTER 14. WHO IS MOST AT RISK?

1. This chapter summarizes and builds on research reported in 2006 by Environment and Human Health, Inc.; see J. Wargo, L. Wargo, and N. Alderman, *The Harmful Effects of Vehicle Exhaust* (North Haven, Conn.: Environment and Human Health, Inc., 2006), http://www.ehhi.org/reports/plastics/ehhi; see also California Environmental Protection Agency, Air Resources Board and American Lung Association of California, "Recent Research Findings: Health Effects of Particulate Matter and Ozone Air Pollution," January 2004, http://www.arb.ca.gov/research/health/fs/pm-03fs.pdf; J. Schwartz, "Harvesting and Long-Term Exposure Effects in the Relation between Air Pollution and Mortality," *American Journal of Epidemiology* 151 (2000): 440–448; EPA, "PM-2.5 Composition and Variability," http://www.epa.gov/ttn/oarpg/naaqsfin; see also J. C. Chow et al., "PM2.5 Chemical Composition and Spatiotemporal Variability during the California Regional PM10/PM2.5 Air Quality Study (CRPAQS)," *Journal of Geophysical Research* 111 (May 25, 2006), http://www.agu.org/pubs/crossref/2006/2005JD006457.shtml.

2. E. L. Avol et al., "Respiratory Effects of Relocating to Areas of Differing Air Pollution Levels," *American Journal of Respiratory and Critical Care Medicine* 164 (2001): 2067–2072; J. D. Schwartz et al., "Analysis of Spirometric Data from a National Sample of Healthy Six- to Twenty-four-year-olds (NHANES II)," *American Review of Respiratory Disorders* 138 (1988): 1405–1414.

3. EPA, Exposure Factors Handbook, 2007; EPA, *Child-Specific Exposure Factors Handbook*, 2002.

4. National Academy of Science, *Clearing the Air: Asthma and Indoor Air Exposures* (Washington, D.C.: National Academy Press, 1999).

5. CDC, "Summary Health Statistics for U.S. Children: National Health Interview Survey, 2003," March 2005, http://www.cdc.gov/nchs/data/series/sr_10/sr10_223.pdf.

6. National Institutes of Health, National Heart, Lung, Blood Institute, "Data Fact Sheet: Asthma Statistics," http://www.nhlbi.nih.gov/health/prof/lung/asthma/asthstat.pdf, 1999; CDC, "Measuring Childhood Prevalence before and after the 1997 Redesign of the National Health Interview Survey," *U.S. Morbidity and Mortality Weekly Report* 49 (2000): 40.

7. S. Brim et al., "Asthma Prevalence among U.S. Children in Under-represented Minority Populations, *Pediatrics* 122 (July 2008): 217–222.

8. C. Aligne et al., "Risk Factors for Pediatric Asthma: Contributions of Poverty, Race, and Urban Residence," *American Journal of Respiratory and Critical Care Medicine* 162 (2000): 873–877; E. F. Crain et al., "An Estimate of the Prevalence of Asthma and Wheezing among Inner-City Children," *Pediatrics* 94 (1994): 356–362; L. Claudio et al., "Environmental Health Sciences Education: A Tool for Achieving Environmental Equity and Protecting Children," *Environmental Health Perspectives* 106 (1998): supplement 3.

9. M. Aubier, "Air Pollution and Allergic Asthma," *Revue Maladies Respiratoires* 17 (2000): 159–165; G. D'Amato, "Outdoor Air Pollution in Urban Areas and Allergic Respiratory Diseases," *Monaldi Archives of Chest Disease* 54 (1999): 470–474; B. Peterson and A. Saxon, "Global Increases in Allergic Respiratory Disease: The Possible Role of Diesel Exhaust Particles," *Annals of Allergy Asthma Immunology* 77 (1996): 263–268.

10. EPA, "Biological Pollutants in Your Home," 1990, http://www.epa.gov/iaq/pubs/bio_1.html.

11. C. S. Kim and T. C. Kang, "Comparative Measurement of Lung Deposition of Inhaled Fine Particles in Normal Subjects and Patients with Obstructive Airway Disease," *American Journal Respiratory Critical Care Medicine* 155 (1997): 899–905; K. D. Yang, "Childhood Asthma: Aspects of Global Environment, Genetics and Management," *Changgeng Yi Xue Za Zhi* (Taiwan) 23 (2000): 641–661; F. P. Perera et al., "Effect of Prenatal Exposure to Airborne Polycyclic Aromatic Hydrocarbons on Neurodevelopment in the First Three Years of Life among Inner-City Children," *Environmental Health Perspectives* 114 (2006): 1287–1292; W. A. Jedrychowski et al., "Variability of Total Exposure to PM2.5 Related to Indoor and Outdoor Pollution Sources, Krakow Study in Pregnant Women," *Science of the Total Environment* 366 (2006): 47–54; F. P. Perera et al., "Biomarkers in Maternal and Newborn Blood Indicate Heightened Fetal Susceptibility to Procarcinogenic DNA Damage," *Environmental Health Perspectives* 112 (2004): 1133–1136.

12. R. J. Delfino et al., "Symptoms in Pediatric Asthmatics and Air Pollution: Differences in Effects by Symptom Severity, Anti-Inflammatory Medication Use, and Particulate Averaging Time," *Environmental Health Perspectives* 106 (1998): 751–761.

13. A. Oosterlee et al., "Chronic Respiratory Symptoms in Children and Adults Living along Streets with High Traffic Density," *Occupational and Environmental Medicine* 53 (1996): 241–247; G. Ciccone et al., "Road Traffic and Adverse Respiratory Effects in Children," *Occupational and Environmental Medicine* 55 (1998): 771–778.

14. N. Kunzli et al., "Public-Health Impact of Outdoor and Traffic-Related Air Pollution: A European Assessment," *Lancet* 356 (2000): 795–801.

15. W. J. Gauderman et al., "Association between Air Pollution and Lung Function Growth in Southern California Children," *American Journal Respiratory Critical Care Medicine* 162 (2000): 1383–1390. In one study, children in communities with higher levels of urban air pollution in Southern California were found to have decreased lung function, and children spending more time outdoors had larger deficits in the growth rate of lung function. The Children's Health Study at the University of Southern California recently found that children who moved to areas with lower PM-10 levels had increased lung function growth rates, while those moving to areas with higher PM-10 levels had reduced lung function growth rates; see E.L. Avol et al., "Respiratory Effects of Relocating to Areas of Differing Air Pollution Levels," *American Journal Respiratory Critical Care Medicine* 164 (2001): 2067–2072; W. J. Gauderman et al., "Association between Air Pollution and Lung Function Growth in Southern California Children," *American Journal of Respiratory Critical Care Medicine* 162 (2000): 1383–1390; W. J. Gauderman et al., "Association between Air Pollution and Lung Function Growth in Southern California Children: Results from a Second Cohort," *American Journal of Respiratory Critical Care Medicine* 166 (2002): 76–84. See also M. Brauer et al., "Air Pollution and Retained Particles in the Lung," *Environmental Health Perspectives* 109 (2001): 1039–1043.

16. B. Brunekreef et al., "Air Pollution from Truck Traffic and Lung Function in Children Living Near Motorways," *Epidemiology* 8 (1997): 298–303. See also P. van Vliet et al., "Motor Vehicle Exhaust and Chronic Respiratory Symptoms in Children Living Near Freeways," *Environmental Research* 74 (1997): 122–132; O. Yu et al., "Effects of Ambient Air Pollution on Symptoms of Asthma in Seattle-Area Children Enrolled in the CAMP Study," *Environmental Health Perspectives* 108 (2000): 1209–1214; R. J. Delfino et al., "Symptoms in Pediatric Asthmatics and Air Pollution: Differences in Effects of Asthma Severity, Anti-Inflammatory Medication Use and Particulate Averaging Time," *Environmental Health Perspectives* 106 (1998): 751–761.

17. G. D. Thurston et al., "Summertime Haze Air Pollution and Children with Asthma," *American Journal Respiratory Critical Care Medicine* 155 (1997): 654–660; R. McConnell et al., "Asthma in Exercising Children Exposed to Ozone: A Cohort Study," *Lancet* 359 (2002): 386–391.

18. Oosterlee et al., "Chronic Respiratory Symptoms in Children and Adults"; Ciccone et al., "Road Traffic and Adverse Respiratory Effects in Children"; M. H. Gielen et al., "Acute Effects of Summer Air Pollution on Respiratory Health of Asthmatic Children," *American Journal of Respiratory Critical Care Medicine* 155 (1997): 2105–2108.

19. C. Nordenhäll et al., "Diesel Exhaust Enhances Airway Responsiveness in Asthmatic Subjects," *European Respiratory Journal* 17 (2001): 909–915; R. J. Delfino et al., "Symptoms in Pediatric Asthmatics and Air Pollution"; Yu et al., "Effects of Ambient Air Pollution on Symptoms of Asthma"; R.

Jorres and H. Magnussen Krankenhaus, "Airways Response of Asthmatics after a Thirty-Minute Exposure, at Resting Ventilation, to 0.25 ppm NO2, or 0.5 ppm SO2," *European Respiratory Journal* 3 (1990): 132–137.

20. CDC National Center for Health Statistics, "State of Childhood Asthma, United States: 1980-2005," December 12, 2006, http://www.cdc.gov/media/pressrel/r061212.htm.

21. P. Lozano et al., "Health Care Utilization and Cost among Children with Asthma Who Were Enrolled in a Health Maintenance Organization," *American Academy of Pediatrics* 99 (1997): 757–764.

22. CDC, "Facts about Chronic Obstructive Pulmonary Disease (COPD)," February 11, 2005, http://www.cdc.gov/nceh/airpollution/copd/copdfaq.htm.

23. See review in W. MacNee and K. Donaldson, "Exacerbations of COPD Environmental Mechanisms," *Chest* 117 (2000): 390–397.

24. Joan Foland, "Cardiovascular Disease, Connecticut's Leading Killer," Connecticut Department of Public Health, issue brief #2002–1, February 2002, http://www.dph.state.ct.us/OPPE/pdfs/cvd.pdf.

25. J. Kaiser, "Epidemiology: Mounting Evidence Indicts Fine-Particle Pollution," *Science* 307 (2005): 1858–1859; P. S. Gilmour et al., "The Procoagulant Potential of Environmental Particles (PM10)," *Occupational and Environmental Medicine* 62 (2005): 164–171; C. A. Pope et al., "Ambient Particulate Air Pollution, Heart Rate Variability, and Blood Markers of Inflammation in a Panel of Elderly Subjects," *Environmental Health Perspectives* 112 (2004): 339–345.

26. C. A. Pope et al., "Lung Cancer, Cardiopulmonary Mortality, and Long-Term Exposure to Fine Particulate Air Pollution," *Journal of the American Medical Association* 287 (2002): 1132–1141; K. Donaldson et al., "Role of Inflammation in Cardiopulmonary Health Effects of PM," *Toxicology and Applied Pharmacology* 207, no. 2 (suppl.) (September 2005): 483–488; C. A. Pope et al., "Cardiovascular Mortality and Long-Term Exposure to Particulate Air Pollution: Epidemiological Evidence of General Pathophysiological Pathways of Disease," *Circulation* 109 (2004): 71–77.

27. American Heart Association, "Air Pollution, Heart Disease, and Stroke: Exposure to Air Pollution Contributes to the Development of Cardiovascular Diseases (Heart Disease and Stroke)," November 2005, http://www.americanheart.org/presenter.jhtml?identifier=4419; D. Krewski et al., "Mortality and Long-Term Exposure to Ambient Air Pollution: Ongoing Analyses Based on the American Cancer Society Cohort," *Journal of Toxicology and Environmental Health* 68 (2005): 1093–1109.

28. A. Peters et al., "Exposure to Traffic and the Onset of Myocardial Infarction," *New England Journal of Medicine* 351 (2004): 1721–1730; G. Hoek et al., "Association between Mortality and Indicators of Traffic-Related Air Pollution in the Netherlands: A Cohort Study," *Lancet* 360 (2002): 1203–1209.

29. EPA, "Final Rule: Control of Hazardous Air Pollutants from Mobile

Sources, Early Credit Technology Requirement," October 9, 2008, http://www.epa.gov/OMS/toxics.htm#mobile.

30. Department of Health and Human Services, "Eleventh Report on Carcinogens," revised January 2001, Public Health Service, National Toxicology Program, http://ntp.niehs.nih.gov/?objectid=035E5806-F735-FE81-FF769DFE5509AF0A.

31. A. J. Thompson, M. D. Shields, and C. C. Patterson, "Acute Asthma Exacerbations and Air Pollutants in Children Living in Belfast, Northern Ireland," *Archives of Environmental Health* 56 (2001): 234–241.

32. World Health Organization, International Agency for Research on Cancer, *Monographs on the Evaluation of Carcinogenic Risk to Humans,* vol. 46: *Diesel and Gasoline Engine Exhaust and Some Nitroarenes,* 1989, http://monographs.iarc.fr/ENG/Monographs/vol46/volume46.pdf. See also World Health Organization, International Programme on Chemical Safety, "Diesel Fuel and Exhaust Emissions," *Environmental Health Criteria* 171 (1996).

33. State of California, "Findings of the Scientific Review Panel on the Report on Diesel Exhaust," April 22, 1999, http://www.arb.ca.gov/toxics/dieseltac/de-fnds.htm.

34. South Coast Air Quality Management District (SCAQMD), "Multiple Air Toxics Exposure Study in the South Coast Air Basin (MATES-II)," March 2000, http://www.aqmd.gov/matesiidf/matestoc.htm.

35. Connecticut Department of Public Health, "Diabetes Fact Sheet," 2005, http://www.dph.state.ct.us/BCH/HEI/diabetes.htm.

36. H. Kan, J. Jia, and B. Chen, "The Association of Daily Diabetes Mortality and Outdoor Air Pollution in Shanghai, China," *Journal of Environmental Health* 67 (2004): 21–26; A. Zanobetti and J. Schwartz, "Are Diabetics More Susceptible to the Health Effects of Airborne Particles?" *American Journal of Respiratory Critical Care Medicine* 164 (2001): 831–833.

37. American Lung Association, "State of the Air 2005," April 28, 2005, http://lungaction.org/reports/stateoftheair2005.html.

38. M. S. O'Neill et al., "Diabetes Enhances Vulnerability to Particulate Air Pollution-Associated Impairment in Vascular Reactivity and Endothelial Function," *Circulation* 111 (2005): 2913–2920.

39. National Institute of Diabetes and Digestive and Kidney Diseases, *National Diabetes Statistics Fact Sheet: General Information and National Estimates on Diabetes in the United States, 2003,* http://diabetes.niddk.nih.gov/dm/pubs/statistics/#7.

40. CDC, "Number of Americans with Diabetes Continues to Increase," 2005, http://www.cdc.gov/od/oc/media/pressrel/fs051026.htm; ibid.

41. G. Dahlquist and L. Hustonen, "Analysis of a Fifteen-Year Prospective Incidence Study of Childhood Diabetes Onset: Time Trends and Climatological Factors," *International Journal of Epidemiology* 23 (1994): 1234–1241; J. Tuomilehto et al., "Evidence for Importance of Gender and Birth Cohort

344 NOTES TO PAGES 227–230

for Risk of IDDM in Offspring of IDDM Parents," *Diabetologia* 38 (1995): 975–982.

42. Delfino et al., "Symptoms in Pediatric Asthmatics and Air Pollution."

43. The odds of symptoms increased 18 percent for a 10 micrograms/cubic meter of air increase in PM-1, and an 11 percent increase in symptoms for a 10 micrograms/cubic meter of air increase in PM-10. The authors concluded: "There is an association between change in short-term air pollution levels, as indexed by PM and CO, and the occurrence of asthma symptoms among children in Seattle." Yu et al., "Effects of Ambient Air Pollution on Symptoms of Asthma."

44. The subjects who had the increased sensitivity were those exposed to PM-2.5 at levels above 100 micrograms per cubic meter of air. See M. Svartengren et al., "Short-Term Exposure to Air Pollution in a Road Tunnel Enhances the Asthmatic Response to Allergen," *European Respiratory Journal* 15 (2000): 716–724.

45. T. Nicolai, "Environmental Air Pollution and Lung Disease in Children," *Monaldi Archives for Chest Disease* 54 (1999): 475–478.

46. Brunekreef et al., "Air Pollution from Truck Traffic." See also Van Vliet et al., "Motor Vehicle Exhaust and Chronic Respiratory Symptoms in Children."

47. H. Gong, Jr., et al., "Comparative Short-Term Health Responses to Sulfur Dioxide Exposure and Other Common Stresses in a Panel of Asthmatics," *Toxicology and Industrial Health* 11 (1995): 467–487.

48. The children with respiratory infections experienced declines in peak respiratory flow associated with exposure to fine particulates during air pollution episodes. See A. Peters et al., "Short-Term Effects of Particulate Air Pollution on Respiratory Morbidity in Asthmatic Children," *European Respiratory Journal* 10 (1997): 872–879.

49. EPA, "Health Assessment Documents for Diesel," http://cfpub.epa.gov/ncea/cfm/recordisplay.cfm?deid=29060; see also D. Diaz-Sanchez, "The Role of Diesel Exhaust Particles and Their Associated Polyaromatic Hydrocarbons in the Induction of Allergic Airway Disease," *Allergy* 52 (1997): 52–56.

50. B. Rudell et al., "Effects on Symptoms and Lung Function in Humans Experimentally Exposed to Diesel Exhaust," *Occupational and Environmental Medicine* 53 (1996): 658–662.

51. These estimates were developed by the American Lung Association based on national rates and applied to county population estimates.

CHAPTER 15. THE TROUBLE WITH DIESEL

1. L. Claudio et al., "Environmental Health Sciences Education: A Tool for Achieving Environmental Equity and Protecting Children," *Environmental Health Perspectives* 106, no. 3 (suppl.) (1998): 849–856; C. Andrew Aligne et al., "Risk Factors for Pediatric Asthma: Contributions of Poverty, Race,

and Urban Residence, *American Journal of Respiratory and Critical Care Medicine* 162 (2000): 873–877; E. F. Crain et al., "An Estimate of the Prevalence of Asthma and Wheezing among Inner-city Children," *Pediatrics* 94 (1994): 356–362; CDC, "CDC's Asthma Prevention Program," http://www.cdc.gov/asthma/NACP.htm, 2008; EPA, "PM2.5 Composition and Sources," 2001, http://www.epa.gov/ttn/oarpg/naaqsfin/pie_txt.pdf; CDC, "Surveillance for Asthma: U.S., 1960–1995," *Morbidity and Mortality Weekly Report* 47 (1998): 1–28; National Center for Health Statistics, "Asthma Prevalence, Health Care Use and Mortality," 2002, http://www.cdc.gov.milli.sjlibrary.org/nchs/products/pubs/pubd/hestats/asthma/asthma.htm.

2. P. J. Landrigan et al., "Environmental Pollutants and Disease in American Children: Estimates of Morbidity, Mortality, and Costs for Lead Poisoning, Asthma, Cancer, and Developmental Disabilities," *Environmental Health Perspectives* 111, no. 3 (March 2003): A144–A145.

3. CDC, "Healthy Youth: Asthma," http://www.cdc.gov/healthyyouth/asthma.

4. J. Wargo and D. Brown, "Children's Exposure to Diesel Exhaust on School Buses," report prepared for Environment and Human Health, Inc., North Haven, Conn., 2000, 76, http://www.ehhi.org/reports/asthma04; see also E. Story et al., "Asthma and Connecticut School Children," report prepared for Environment and Human Health, Inc., North Haven, Conn., 2000, 55, http://www.ehhi.org/reports/asthma03.

5. P. R. Nader et al., "Moderate-to-Vigorous Physical Activity from Ages Nine to Fifteen Years," *JAMA* 300, no. 3 (2008): 295–305; EPA, *Child-Specific Exposure Factors Handbook* (Washington, D.C.: National Center for Environmental Assessment, 2002), http://cfpub.epa.gov/ncea/cfm/recordisplay.cfm?deid=55145.

6. EPA, "Health Assessment Document for Diesel Exhaust," 2002, http://cfpub.epa.gov/ncea/cfm/recordisplay.cfm?deid=29060.

7. A. Lloyd and T. Cackette, "Diesel Engines: Environmental Impact and Control," *Journal of the Air and Waste Management Association* 51 (2001): 818. Almost 94 percent of diesel particulate matter is composed of elemental and organic carbon. Within-vehicle concentrations of black carbon were measured in Sacramento (0–10 micrograms/cubic meter of air, or ug/m3), and Los Angeles (3–40 ug/m3); see California Air Resources Board, "Measuring Concentrations of Selected Air Pollutants inside California Vehicles," 1998, http://www.arb.ca.gov/research/abstracts/95-339.htm. Black carbon within vehicles was detected at 5 ug/m3 in background Los Angeles air; at 15 ug/m3 when following a diesel vehicle with a high exhaust pipe; at 50 ug/m3 when following a truck with a low exhaust pipe; and at 130 ug/m3 when following an urban transit bus; see S. A. Fruin et al., "Fine Particle and Black Carbon Concentrations inside Vehicles," Tenth Annual Conference on International Society of Exposure Analysis, October 25, 2000, cited in Lloyd and Cackette, "Diesel Engines."

8. ATSDR, "Medical Management Guidelines for Formaldehyde," February 7, 2008, http://www.atsdr.cdc.gov/mhmi/mmg111.html.

9. Each data point represents a ten-second average of recorded data. Unless otherwise noted in the analyses and charts that follow, summary statistics and distributions are presented for data averaged over ten-second intervals.

10. These data were collected within a very rural environment with few detectable additional sources of carbon or PM-2.5. The relatively flat readings at the beginning and end of each run are consistent with State of Connecticut ambient outdoor particulate readings for the same period.

11. Levels of PM-2.5 found within diesel-powered school buses were higher than levels measured at fixed monitoring facilities maintained to comply with federal law. Differences between state averages and our findings may be explained in part by the location of sampling equipment. Also, the State of Connecticut averages its findings over 24 hours for 365 days, and then averages these results over three years. This ensures that nights and weekends (when traffic and industrial activity are minimal) will reduce reported levels of particulates.

12. Oak Ridge National Laboratory, *Transportation Energy Data Book*, 27th ed., 2008, 5–20, http://cta.ornl.gov.

13. International Center for Technology Assessment, "In-Car Air Pollution: The Hidden Threat to Automobile Drivers," 2000, http://www.icta.org; California Air Resources Board, "Measuring Concentrations of Selected Air Pollutants inside California Vehicles."

14. EPA, "Amendments to Regulations for Heavy-Duty Diesel Engines," *Federal Register* 71, no. 168 (2006): 51481–51489, http://www.epa.gov/otaq/diesel.htm.

15. State of California Air Resources Board, "Characterizing the Range of Children's Pollutant Exposure during School Bus Commutes," http://www.arb.ca.gov/research/schoolbus/schoolbus.htm 2003; Health Canada, "Evaluation of the Levels of Diesel-Related Pollutants on School Buses during the Transportation of Children," 2006, http://www.hc-sc.gc.ca/ewh-semt/pubs/air/bus-autobus/index_e.html.

16. EPA, "U.S. Settles Clean Air Case against Toyota—Toyota Will Convert School Buses to Run Cleaner as Part of Settlement," March 7, 2003, http://yosemite.epa.gov/opa/admpress.nsf/b1ab9f485b098972852562e7004dc686/69f88fa92e75411585256ce200676104?OpenDocument; State of Connecticut, Public Act 02–56, *An Act Concerning the Idling of School Buses*, http://dep.state.ct.US/whatshap/press/2005/cr052305.htm; State of California, California Air Resources Board, "Summary of Anti-Idling Initiatives in Other States, 2002," http://www.epa.gov/cleanschoolbus/antiidling.htm.

17. *Whitman vs. American Trucking Associations, Inc.*, 531 U.S. 437 (2001), 99–1257.

CHAPTER 16. FORGOTTEN LESSONS

1. National Research Council, *The Biological Effects of Atomic Radiation: A Report to the Public from a Study by the National Academy of Sciences* (Washington, D.C.: National Research Council, 1956).

2. Eve Edstrom, "Study Group Urged on Radiation Perils," *Washington Post and Times Herald*, November 8, 1956.

3. "Stevenson Sees a Cover-up on Bomb," *New York Times*, November 3, 1956.

4. National Research Council, *Biological Effects of Atomic Radiation*.

5. Richard Rhodes, *Arsenals of Folly: The Making of the Nuclear Arms Race* (New York: Knopf, 2007).

6. U.S. Congress, HR 94–69, "Oversight Hearings: Radiological Contamination of the Oceans," 1977, cited in L. Evenson, "The Ethical Implications of Ocean Dumping," senior thesis, University of California, Santa Barbara, 1981.

7. U.S. Department of Defense, *Defense Environmental Programs Annual Report to Congress, Fiscal Year 2007:* "In FY2007 alone, DoD obligated approximately $3.7 billion for the four environmental programs—$299.6 million for conservation; $1.4 billion for environmental restoration activities at active installations and formerly used defense sites; $1.4 billion for compliance; and $130.2 million for pollution prevention. DoD also obligated $492.7 million for environmental restoration and closure related compliance requirements at BRAC installations, including those identified for closure under BRAC 2005." See https://www.denix.osd.mil/portal/page/portal/denix/environment/ARC/FY2007.

8. EPA, "PM-2.5 Composition and Variability," June 1997, http://www.epa.gov/ttn/oarpg/naaqsfin.

CHAPTER 17. THE QUIET REVOLUTION IN PLASTICS

1. K. Weiss, "Plague of Plastic Chokes the Seas," *Los Angeles Times*, August 2, 2006. This chapter builds on research reported in 2008 by Environment and Human Health, Inc. For more detailed citations see J. Wargo et al., "Plastics That May Be Harmful to Children and Reproductive Health," Environment and Human Health, Inc., North Haven, Conn., 2008, http://www.ehhi.org/reports/plastics/ehhi_plastics_report_2008.pdf; Society of the Plastics Industries, *Economic Statistics*, 2008, http://www.plasticsindustry.org/aboutplastics/?navItemNumber=100.

2. EPA, "2000–2001 Pesticide Market Estimates: Sales," 2008, http://www.epa.gov/oppbead1/pestsales/01pestsales/sales2001.htm.

3. "Packaging for Snack Foods Forecast to Grow by 3.7 Percent," *FoodProductionDaily.com*, May 18, 2006; "Dairy Packaging Demand Forecast to Rise 4 Percent," *FoodProductionDaily.com*, January 18, 2007.

4. R. Lipton and D. Barboza, "As More Toys Are Recalled, Trail Ends in China," *New York Times*, June 19, 2007; European Council of Vinyl Manu-

facturers, "Where Is PVC Used? PVC the Right Choice for Toys," http://www.ecvm.org/code/page.cfm?id_page=128, accessed October 2008.

5. The Society of the Plastics Industry Inc., "SPI Material Container Coding System," October 2008, http://www.astm.org/SNEWS/SO_2008/wilhelm_so08.html; Federal Trade Commission, "Guides for the Use of Environmental Marketing Claims," 2008, http://www.ftc.gov/bcp/grnrule/guides980427.htm.

6. EPA, "Municipal Solid Waste in the United States: Facts and Figures," 2008, http://www.epa.gov/epaoswer/non-hw/muncpl/pubs/mswchar05.pdf.

7. National Association for PET Container Resources, "2006 Report on Post Consumer PET Container Recycling Activity," 2006, http://www.napcor.com/pdf/2006PET_Report.pdf.

8. FDA, Center for Food Safety and Applied Nutrition, "Guidance for Industry: Use of Recycled Plastics in Food Packaging: Chemistry Considerations," August 2006, http://www.foodsafety.gov/~dms/opa2cg3b.html.

9. American Chemistry Council, "Packaging and Consumer Products," 2007, http://www.americanchemistry.com/s_plastics/sec_content.asp?CID=1078&DID=4232.

10. Pacific Institute, "Bottled Water and Energy: A Fact Sheet," 2008, http://www.pacinst.org/topics/water_and_sustainability/bottled_water/bottled_water_and_energy.html; J. Larsen, "Bottled Water Boycotts: Back-to-the-Tap Movement Gains Momentum," *Earth Policy Institute*, December 7, 2007; http://www.pacinst.org/topics/water_and_sustainability/bottled_water/bottled_water_and_energy.html.

11. T. Colborn and C. Clement, *Chemically Induced Alterations in Sexual and Functional Development: The Wildlife/Human Connection* (Princeton, N.J.: Princeton Scientific Publishing, 1992).

12. D. A. Crain et al., "Alteration in Steroidogenesis in Alligators (*Alligator mississippiensis*) Exposed Naturally and Experimentally to Environmental Contaminants," *Environmental Health Perspective* 105 (1997): 528–533; M. P. Gunderson, E. Oberdörster, and L. J. Guillette, Jr., "EROD, MROD, and GST Activities in Juvenile Alligators Collected from Three Sites in the Kissimmee-Everglades Drainage, Florida (USA)," *Comparative Biochemistry and Physiology—Part C: Toxicology and Pharmacology,* 139 (2004): 39–46; G. A. Fox, "Epidemiological and Pathobiological Evidence of Contaminant-Induced Alterations in Sexual Development in Free-Living Wildlife," in *Chemically-Induced Alterations in Sexual and Functional Development: The Wildlife/Human Connection,* ed. T. Colborn and C. Clement (Princeton, N.J.: Princeton Scientific Publications, 1992), 147–158.

13. A. L. Herbst, H. Ulfelder, and D. C. Poskanzer, "Adenocarcinoma of the Vagina: Association of Maternal Stilbestrol Therapy with Tumor Appearance in Young Women," *New England Journal of Medicine* 284 (1971): 878–881, as cited in G. Schonfelder et al., "*In Utero* Human Exposure to Low

Doses of Bisphenol A Lead to Long-Term Deleterious Effects in the Vagina," *Neoplasia* 4, no.2 (2002): 98–102.

14. CDC, "Potential Health Risks for Third Generation (Offspring of DES
Daughters and Sons)," 2008, http://www.cdc.gov/DES/consumers/about/
concerns_offspring.html. DES has long been presumed to be between
1,000 and 10,000 more potent than BPA, even though the two compounds
have molecular similarity. See also T. Schettler et al., *Generations at Risk:
Reproductive Health and the Environment* (Cambridge, Mass.: MIT Press,
1996). In 1938, DES became the first manufactured estrogen. The drug
was recommended to pregnant women at risk of miscarriage and was sold
under more than two hundred different brand names. CDC, "DES Update:
Health Care Providers," http://www.cdc.gov/des/hcp/index.html.

15. A. M. Soto et al., "P-Nonyl-phenol: An Estrogenic Xenobiotic Released
from 'Modified' Polystyrene," *Environmental Health Perspectives* 92 (1991):
167–173.

16. National Academy of Sciences, *Pesticides in the Diets of Infants and Children*
(Washington, D.C.: National Academies Press, 1993), 483, http://books.nap
.edu/openbook.php?isbn=0309048753&page=1; *Food Quality Protection
Act of 1996*, U.S. Public Law 104–170, http://www.epa.gov/pesticides/
regulating/laws/fqpa, section 408, "Estrogenic Substances Screening Pro-
gram"; *Safe Drinking Water Act, as Amended through Public Law 107–377*,
U.S. Code 42 (December 31, 2002), "Estrogenic Substances Screening
Program," http://www.epa.gov/lawsregs/laws/sdwa.html.

17. National Research Council, Commission on Life Sciences, Board on
Environmental Studies and Toxicology, *Hormonally Active Agents in the
Environment* (Washington, D.C.: National Academy Press, 1999). See also
National Research Council et al., "Medical Hypothesis: Xenoestrogens as
Preventable Causes of Breast Cancer," *Environmental Health Perspectives*
101, no. 5 (1993): 372–377; M. E. Herman-Giddens et al., "Secondary Sexual
Characteristics and Menses in Young Girls Seen in Office Practice: A Study
from the Pediatric Research in Office Settings Network," *Pediatrics* 99
(1997): 505–512; R. M. Sharpe and N. E. Skakkebaek, "Are Oestrogens Involved in Falling Sperm Counts and Disorders of the Male Reproductive
Tract?" *Lancet* 341 (1993): 1392–1395; N. E. Skakkebaek et al., "Germ Cell
Cancer and Disorders of Spermatogenesis: An Environmental Connection?" *Acta Pathologica, Microbiologica et Immunologica* 106, no. 1 (1998):
3–11; J. Toppari et al., "Male Reproductive Health and Environmental Xeno-
estrogens," *Environmental Health Perspectives* 104 (1996): 741–803; E.
Carlsen et al., "Evidence for Decreasing Quality of Semen during Past Fifty
Years," *British Medical Journal* 305 (1992): 609–613; L. E. Gray and W. R.
Kelce, "Latent Effects of Pesticides and Toxic Substances on Sexual Differ-
entiation of Rodents," *Toxicology and Industrial Health* 12 (1996): 515–553;
W. R. Kelce and L. E. Gray, "Endocrine Disruptors: Effects on Sex Steroid
Hormone Receptors and Sex Development," *Handbook Experimental*

Pharmacology 124 (1997): 435–474, as referenced in G. Lemasters et al., "Workshop to Identify Critical Windows of Exposure for Children's Health: Reproductive Health in Children and Adolescents Work Group Summary," *Environmental Health Perspectives* 108, no. 3 (suppl.) (2000); B. Hileman, "Chemical Exposures: Unusual Cross-Disciplinary Meeting Explores Effects of Environmental Compounds on Human Development and Reproduction," *Chemical and Engineering News* 85, no. 11 (2007): 29–32.

18. Y. Sun et al., "Determination of Bisphenol A in Human Breast Milk by HPLC with Column-Switching and Fluorescence Detection," *Biomedical Chromatography* 18, no. 8 (2004): 501–507; R. Kuruto-Niwa et al., "Measurement of Bisphenol A Concentrations in Human Colostrum," *Chemosphere* 66, no. 6 (2007): 1160–1164; Y. Ikezuki et al., "Determination of Bisphenol A Concentrations in Human Biological Fluids Reveals Significant Early Prenatal Human Exposure," *Human Reproduction* 17, no. 11 (2002): 2839–2841; O. Takahashi and S. Oishi, "Disposition of Orally Administered 2,2-Bis(4-hydroxyphenyl)propane (Bisphenol A) in Pregnant Rats and the Placental Transfer to Fetuses," *Environmental Health Perspectives* 108, no. 10 (2000): 931–935; G. Schonfelder et al., "Parent Bisphenol A Accumulation in the Human Maternal-Fetal-Placental Unit," *Environmental Health Perspectives* 110, no. 11 (2002): A703–A707; Kuruto-Niwa et al., "Measurement of Bisphenol A Concentrations in Human Colostrum"; A. M. Calafat et al., "Urinary Concentrations of Bisphenol A and 4-Nonylphenol in a Human Reference Population," *Environmental Health Perspectives* 113, no. 4 (2005): 5; A. M. Calafat et al., "Exposure of the U.S. Population to Bisphenol A and 4-tertiary-Octylphenol: 2003–2004," *Environmental Health Perspectives* 116, no. 1 (2008): 39–44.

19. European Union, "Risk Assessment Report: 4,4'-isopropylidenediphenol (Bisphenol A)," http://jama.ama-assn.org/cgi/content/full/300/11/1303, 2003; N. K. Wilson et al., "An Observational Study of the Potential Human Exposures of Preschool Children to Pentachlorophenol, Bisphenol-A, and Nonylphenol at Home and Daycare," *Environmental Research* 103, no. 1 (2007): 9–20; K. Miyamoto and M. Kotake, "Estimation of Daily Bisphenol-A Intake of Japanese Individuals with Emphasis on Uncertainty and Variability," *Environmental Science* 13 (2006): 15–29; J. Sajiki et al., "Bisphenol A (BPA) and Its Source in Foods in Japanese Markets," *Food Additives and Contaminants* 24, no. 1 (2007): 103–112.

20. J. E. Biles et al., "Determination of Bisphenol-A in Reusable Polycarbonate Food-Contact Plastics and Migration to Food-Simulating Liquids," *Journal of Agricultural and Food Chemistry* 45 (1997): 3541–3544; H. H. Le et al., "Bisphenol A is Released from Polycarbonate Drinking Bottles and Mimics the Neurotoxic Actions of Estrogen in Developing Cerebellar Neurons," *Toxicology Letters* 176, no. 2 (2008): 149–156.

21. C. Brede et al., "Increased Migration Levels of Bisphenol A from Polycar-

bonate Baby Bottles after Dishwashing, Boiling and Brushing," *Food Additives and Contaminants* 20, no. 7 (2003): 684–689.

22. E. L. Bradley, W. A. Read, and L. Castle, "Investigation into the Migration Potential of Coating Materials from Cookware Products," *Food Additives and Contaminants* 24, no. 3 (2007): 326–335; J. López-Cervantes and P. Paseiro-Losada, "Determination of Bisphenol A in, and Its Migration from, PVC Stretch Film Used for Food Packaging," *Food Additives and Contaminants* 20, no. 6 (2003): 596–606; A. Ozaki et al., "Migration of Bisphenol A and Benzophenones from Paper and Paperboard Products Used in Contact with Food," *Shokuhin Eiseigaku Zasshi* (Journal of the Food Hygenic Society of Japan) 47, no. 3 (2006): 99–104; C. Thomsen, E. Lundanes, and G. Becher, "Brominated Flame Retardants in Archived Serum Samples from Norway: A Study on Temporal Trends and the Role of Age," *Environmental Science and Technology* 36, no. 7 (2002): 1414–1418; K. Jakobsson et al., "Exposure to Polybrominated Diphenyl Ethers and Tetrabromobisphenol A among Computer Technicians," *Chemosphere* 46, no. 5 (February 2002): 709–716.

23. F. vom Saal et al., "Chapel Hill Bisphenol A Expert Panel Consensus Statement: Integration of Mechanisms, Effects in Animals and Potential to Impact Human Health at Current Levels of Exposure," *Reproductive Toxicology* 24, no. 2 (2007): 131–138.

24. Ibid.; A. L. Wozniak, N. N. Bulayeva, and C. S. Watson, "Xenoestrogens at Picomolar to Nanomolar Concentrations Trigger Membrane Estrogen Receptor-a mediated Ca++ Fluxes and Prolactin Release in GH3/B6 Pituitary Tumor Cells," *Environmental Health Perspectives* 113 (2005): 431–439.

25. Wargo et al., "Plastics That May Be Harmful to Children and Reproductive Health."

26. Vom Saal et al. "Chapel Hill Bisphenol A Expert Panel Consensus Statement."

27. National Institute of Environmental Health Sciences, "Draft NTP brief on Bisphenol A, CAS NO. 80–05–7," April 14, 2008, http://cerhr.niehs.nih.gov/chemicals/bisphenol/BPADraftBriefVF_04_14_08.pdf.

28. American Chemistry Council, Phthalate Esters Panel, "Phthalate Esters Panel, American Chemistry Council, Provides a Review of Leading Human Studies on Phthalates," July 27, 2007, http://www.biomatrix.co.uk/article.cfm/id/191226.

29. H. M. Koch et al., "Internal Exposure of the General Population to DEHP and Other Phthalates: Determination of Secondary and Primary Phthalate Monoester Metabolites in Urine," *Environmental Research* 93, no. 2 (2003): 177–185; M. J. Silva et al., "Detection of Phthalate Metabolites in Human Amniotic Fluid," *Bulletin of Environmental Contamination and Toxicology* 72 (2004): 1226–1231; A. M. Calafat et al., "Exposure to Di-(2-ethylhexyl) phthalate among Premature Neonates in a Neonatal Intensive Care Unit," *Pediatrics* 113 (2004): 429–434; M. J. Silva et al., "Urinary Levels of Seven

Phthalate Metabolites in the U.S. Population from the National Health and Nutrition Examination Survey (NHANES), 1999–2000," *Environmental Health Perspectives* 112 (2004): 331–338; M. J. Silva et al., "Analysis of Human Urine for Fifteen Phthalate Metabolites Using Automated Solid-Phase Extraction," *Journal of Chromatography B* 805 (2004): 161–167; D. B. Barr et al., "Assessing Human Exposure to Phthalates Using Monoesters and Their Oxidized Metabolites as Biomarkers," *Environmental Health Perspectives* 111 (2003), 1148–1151; H. M. Koch, H. Drexler, and J. Angerer, "An Estimation of the Daily Intake of Di(2-ethylhexyl)phthalate (DEHP) and Other Phthalates in the General Population," *International Journal of Hygiene and Environmental Health* 206 (2003): 77–83; CDC, *Third National Report on Human Exposure to Environmental Chemicals* (Atlanta: CDC, 2005), http://www.cdc.gov/exposurereport; M. Wittassek, W. Heger, and H. M. Koch, "Daily Intake of Di(2-ethylhexyl)phthalate (DEHP) by German Children: A Comparison of Two Estimation Models Based on Urinary DEHP Metabolite Levels," *International Journal of Hygiene and Environmental Health* 210 (2007): 35–42; K. M. Shea, "Pediatric Exposure and Potential Toxicity of Phthalate Plasticizers: Technical Report," *Pediatrics* 111, no.6 (2003): 1467–1474; B. Blount, M. Silva, and S. Caudill, "Levels of Seven Urinary Phthalate Metabolites in a Human Reference Population," *Environmental Health Perspectives* 108 (2000): 979–982; J. J. Adibi et al., "Prenatal Exposures to Phthalates among Women in New York City and Krakow, Poland," *Environmental Health Perspectives* 11, no. 14 (2003): 1719–1722; G. Latini et al., "Exposure to Di(2-ethylhexyl)phthalate in Humans during Pregnancy," *Biology of the Neonate* 83 (2003): 22–24.

30. FDA, *Code of Federal Regulations*, 1999, in particular 21 *CFR* 175.300, 175.105, 176.210, 177.1010, 177.1200, 178.3910, all at http://www.accessdata.fda.gov/SCRIPTs/cdrh/cfdocs/cfcfr/CFRSearch.cfm; see also ATSDR, "Toxicological Profile, Di(2-Ethylhexyl)Phthalate: Potential for Human Exposure," 2003, http://www.atsdr.cdc.gov/toxprofiles/tp9.html; J. H. Petersen and T. Breindahl, "Plasticizers in Total Diet Samples, Baby Food and Infant Formulae," *Food Additives and Contaminants* 17, no. 2 (2000): 133–141; ATSDR, *Toxicological Profile for Di(2-ethylhexyl)phthalate* (Atlanta: Public Health Service, U.S. Department of Health and Human Services, 1993); Y. Tsumura et al., "Eleven Phthalate Esters and Di(2-ethylhexyl) Adipate in One-Week Duplicate Diet Samples Obtained from Hospitals and Their Estimated Daily Intake," *Food Additives and Contaminants* 18, no. 5 (2001): 449–460; G. Latini, C. De Felice, and A. Verrotti, "Plasticizers, Infant Nutrition and Reproductive Health," *Reproductive Toxicology* 19 (2004): 27–33; G. K. Mortensen, K. M. Main, and A. M. Andersson, "Determination of Phthalate Monoesters in Human Milk, Consumer Milk, and Infant Formula by Tandem Mass Spectrometry (LC-MS-MS)," *Analytical and Bioanalytical Chemistry* 382, no.4 (2005): 1084–1092; Shea, "Pediatric Exposure and Potential Toxicity of Phthalate Plasticizers"; A. M. Calafat et al.,

"Automated Solid Phase Extraction and Quantitative Analysis of Human Milk for Thirteen Phthalate Metabolites," *Journal of Chromatography B Analytical Technologies in the Biomedical and Life Sciences* 805 (2004): 49–56; J. Zhu et al., "Phthalate Esters in Human Milk: Concentration Variations over a Six-Month Postpartum Time," *Environmental Science and Technology* 40, no. 17 (September 1, 2006): 5276–5528; K. M. Main, G. K. Mortensen, and M. M. Kaleva, "Human Breast Milk Contamination with Phthalates and Alterations of Endogenous Reproductive Hormones in Infants Three Months of Age," *Environmental Health Perspectives* 114 (2006): 270–276; S. Srivastava et al., "Biochemical Alterations in Rat Fetal Liver Following in Utero Exposure to Di(2-ethylhexyl)phthalate (DEHP)," *Indian Journal of Experimental Biology* 27 (1989): 885–888; L. A. Dostal, R. P. Weaver, and B. A. Schwetz, "Transfer of di(2-ethylhexyl) Phthalate through Rat Milk and Effects on Milk Consumption and the Mammary Gland," *Toxicology and Applied Pharmacology* 91 (1987): 315–325; D. Parmar, S. P. Srivastava, and P. K. Seth, "Hepatic Mixed Function Oxidases and Cytochrome P-450 Contents in Rat Pups Exposed to Di-(2-ethylhexyl)phthalate through Mother's Milk," *Drug Metabolism and Disposition* 13 (1985): 368–370.

31. NRDC, "Bottled Water: Pure Drink or Pure Hype?" 1999, http://www.nrdc.org/water/drinking/nbw.asp.

32. P. A. Clausen et al., "Simultaneous Extraction of Di(2-ethylhexyl) phthalate and Nonionic Surfactants from House Dust: Concentrations in Floor Dust from Fifteen Danish Schools," *Journal of Chromatography* 986 (2003): 179–190; H. Fromme et al., "Occurrence of Phthalates and Musk Fragrances in Indoor Air and Dust from Apartments and Kindergartens in Berlin (Germany)," *Indoor Air* 14 (2004): 188–195; R. A. Rudel et al., "Phthalates, Alkylphenols, Pesticides, Polybrominated Diphenyl Ethers, and Other Endocrine-Disrupting Compounds in Indoor Air and Dust," *Environmental Science and Technology* 37 (2003): 4543–4553; M. Wensing, E. Uhde, and T. Salthammer, "Plastics Additives in the Indoor Environment: Flame Retardants and Plasticizers," *Science of the Total Environment* 339 (2005): 19–40; C. J. Weschler, "Indoor-Outdoor Relationships for Nonpolar Organic Constituents or Aerosol Particles," *Environmental Science and Technology* 18 (1984): 648–652; R. Rudel et al., "Phthalates, Alkylphenols, Pesticides, Polybrominated Diphenyl Ethers, and Other Endocrine Disrupting Compounds in Indoor Air and Dust," *Environmental Science and Technology* 37, no. 20 (2003): 4543–4553; "What You Should Know about Chemicals in Your Cosmetics," *Consumer Reports*, 2007, http://money.aol.com/consreports/smartshopping/home_garden/_a/what-you-should-know-aboutchemicals-in/20070117090309990001; N. Singer, "Looking at the Bottle and What's in It," *New York Times*, February 15, 2007.

33. K. Bouma and D. J. Schakel, "Migration of Phthalates from PVC Toys into Saliva Stimulant by Dynamic Extraction," *Food Additives and Contaminants*

19 (2002): 602–610; T. Niino et al., "Monoester Formation by Hydrolysis of Dialkyl Phthalate Migrating from Polyvinyl Chloride Products in Human Saliva," *Journal of Health Science* 47 (2001): 318–322; National Toxicology Program, U.S. Department of Health and Human Services, "NTP-CERHR Expert Panel: Di(2-ethylhexyl)phthalate," *Journal of Health Science* 47 (2001): 318–322; "U.S. Agency, Lawmakers Seek Tests of China-Made Toys," *Reuters,* August 14, 2007.

34. J. Barrett, "Draft Brief on DEHP," *Environmental Health Perspectives* 114 (2006): A580–A581; A. Calafat et al., "Exposure to Di-(2-Ethylhexyl) Phthalate among Premature Neonates"; FDA, "FDA Public Health Notification: PVC Devices Containing the Plasticizer DEHP," July 12, 2002, http://www .fda.gov/cdrh/safety/dehp.html; Shea, "Pediatric Exposure and Potential Toxicity of Phthalate Plasticizers"; NIEHS News, "NIEHS Investigates Links between Children, the Environment, and Neurotoxicity," *Environmental Health Perspectives* 109 (2001): 260–261; National Toxicology Program, U.S. Department of Health and Human Services, "NTP-CERHR Monograph on the Potential Human Reproductive and Developmental Effects of Di-(2-ethylhexyl) phthalate (DEHP)," 2006, http://cerhr.niehs .nih.gov/chemicals/phthalates/didp/DIDP_Monograph_Final.pdf.

35. Y. Nakamura, "Teratogenicity of Di-(2-ethylhexyl) Phthalate in Mice," *Toxicology Letters* 4 (1979): 113–117; D. Parmar et al., "Hepatic Mixed Function Oxidases and Cytochrome P-450 Contents in Rat Pups Exposed to Di-(2-ethylhexyl)phthalate through Mother's Milk," *Drug Metabolism and Disposition* 13, no. 3 (1985): 368–370; L. A. Dostal et al., "Transfer of di(2-ethylhexyl) phthalate through Rat Milk and Effects on Milk Composition and the Mammary Gland," *Toxicology and Applied.Pharmacology* 91 (1987): 315–325; L. A. Dostal et al., "Hepatic Peroxisome Proliferation and Hypolipidemic Effects of di(2-ethylhexyl)phthalate in Neonatal and Adult Rats," *Toxicology and Applied Pharmacology* 87 (1987): 81–90; EPA, "Toxicity and Exposure Assessment for Children's Health, Phthalates, TEACH Chemical Summary," November 2008, http://www.epa.gov/teach.

36. L. E. Gray, C. Wolf, and C. Lambright, "Administration of Potentially Antiandrogenic Pesticides (procymidone, linuron, iprodione, chlozolinate, p,p'-DDE, and ketoconazole) and Toxic Substances (dibutyl-and diethylhexyl phthalate, PCB 169, and ethane dimethane sulphonate) during Sexual Differentiation Produces Diverse Profiles of Reproductive Malformations in the Male Rat," *Toxicology and Industrial Health* 15 (1999): 94–118.

37. L. G. Parks et al., "The Plasticizer Diethylhexyl Phthalate Induces Malformations by Decreasing Fetal Testosterone Synthesis during Sexual Differentiation in the Male Rat," *Toxicological Sciences* 58 (2000): 339–349; J. Borch et al., "Steroidogenesis in Fetal Male Rats Is Reduced by DEHP and DINP, But Endocrine Effects of DEHP Are Not Modulated by DEHA in Fetal, Prepubertal and Adult Male Rats," *Reproductive Toxicology* 18 (2004):

53–61; L. E. Gray et al., "Perinatal Exposure to the Phthalates DEHP, BBP, and DINP, but not DEP, DMP, or DOTP, Alters Sexual Differentiation of the Male Rat," *Toxicological Sciences* 58 (2000): 350–365; National Toxicology Program, Center for the Evaluation of Risks to Human Reproduction, *Expert Panel Report on DEHP*, 2000, http://www.toxicologysource.com/scitox/deh-phthalatetox.html; M. Ablake et al., "Di-(2-ethylhexyl) Phthalate Induces Severe Aspermatogenesis in Mice: However, Subsequent Antioxidant Vitamins Supplementation Accelerates Regeneration of the Seminiferous Epithelium," *International Journal of Andrology* 27, no. 5 (2004): 274–281.

38. S. M. Duty et al., "Phthalate Exposure and Human Semen Parameters," *Epidemiology* 14 (2003): 269–277; R. Hauser et al., "DNA Damage in Human Sperm Is Related to Urinary Levels of Phthalate Monoester and Oxidative Metabolites," *Human Reproduction* 22, no. 3 (2007): 688–695.

39. S. H. Swan et al., "Decrease in Anogenital Distance among Male Infants with Prenatal Phthalate Exposure," *Environmental Health Perspectives* 113 (2005): 1056–1061; G. Lottrup et al., "Possible Impact of Phthalates on Infant Reproductive Health," *International Journal of Andrology* 29 (2006): 172–180.

40. NIEHS News, "NIEHS Investigates Links between Children, the Environment, and Neurotoxicity"; National Toxicology Program, U.S. Department of Health and Human Services, "NTP-CERHR Monograph on the Potential Human Reproductive and Developmental Effects of Di-(2-ethylhexyl) phthalate (DEHP)."

41. L. Øie, L-G Hersoug, and J. O. Madsen, "Residential Exposure to Plasticizers and Its Possible Role in the Pathogenesis of Asthma," *Environmental Health Perspectives* 105 (1997): 972–978; C. G. Bornehag et al., "The Association between Asthma and Allergic Symptoms in Children and Phthalates in House Dust: A Nested Case-Control Study," *Environmental Health Perspectives* 112, no. 14 (2004): 1393–1397; J. J. Jaakkola, A. Ieromnimon, and M.S. Jaakkola, "Interior Surface Materials and Asthma in Adults: A Population-Based Incident Case-Control Study," *American Journal of Epidemiology* 164, no. 8 (2006): 742–749; S. Caress and A. Steinemann, "National Prevalence of Asthma and Chemical Hypersensitivity: An Examination of Potential Overlap," *Journal of Occupational and Environmental Medicine* 47 (2005): 518–522; R. B. May et al., "D-(2-ethyl)-phthalate as Plasticizer in PVC Respiratory Tubing Systems: Indication of Hazardous Effects on Pulmonary Function in Mechanically Ventilated, Preterm Infants," *European Journal of Pediatrics* 147 (1998): 41–46.

42. *U.S. Toxic Substance Control Act*, Public Law 94–469, 90 *U.S. Statutes at Large* (1976): 2003; General Accountability Office, "Chemical Regulation: Options Exist to Improve EPA's Ability to Assess Health Risks and Manage Its Chemicals Review Program," 2005, GAO-05–458, http://www.gao.gov/new.items/do5458.pdf; GAO, "Toxic Chemicals: EPA's Toxic Release Inven-

tory Is Useful But Can Be Improved," June 1991, RCED-91–121, 26, http://
archive.gao.gov/d2ot9/144255.pdf.

43. J. Duffy, "EU Investigates Chemical That May Harm," *Sunday Herald*, November 11, 2007; Europa, "Summaries of Legislation: Phthalate-Containing
Soft PVC Toys and Childcare Articles," http://europa.eu/scadplus/leg/en/
lvb/l32033.htm.

44. FDA Center for Devices and Radiological Health, "FDA Public Health
Notification: PVC Devices Containing the Plasticizer DEHP," July 12, 2002,
http://www.fda.gov/cdrh/safety/dehp.html.

CHAPTER 18. GREEN INTELLIGENCE

1. *Children's Health Act of 2000*, Public Law 106–310, 106th Cong., 2d sess.
(October 17, 2000), sec. 1004, authorizing the planning and implementation of the National Children's Study by a consortium of the U.S. Department of Health and Human Services, National Institutes of Health,
NIEHS, CDC, EPA, and U.S. Department of Education, http://www
.nationalchildrensstudy.gov/about/pages/funding.aspx.

2. U.S. National Institutes of Health, *Women's Health Study*, 2006, http://
www.clinicaltrials.gov/ct/show/NCT00000479.

3. *Pollution Prevention Act*, codified at *U.S. Code* 42 (1990): 13101 et seq.;
Emergency Planning and Community Right to Know Act, U.S. Code 42
(1986): 11001 et seq.

4. Paul T. Anastas and John C. Warner, *Green Chemistry: Theory and Practice*
(New York: Oxford University Press, 2000).

EPILOGUE

1. These suggestions build on a paper written by John Wargo and Linda
Wargo, "The State of Children's Health and Environment, 2002,"
Children's Health Environmental Coalition, Los Angeles. Out of print.

INDEX